Introduction to Aviation

Geoffrey Whitehurst

Revised Printing

Kendall Hunt
publishing company

Kendall Hunt
publishing company

www.kendallhunt.com
Send all inquiries to:
4050 Westmark Drive
Dubuque, IA 52004-1840

Revised Printing 2014

ISBN 978-1-4652-4842-8

Printed in the United States of America
10 9 8 7 6 5 4 3

CONTENTS

List of Figures *vii*

CHAPTER 1 **Man's Efforts to Fly 1**

Leonardo da Vinci 3
Development of the Lighter-than-Air Ship 5
The Dirigible 7
The Future of Zeppelins 10

CHAPTER 2 **Development of the Airplane 13**

George Cayley 14
William Henson and John Stringfellow 14
Jean-Marie Le Bris 16
Francis Wenham 16
Alphonse Penaud 17
H. F. Phillips 17
Clement Ader 17
Hiram Maxim 18
Otto Lilienthal 19
Percy Pilcher 22
John J. Montgomery 23
Octave Chanute 24
Samuel P. Langley 24
The Wright Brothers 28
Tuskegee Airmen 38
Female Aviators 41
Claude Dornier 41
DO-X 42
Hercules HK-1 SPRUCE GOOSE 43

CHAPTER 3 **World War I 45**

The National Advisory Committee for Aeronautics (NACA) 47

CHAPTER 4 **Commercial Aviation in the U.S. pre-WW II 49**

A Rough Beginning 49
Scheduled Airmail Service 50
Transcontinental Airmail 52
The Lighted Airway 53
The Privatization of Airmail 54
The Airmail Act of 1925 (The Kelly Act) 55
The Founding of the Airlines 58
The Morrow Board 58
The Air Commerce Act of 1926 58
Lindbergh 59
The Daniel Guggenheim Fund for the Promotion of Aeronautics 61
Airways—from Lighted Beacons to Radio Navigation 62
Amelia Earhart 63
From Mail Carriers to Airlines 66
United Air Transport 67
North American Aviation 67
Aviation Corporation 67
Transcontinental Air Transport—A Novel Idea 67
Progenitors of the Big Four 68
The Airmail Act of 1930 (McNary Watres Act) 68
New Deal—The Roosevelt Administration 73
 A New Broom Sweeps Clean 73
The Airmail Act of 1934 (The Black-McKellar Act) 74
State of the Airlines before the Civil Aeronautics Act 75
The National Advisory Committee on Aeronautics (NACA) 75
The Big Four 75
 American Airlines 75
 Eastern Air Lines 77
 TWA 77
 United Airlines 81
The Lesser Lines 82
Pan American Airways 82
Summary of Airlines' Condition 88
The Civil Aeronautics Act of 1938 (McCarran-Lea Act) 88
Endnotes 91

CHAPTER 5 **World War II 93**

The Lend Lease Act 94
The Chosen Instrument 95
The Airlines at War 95

CHAPTER 6 **Commercial Aviation in the U.S. post-WW II 99**

On the Way to the Jet Age 102
Jets 103
The British Comet 105
America Catches Up 106
The Next Jets 109
The Really Big Jets 111

CHAPTER 7 **Aviation and Federal Legislation 115**

Airmail Act of 1925 (The Kelly Act) 116
Air Commerce Act of 1926 118
Airmail Act of 1930 (McNary-Watres Act) 119
Airmail Act of 1934 (Black-McKellar Act) 124
Civil Aeronautics Act of 1938 (McCarran-Lea Act) 125
 The Grandfather Clause 126
The Federal Aviation Act—1958 127
Section 101—Definitions 129

Section 102—Declaration of Policy: The Board Factors for Interstate, Overseas and Foreign Air Transportation 132
Section 103—Declaration of Policy: The Secretary of Transportation 133
Section 104—Public Right of Transit 134
Section 105—Federal Preemption 134
Section 106—Report on Subsidy Cost-Sharing 134
Section 107—Safety Study 135
Section 108—Report on Air Carrier Marketing of Tours 135
Civil Aeronautics Board 135
Functions of the Board Members 136
Federal Airport Act of 1946 137
Airport and Airway Development Act of 1970 (Airport Development Aid Program) 138
Airport Improvement Program 140
Most Recent Developments 141
Background and Analysis 141
Founding Legislation 142
Airport and Airway Development and Revenue Acts of 1970 (P.L. 91-258) 142
Airport and Airway Improvement Act of 1982 (P.L. 97-248) 142
The Airport and Airway Trust Fund 142
Airport Improvement Program Funding 143
Airline Deregulation Act of 1978 144
Deregulation—What Is It? 146
Comments 147
Labor Relations 148
Essential Air Service 148
Airports 149
Postal Service 149
International 149
Summary of Deregulation Changes 150
What Remains Regulated 154
Effects of Deregulation 155
Endnotes *159*

CHAPTER 8 **The Federal Aviation Act** **161**
The Department of Transportation 163
The Federal Aviation Administration 164
Regulation 166
Certification 166
Airmen 166
Aircraft 166
Air Carriers 167
Air Navigation Facilities 167
Air Agencies 167
Airports 167
Designees 167
Investigation 167
Enforcement 168
Operations 168
Air Traffic Control 169
Radio Aids to Navigation 169
National Airports 169
Monroney Aeronautical Center 169
William J. Hughes Technical Center 169
Education 170
Funding 170
Registration and Recordation 170
Commercial Space Transportation 170
Dual Mandate 170
The National Transportation Safety Board 171
Endnote *173*

CHAPTER 9 **Prelude to Deregulation** **175**
The Economic Nature of Transportation 175
The Nature of Regulated Transportation 176
The Airline Deregulation Act of 1978 179
Braniff—A Case History Under Deregulation 179
The United States Bankruptcy Act 180
The Progression of Deregulation 181
Airline Fares 181
Passenger Travel 182
Number of Carriers 182
Market Share 183
Safety 183
Employment 183
The Evolution of Operating Practices 184
Hub and Spoke 184
Low Fare, Point-to-Point Service 185
The Regional Jet 186
The Fourth Wave—Demise of the Legacy Airlines Business Model 188
Endnotes *189*

CHAPTER 10 **Airports** **191**
Federal Airport Act of 1946 193
Airport Security 194
Evolution of a New National Airport Policy 195
Airports and Deregulation 195
Airport Ownership 197
Airport Funding for Operations and Capital Improvements 197
Airport Revenue Bonds 198
Airport Improvement Program Funds (AIP) 198
Airport User Charges 198
Passenger Facility Charges (PFCs) 198
State and Local Government Programs 198

Airport Facilities Use Arrangements 199
Residual Use and Lease Agreements 199
Compensatory Use and Lease Agreements 199
Hybrid Use and Lease Agreements 199
Gate Leasing Arrangements 200
Exclusive-Use Arrangements 200
Preferential-Use Arrangements 200
Airport-Controlled or Common-Use Arrangements 200
DOT Interest in Airport Practices—Unfair Competition 201
Airport Privatization 201
Wendell H. Ford Aviation Investment and Reform Act for the 21st Century (AIR-21) 202
Airports and the Environment 204
The Air Pollution Act and The Clean Air Act 204
Aviation Impacts 204
Aviation-Generated Air Pollution 205
Aircraft Noise 207
Aircraft Noise Abatement Act of 1968 207
Noise Control Act of 1972 and Aviation Safety and Noise Abatement Act of 1979 207
Airport Noise and Capacity Act of 1990 208
Endnotes 210

CHAPTER 11 Human Factors in Aviation 211

Basic Concepts 211
Human Factors 211
Aircraft Accidents 211
SHELL Model 212
Liveware-Software 212
Liveware-Hardware 212
Liveware-Environment 212
Liveware-Liveware 213
Basic Physiology 213
The Sensory System 213
Central and Peripheral Nervous System 213
Human Information Processing 214
Selection and Execution 214
Attention 214
Vigilance 216
Perception 216
Memory 216
Working Memory (Short-term Memory) 217
Long-term Memory 217
Behavior 217
Rule-based behavior 217
Knowledge-based Behavior 217
Skill-based Behavior 218
Phases of Motor Memory 218
Receptors and Sensory Stores 218
Learning 218
Response Selection 218
Behavioristic Approach 219
Cognitive Approach 219
Learning Techniques 220
Memory Techniques 220
Skill Development 220
Motivation 220
Response Execution (Actions) 220
Response Times 221
Feedback 221
Influence of Stress and Arousal 221
Integration of Sensory Inputs 221
Visual Perception 221
Visual Perception in Flight 223
Approach and Landing 223
Height Judgment 223
Constant Relative Bearing 223
Illusions 224
Visual Illusions in Flight 224
Vestibular Illusions 225
Gravic Illusions 225
Gyral Illusions 225
Coriolis Illusion 225
The Leans 226
Get on Gauges 226
Central Illusions 226
Flicker Vertigo 226
Decision Making 226
Rational and Intuitive 226
Decision-making Process 227
Decision-making Styles 227
Democratic 227
Autocratic 227
Consensus 228
Collective-Participative 228
Decision-making Concepts 228
Error Sources and Limits in Decision Making 228
Decision-making Models 229
Decision-making Processes 229
Skill-based Decision Making 229
Rule-based Decision Making 229
Knowledge-based Decision Making 230
Risk Assessment and Options 230
Decision and Response 230

CHAPTER 12 **Principles of Navigation 233**
Form of the Earth 233
The Solar System 233
The Earth 233
Direction 234
Sexagesimal System 234
Earth Reference Systems 234
Circles on the Earth 235
The Graticule 236
Latitude 236
Longitude 237
Positions in Latitude and Longitude 237
Convergency 238
Rhumb Line 238
Distance on the Earth 238
Metric and Imperial Measures 238
The Kilometer (km) 238
The Statute Mile (sm) 239
The Nautical Mile (nm) 239
Direction on the Earth 239
The Compass 239
The Earth's Magnetic Poles 240
Magnetic Variation 241
Deviation 241
Heading 241
Track (or Course) 241
Bearing 242
Relative Bearings 242

CHAPTER 13 **International Aviation 243**
Conferences and Conventions 243
Paris Conference—1910 243
Paris Convention—1919 244
Madrid Convention—1926 244
Pan American Convention on Commercial Aviation—1928 245
Warsaw Convention—1929 245
International Air Transportation of Goods and Litigation Under the Warsaw Convention 247
Convention on Damage Caused by Foreign Aircraft to Third Parties on the Surface—1933 248
Chicago Convention on International Civil Aviation—1944 (Chicago Conference) 249
Outcomes of Chicago Conference—1944 250
Bilateral Agreements 250
Provisional International Civil Aviation Organization 251
Permanent International Civil Aviation Organization (ICAO) 251
International Air Services Transit Agreement 251
International Air Transport Agreement (Five Freedoms Agreement) 252
Bermuda Agreement—1946 (U.N. Treaty Series 36, Vol. 3, Page 253) 252
Rights in Aircraft Convention 253
Ground Damage Convention 253
Air Offenses Convention 253
Hijacking Convention 254
Sabotage Convention 254
International Air Transportation Competition Act of 1979 254
International Deregulation 255
The United States Policy on International and Foreign Air Transportation 256
United States Policy for the Conduct of International Air Transportation Negotiations (August 21, 1978) 257
Introduction 257
Goals of U.S. International Air Transportation Policy 257
Translating Goals into Negotiating Objectives 258
Explanation of Objectives 258
Negotiating Principles 259
Interagency Committee on International Air Transportation Policy 259
Free Trade 262
Bibliographic Note 263
Endnotes 263

FIGURES

1-1 Leonardo da Vinci's Sketches.
1-2 Leonardo da Vinci's sketches of muscle-powered flying machines.
1-3 Balloon which carried de Rozier and the Marquis de' Arlandes over Paris on November 21, 1783.
1-4 Explosion and crash of the *Hindenburg*, Lakehurst, New Jersey, May 6, 1937.
2-1 Sir George Cayley.
2-2 Cayley's Aerial Carriage.
2-3 Flights by Lilienthal.
2-4 Manly and Langley.
2-5 Launching the aerodrome.
2-6 The Wright Brothers (top). The Wright Flyer (middle). December 17, 1903; the flight the world remembers (bottom).
2-7 The Wright Flyer.
2-8 Glenn Curtiss in his *June Bug*.
2-9 Alberto Santos-Dumont, 14 BIS, France 1906. Wings and engine were at the rear of the plane.
2-10 Harriet Quimby.
2-11 Wilbur Wright and the first female pilot in America, Harriet Quimby.
4-1 President Woodrow Wilson at the inauguration of airmail—May 15, 1918.
4-2 Major Reuben Fleet (on the left) briefs airmail pilot Lt. George Boyle before he begins his flight on May 15, 1918.
4-3 Lt. George Boyle takes off for Philadelphia.
4-4 Original Transcontinental Airmail Route—compare to the Transcontinental Railroad, Figure 3-3.
4-5 Charles Lindbergh on CAM 2, flown between Chicago and St. Louis.
4-6 Ford Trimotor.
4-7 Charles Lindbergh pays a visit to Orville Wright at Wright Field, Dayton, OH, June 22, 1927.
4-8 Juan Trippe and Charles Lindbergh.
4-9 Schematic of the four-course radio range.
4-10 Amelia Earhart.
4-11 Route of Transcontinental Air Transport—depicting night and day portions of the route.
4-12 Airmail airways.
4-13 The Boeing 247 was to be a great leap forward with its low mono-wing, and two engines instead of three that were mounted into wings in nacelles that greatly reduced drag.
4-14 DC-1, designation for the Douglas Commercial Number 1. This was the only one ever built.
4-15 DC-3—The plane that changed the world.
4-16 Boeing 307.
4-17 Progress in airmail transport service, 1926–1936.
4-18 Source of revenues of the airline companies.
4-19 The S-42 had a range of 2,520 miles.

4-20 The B-314 Was the largest aircraft to be used in scheduled service until the arrival of jumbo jets in the late 1960s.
4-21 The beginnings of air traffic control—Earl Ward (left) organized the Newark, New Jersey air traffic facility in the mid-1930's. Here he tracks a flight with the aid of a calipar as R. A. Eccles watches. The pointed markers representing aircraft were moved across the map as flights progressed.
6-1 The DC-4 became available to the domestic fleet in 1946.
6-2 The Constellation was much more expensive to maintain than other comparable aircraft.
6-3 The DC-6 was launched in coast-to-coast service on April 27, 1947 with one stop enroute for fuel.
6-4 Boeing 377, Constellation 049, and DC-4.
6-5 Germany's jumo-004 engine.
6-6 The Comet entered into service on British Overseas Airways Corporation in May 1952.
6-7 A KC-97 refueling a B-47.
6-8 The Caravelle was a prototype with the aircraft's engines on the side of the fuselage near the tail of the aircraft.
6-9 Boeing 727.
6-10 Douglas DC-9.
6-11 Boeing 737.
6-12 The Boeing 747 flew faster than previous models.
6-13 Lockheed L-1011.
6-14 DC-10.
7-1 Three-engined airplane of the 1930s meets the three-engined airplane of the 1960s.
7-2 A DC-3 of the 1930s superimposed on the outline of a Boeing 747 of today.
7-3 Major Hub Cities in the United States.
8-1 Estimated controller losses and planned hires, fiscal years 2006–2015.
8-2 NTSB's Organization.
8-3 Number of accident investigations completed by NTSB by mode, fiscal years 2002–2005.
8-4 Fatalities by transportation mode, 2005.
9-1 Median fares have declined almost 40 percent since 1980 as measured in 2005 dollars.
9-2 Safety record of U.S. airlines—fatal accidents per million aircraft miles.
9-3 Accident rates.
9-4 A wide array of "regional jets."
9-5 Uses of regional jets.
9-6 U.S. regional airline growth, 1978–1996.
10-1 98% of airport revenue comes from airport users: U.S. Airport Sources of Revenue, 2001.
10-2 Significant progress in reduction of aircraft noise.
11-1 Leading causes of aircraft accidents over past 40 years.
11-2 The hawkins SHELL.
11-3 Brain and section of spinal cord.
11-4 Simple human information processing.
11-5 Human information processing.
11-6 Some common visual Illusions.
11-7 Rate of change of apparent size.

12-1 Solar system.
12-2 Earth's inclination.
12-3 Earth's shape.
12-4 Basic directions.
12-5 Compass rose.
12-6 Great circle and small circle.
12-7 Latitude and longitude.
12-8 Geocentric and geodetic latitude.
12-9 Latitude and longitude of New York.
12-10 Convergency and great circle direction.
12-11 A simple magnetic compass.
12-12 The earth's magnetic field and magnetic poles.
12-13 Changing positions of the north and south magnetic poles.
12-14 Variation.
12-15 Heading, track and drift.
12-16 Basic clock code.

CHAPTER 1

Man's Efforts to Fly

Throughout humankind's early years on this earth progress was limited by means of transportation. We progressed successfully through various evolutionary stages and found ourselves capable of covering greater distances. It was always our fondest desire to break away from the bonds of the earth and become airborne. For centuries humans had looked at the birds and envied their freedom of flight. Ages passed while humans enviously watched the birds soaring in seemingly effortless flight. Archaeological findings of civilizations dating back to 3000–4000 B.C. indicate humankind's early interest in flight.

For centuries people had visions of magical powers and other occult devices that would unlock the key to the secrets of flying. Many of these strange ideas were recorded and are available today in historical archives.

The beginning of the story of the earliest known evidence of human efforts to fly goes back to the ancient literature of Greek and Roman mythology. In this world of legend the ways of flight were many.

Apollo's son **Phaethon** rode the skies in a sun chariot. There were winged men of Egypt and the winged bull of Assyria, spoken of in the Bible, who protected the palace of King Sargon in 708 B.C. Well remembered is **Sinbad** the Sailor who flew on his bird "Roc." The Arabs had their magical flying carpets. The pre-Christian settlers of the Middle East dreamed of traveling by smoke.

One of the oldest tales is about the Greek mythical character **Icarus**. He was the son of **Daedalus** who was a sculptor, architect, and inventor. When he feared that his nephew Talus would surpass him in originality he threw him from the Acropolis in Athens and fled to Crete. There he built a labyrinth to house the Minotaur for King Minos. The king refused to allow him to leave and put him in prison. Daedalus then constructed wings made of wax feathers for himself and his son Icarus. They flew away but Icarus flew too close to the sun; his wings melted and he fell to his death into the sea. Daedalus flew on and successfully escaped to live alone in Sicily.

Nike, the winged goddess of victory whose statue appeared on an island in the Aegean Sea in 300 B.C. Hermes, the Greek god called **Mercury** by the Romans, was the winged messenger of the gods. There is the story of the flying horse **Pegasus**, the famous ancestor of our present-day horse, who carried Perseus whenever he chose.

A Persian legend from 1500 B.C. says that **King Kai Ka'us** or **Koos/Kaoos**, who built the tower of Babylon, was tempted by evil spirits to invade the heavenly realm. To his gold and wood throne he had attached four poles. Four large eagles were tied to the poles. The birds, flapped furiously and as the story goes, lifted the throne a short distance before becoming exhausted, causing the king and throne to crash. His fall was taken as a sign that people were not meant to fly. Thereafter the king was known as the Foolish King.

In the Americas, the Incas had their legend of **Ayar Utso**, who grew wings and flew to the sun. From ancient days through the pre-Christian era, until the fifteenth century, a conglomeration of ideas on flying had been presented. All were impossible and none were worthy of fact. Humankind had yet to fly.

In 863, **Bladud**, the King of Britain attempted to fly using wings of feathers. He jumped from a London church tower, broke his neck and died. In Nero's time, Simon the Magician was killed when he attempted to fly in the Roman Forum.

There was a Saracen who, in 1100 B.C. in the presence of his emperor, Comnenus, attempted to fly around the Hippodrome of Constantinople. Legend says that he took off clad in a long white robe spread out to catch the breeze. He jumped from a high tower, but the weight of his body dragged him down and he did not survive.

Spain was the scene of an attempted flight about the year 875 when an Andalusian physician named **Abbas ibn-Firnas** ventured to fly. He covered himself with feathers, attached a couple of wings to his body, climbed onto a ledge and flung himself into the air. According to the testimony of several who had witnessed the performance, he flew to a considerable distance as if he had been a bird. However, when the ambitious physician attempted to land like a bird, he crashed heavily to the ground and severely injured his back. With more sympathy than understanding, the account explained that he had crashed because, "not knowing that birds when they alight come down upon their tails, he forgot to provide himself with one."

The lack of a tail was blamed also for the near undoing of an eleventh century English monk named **Eilmer** who was one of the first tower jumpers. In 1065 he fitted himself with wings and leaped from a tower at Malmesbury Abbey. A bewildered medieval historian reported Eilmer "had by some means fastened wings to his hands and feet so that, mistaking fable for truth, he might fly like Daedalus, and collecting the breeze on the summit of a tower, he flew for the distance of several hundred feet." Historians generally agree that Eilmer did achieve some sort of uncontrolled glide. If he managed to cover an eighth of a mile he could credit luck more than skill. Like Abbas ibn-Firnas, he had a rough landing. He fell hard to the ground, broke his legs and was crippled for life. He said that "the cause of his failure was his forgetting to put a tail on the back part."

Roger Bacon, the celebrated English philosopher and scientist, who lived from 1214 to 1292 was probably the first person to suggest an apparatus might propel a man through the air. He also was the first to commit to paper any scientific speculation about flight. Bacon, a Franciscan monk, was a prolific writer with a deep interest in natural science, alchemy, and mathematics. Around the year 1250 he wrote a book titled *Of the Marvelous Powers of Art and Nature* in which he sought to demonstrate the superiority of reason over the magical powers claimed by the alchemists of his day. Mankind, he said, was fully capable of building "instruments to fly that would be propelled through the air by flapping, bird-like wings." Furthermore, he said: "It is not necessarily impossible for humans to fly, but it so happens that God did not give them the knowledge of how to do it. It follows, therefore, that anyone who claims that he can fly must have sought the aid of the devil. To attempt to fly is therefore sinful."

John Damian de Falcuis was an Italian scientist and court physician for King James IV in Scotland. He promised the king that he could create gold from base metals. When unable to do so he then said that he was able to fly. In September 1507, he had constructed a set of ten-foot wings. He jumped from the castle wall breaking a leg in doing so. He said that his failure was because he had used chicken feathers and chickens do not fly.

As late as 1801, a seventy-two year old French general, **Resnier de Goue**, jumped from a castle rampart while flapping wildly. He landed in the river Charente below and survived with no injury.

Leonardo da Vinci

We next come to the beginning of the sixteenth century and that incredible genius, Leonardo da Vinci, the prophet of flight and a man of the Renaissance. Better known as the painter of such masterpieces as *The Last Supper* and the *Mona Lisa*, Leonardo da Vinci was also known for his insatiable thirst for knowledge. The idea of flight captured his imagination and he spent many hours devoting his fertile mind to the problem of man joining birds in flight. Leonardo, for all his genius, failed to achieve a design that could get a man off the ground, but the brilliance of his fifteenth century sketches pointed the way to eventual development of the parachute and the helicopter, if not the airplane itself.

Born on April 15, 1452, near the Tuscan town of Vinci, Leonardo was the illegitimate son of notary Ser Piero da Vinci and a young woman named Caterina. He was raised by his father and at age sixteen he began his apprenticeship with Master Verrochio of Florence, a master goldsmith and sculptor and a typical Renaissance seeker of truth. Encouraged by Verrochio, Leonardo mastered not only the arts, including music, but all the known physical sciences as well, including mathematics, biology, astronomy, and physics. By the time he reached the age of twenty-one, he had the skilled, expressive hands of an artist and the mind of a scientist-philosopher. Whatever interested him, he pinned down on his drawing board in the form of a sketch or detailed drawing. His painting brought him early fame, since art was a premier commodity in Renaissance Europe. The exploration of the physical world remained his passionate interest and his boyhood fascination with the flight of birds never left him.

As a biologist who had dissected many birds as well as human remains, Leonardo recognized that proportionally man does not have as much muscular power available for wingpower as do birds. To him the solution was simple to supply an adequate mechanical linkage and leverage for a human flying machine. With painstaking care and precise draftsmanship, he calculated the center of gravity of various birds, and he worked out weight/lift ratios and the wing area necessary to support the average man, as well as the leg and arm power required to move the wings in a bird-like manner.

His early designs, beginning in 1486, looked like a surfboard to which the prone aviator was fastened by means of an iron girdle. Wings were operated (a downward and rearward beat) by means of stirrups, with the right foot lowering the wings and the left foot raising them. Later he introduced hand cranks to help the wing elevation and added a rudder attached to a hoop around the pilot's head (a device later used until the time of the Wright brothers first powered flight).

A designer, not a builder, Leonardo cautioned any ambitious craftsman "this machine should be tried over a lake, and you should carry along a wineskin, so that in case you fall you will not be drowned." Thus was proposed the first form of survival gear for pilots, and a very practical idea, considering the weight of his flying machine.

Leonardo later designed a boat-shaped ornithopter, using oars to move the wings; a pedal-powered machine; and various other combinations of hand and leg power with spring devices to assist in the recovery stroke. He also became interested in the helical wheel as a form of airscrew and in 1486 to 1490. He sketched model designs for a vertical lift machine, which appears to be the earliest suggestion for a helicopter. Incidentally, he used the same principle of the helix to suggest a means of harnessing the movement of heated air as a source of power to continuously turn a roasting spit; actually a very practical design.

There is no evidence, however, that any of Leonardo's aeronautical designs were ever built, let alone put to the test.

When he died in 1519 he left his collection of over 5,000 pages of manuscripts and drawings to a friend who never made them public. It was not until the late 1800's that his work received serious notice. By then many of his ideas had

FIGURE 1-1 Leonardo da Vinci's Sketches

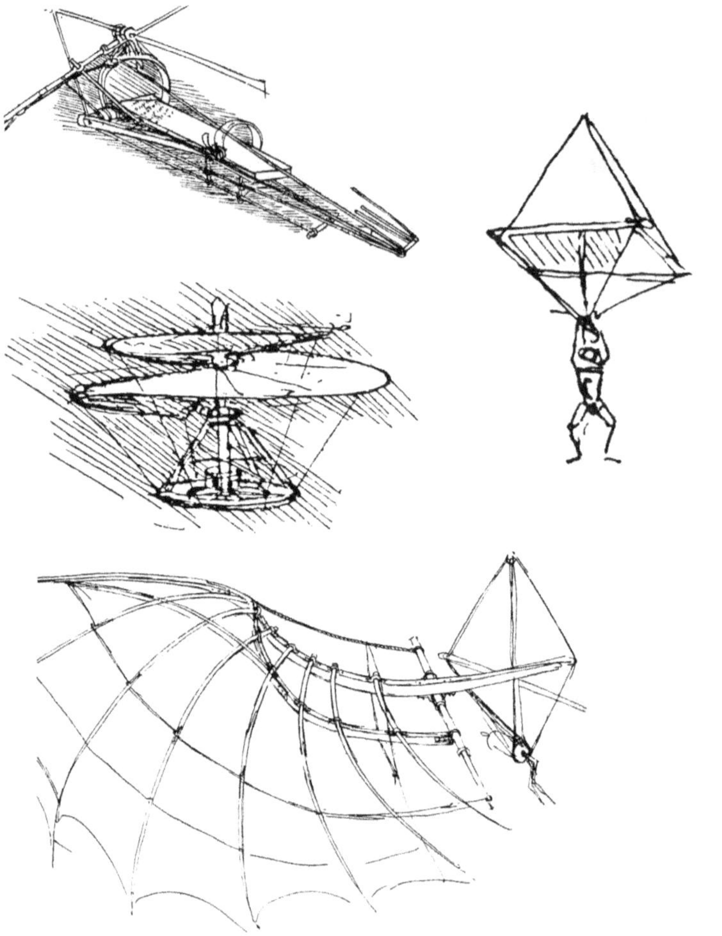

Left top: Self-portrait of da Vinci made in 1514. Right, top to bottom: An ornithopter flying machine to be flown in the prone position was designed in 1490. The first parachute design. da Vinci's Aerial Screw, a spring driven helicopter, designed in the 1480s, is considered the ancestor of the helicopter. Leonardo da Vinci's wing design.

been discovered by others. If his work had been known earlier the progress of flight may well have advanced by centuries. (See Figure 1-1.) We can mark da Vinci's work as the turning point in the people's efforts to become airborne. It was da Vinci, a true scientist, mathematician, and scholar, who applied known scientific knowledge to flight theory. From his notebooks we know that he spent years trying to find the solution to the problem of human flight, and that he knew that someday it would come about. (See Figure 1-2.) Da Vinci's invention of the propeller with which he equipped his small helicopter models was a contribution to the ages. Hardly less important was his invention of the principle of the parachute. It was a tent made of linen that would enable a person to throw him or herself down from a great height without sustaining any injury.

Da Vinci is remembered as a genius of the Renaissance who studied the anatomy of birds and their movement. He wrote the first important paper on the mechanism of flight that was the beginning of scientific research. In his treatise of 1505 he said:

> There is in man the ability to sustain himself in the air by the flapping of wings. A bird is an instrument working according to mathematical law, which instrument is within the capacity of man to reproduce in all its movements.

Da Vinci constructed a flying machine based on the wing structure of the bat. One of his servants supposedly tried to fly it and crashed after a short flight. History will attest to the fact that the bird was indeed man's teacher in the difficult art of flight.

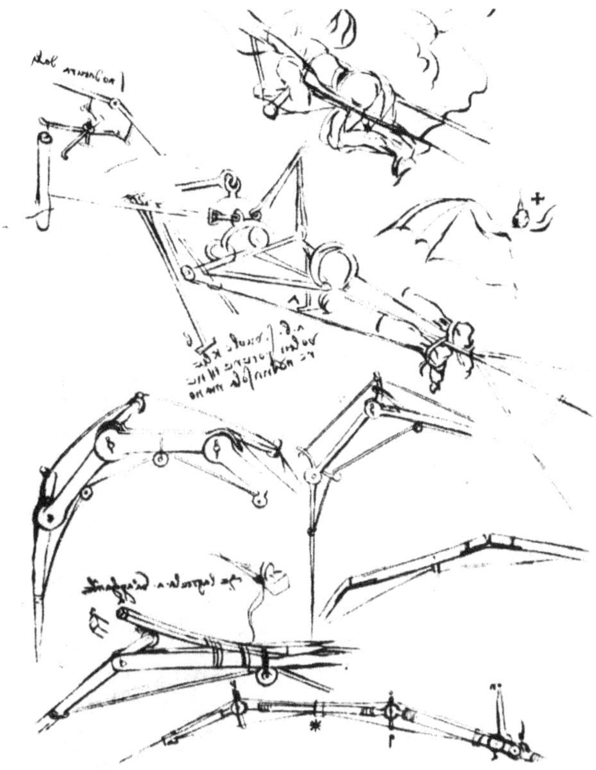

FIGURE 1-2 Leonardo da Vinci's sketches of muscle-powered flying machines.

After da Vinci's work there was little aeronautical progress for many years. In the early seventeenth century, **Gaspar Schott**, a physicist, observed that empty egg shells were extremely light and also that dew rose from the grass in the early morning. He reasoned that the way to fly was to get a sufficient number of egg shells filled with dew and be lifted into the sky.

De Lana in 1670 designed a vacuum balloon airship, which could not fly because he did not know about atmospheric pressure on the globes which he proposed to hold it aloft.

In the middle of the seventeenth century, the Italian scientist **Giovanni Borelli** said that it is impossible that man should be able to fly aircraft by using his own strength. He determined that the key to flight was to learn how birds flew, but not try to imitate them. Borelli's theory was disproved August 23, 1977, when a 70-pound aircraft with a 96-foot wingspread flew by man power only. The feat won for its builder, Paul MacCready, $86,000 awarded by the British Aeronautical Society.

A humorous thought of an unknown writer was that a goose is able to fly because of its feathers and that, if enough goose feathers could be used to cover a man, then he also could fly.

In Portugal in the eighteenth century, **Friar Bartholomew de Gusman** invented an airship which was to be lifted by magnets. Bellows were to provide a draft in the event the wind did not blow the craft hard enough.

The Frenchman, **Besnier**, constructed what some regard as the first glider. It was an ornithopter in which he jumped from a garret window and flopped to the ground below; he immediately decided that he was through flying.

In 1742 the **Marquis de Bacqueville** attached paddle-shaped wings to his arms and legs. He sought to fly across the river Seine, but instead dropped into a washerwoman's boat in the river and broke both his legs.

Development of the Lighter-than-Air Ship

The first successful experiment in air transportation was accomplished with the efforts of two French brothers, **Joseph and Etienne Montgolfier**. They lived near Paris and were papermakers. One evening Joseph Montgolfier threw a piece of lightweight paper into the fireplace and saw it go up the chimney. He thought this occurred because the fire produced a lighter-than-air gas, and reasoned that it could be used as lifting power.

The brothers immediately experimented using paper and then a silk bag filled with smoke from fire. The bag rose to the roof. This aroused their interest and spurred them on to making a bigger paper bag which developed into the first balloon. On June 4, 1783, a balloon made its first official public appearance. It was a 31,784 cubic foot envelope made of cotton sewn on paper coated with alum. It held a gondola with burning straw and wool. The balloon ascended to 6,600 feet for ten minutes. It burned when it landed in a nearby field. There were sufficient witnesses to this flight to register the Montgolfiers' discovery with the French Science Academy. Benjamin Franklin saw this along with several thousand others and said, "This is a discovery of great importance which may possibly give a new turn to human affairs."

In an experiment on September 19, 1783, they sent aloft a sheep, a rooster, and a duck which became the first aerial passengers in a man-made machine.

The Montgolfiers' experiments attracted the head of Louis XVI's natural history collection, **Jean de Rozier**, a noted French scientist. He helped the Montgolfiers design and build a balloon large enough to carry a man. On November 21, 1783, de Rozier became the world's first successful test pilot, as well as the first human to accomplish flight in a man-made device, when he ascended to an altitude over 1,000 feet and stayed aloft 23 minutes. A cousin of Louis XVI, Marquis de'Arlandes, accompanied de Rozier on this flight. (See Figure 1-3.)

FIGURE 1-3 Balloon which carried de Rozier and the Marquis de' Arlandes over Paris on November 21, 1783.

In June 1785, de Rozier fell to his death when his balloon caught fire in an attempt at crossing the English Channel. In England in 1784, Mrs. Tible became the first woman to fly a balloon.

On December 1, 1783, a coating of dissolved rubber was added by Professor **Jacques Charles** to the silk fabric of the balloon, to make an airtight covering. Charles' balloon was filled with hydrogen, a very flammable gas that is seven times lighter than air. Charles, a chemist, made an ascent from Paris which demonstrated the advantage of hydrogen over the heated air. This flight lasted one hour and forty-five minutes, and the distance traveled was 27 miles.

The first known balloon ascent in the United States took place at Baltimore on June 24, 1784. **Peter Carnes** built a Montgolfier-type balloon. Before launching, a thirteen-year-old boy, Edward Warren, who was standing in the crowd asked to make the trip. The boy became the first aerial passenger in the United States.

In England on January 7, 1785, **Jean Pierre Blanchard**, a Frenchman, and **John Jeffries**, a Boston physician who financed the project, made the first balloon flight across the English Channel from Dover to Calais. On January 9, 1793, manned ballooning was only nine years old when Blanchard, who had made 44 flights in Europe, arrived in Philadelphia. President George Washington gave him a letter of passage in case he met difficulty after landing; thousands watched as Blanchard and a companion, a little black dog, ascended in a hydrogen balloon from the Walnut Street prison yard. They soared to 5,812 feet and floated 15 miles in 46 minutes before putting down gently near Woodbury, in what is now Deptford, New Jersey. The temperature was

55°F—"most delightful and quite extraordinary for this season of the year," he wrote.

Blanchard died in 1809 from a heart attack. His second wife, Marie-Madeleine Sophie Armant, was thirty years younger than he. She carried on his work and became well known throughout Europe. She had made her first flight in 1805. She became the first woman to die in a balloon accident in 1819. Her balloon caught fire when it hit rooftops and she fell to the street below.

The earliest use of balloons for military purposes dates back to the French Revolution. On June 26, 1794, two soldiers in a balloon directed artillery fire by using signal flags. Later, Napoleon Bonaparte proposed to send armies via balloons across the English Channel to attack England in 1804. During the U.S. Civil War Abraham Lincoln ordered the use of balloons to observe and map the Confederate positions. Balloons were also used during the Franco-Prussian War in 1870, when letters were sent out of the besieged city of Paris by balloon. Felix Nadar, who organized this service, became involved in the first air battle in history when he encountered a German balloon at 10,000 feet over Paris and had to shoot his way out of the situation. The French balloons were made in the Paris railroad station.

The Dirigible

In order to gain control, balloons became elongated with vertical and horizontal planes at the end for directional control. Engines were soon developed for driving power.

The dirigible was the logical outgrowth of the balloon. In 1852 French engineer **Henri Giffard** constructed the first practical steerable balloon of the elongated design. It was 144 feet long and 40 feet in diameter.

Captain Thomas Baldwin made the first parachute jump in America from a balloon in 1887 at San Francisco. He built the first dirigible for the United States Army, as well as the first dirigible balloon in the United States.

Alberto Santos-Dumont, a Brazilian, was among the early builders and pilots of controlled airships. He flew his airships in Paris beginning in September, 1898, and later flew around the Eiffel Tower. He built and flew a total of 14 airships. Santos-Dumont said, "Trying to propel a dirigible balloon through the air is like pushing a candle through a brick wall."

History may prove that the balloon was of no real assistance to the development of aviation. In fact, it may even be considered an obstacle to the progress of the airplane, because the balloon's development distracted the direction of research, which could have been devoted to the study of the airplane.

Probably the greatest contributor to lighter-than-air craft was **Ferdinand von Zeppelin**. On July 2, 1900, the first rigid airship to fly made history in its 17-minute flight over Lake Constance near Friedrichshafen, Germany. It was the 420 foot LZ-1 designed by Count Ferdinand von Zeppelin. The LZ-1 meant "Luftship Zeppelin One."

Built rigidly with an aluminum framework, inflated with hydrogen gas and propelled by two 16-horsepower engines, the cruising speed of the dirigible was 20 miles per hour.

Count Zeppelin, a retired general of the Prussian Army, made two other test flights in the LZ-1. Although the second flight lasted a full eight minutes, journalists and critics belittled his efforts. Among these critics was a young yachtsman and economist, **Dr. Hugo Eckener**.

Undaunted by this criticism Count Zeppelin designed, experimented, and built three other models. It was the LZ-4 that brought him national acclaim. Successful flights of this lighter-than-airship lasting twelve and twenty-four hours converted Dr. Eckener from a skeptic to an admirer.

When the LZ-4 crashed and burned at Echterdingen, the people of Germany arose to the occasion. They subscribed six million marks so that Count Zeppelin could continue developing airships.

In 1908 Count Zeppelin celebrated his seventieth birthday. German cities named streets in his honor. Prince Henry and the Crown Prince clamored for rides in his airship. The Kaiser, in awarding the decoration of the Order of the Eagle, honored Count Zeppelin as the greatest German of the twentieth century.

Dr. Eckener joined Count Zeppelin in 1909 in forming DERLAG, the world's first passenger airline. They built aerodome facilities at Berlin, Dresden, Frankfurt, and Hamburg; they produced such airships as Deutschland, Hansa, Sachsen, Schwaben, and Viktoria-Luise. By 1913 they boasted of a perfect safety record in over 1,600 flights carrying more than 35,000 passengers.

The German navy under Admiral Alfred von Tirpitz ordered its first airship, the L-1 from Count Zeppelin in 1912. The airship was principally used for training purposes, but it was also teamed up with the fleet on scouting missions. A larger model, the L-2, was produced in 1913. Although both ships were casualties, the German Admiralty ordered another ship and placed Peter Strasser in command of the Naval Airship Service.

Zeppelin designed and built airships for the German war machine in World War I. They were used to fly over England and drop bombs from high altitudes. Not until the Allies were able to destroy the Zeppelin manufacturing works was this threat ended.

When World War I began in August 1914, the German Naval Airship Service had only one Zeppelin, the L-3, and no base facilities. These conditions were soon remedied. Construction started on a main base at Nordholz and on two others, one near the Dutch border at Hage and another near the Danish frontier at Tondern. The war stimulated development and 61 airships saw action during the war.

During the first two years of the war and before England had perfected its defenses, these raiders wrought havoc. The first air raid began the evening of January 19, 1915, when Zeppelins L-3 and L-4 flew over Yarmouth and King's Lynn. Zeppelin raids continued for the next two years doing only sporadic damage. Yet they ultimately forced the Home Defense to tie down 17,341 anti-aircraft troops and 12 R.F.C. squadrons with 110 planes, which were all desperately needed on the Somme.

As the Zeppelins proved vulnerable to bad weather, anti-aircraft fire, and fighter planes, the German High Command turned to long-range bombers. The most damaging raid by Zeppelins occurred on September 8, 1915, when two tons of bombs dropped on London.

Of 125 Zeppelins built by Germany between 1914 and 1918, only six survived the war.

After World War I other nations of the world became interested in dirigibles, no doubt because of impressive use Germany made of them during the war. Great Britian built the R-33 and the R-34. Honors for the first transatlantic crossing went to the R-34 which completed its round trip on July 13, 1919. It took 75 hours for this first crossing. Later both the R-33 and the R-34 were severely damaged during rough weather.

In the early 1920s both England and France built and experimented with dirigibles. Both countries soon became discouraged due to failures. France ceased in 1923 after the Dixmunde disappeared over the Mediterranean and no trace of it was ever found. England continued until 1930 when their R-101 crashed and killed 47 persons.

In 1923 the United States Navy began operating one rigid airship, the *Shenandoah.* It crashed in Ohio in 1925 after breaking into three parts killing fourteen crew members. The *Shenandoah* was built from Germany's World War I design.

Another dirigible arrived at the Lakehurst, New Jersey, Naval Air Station on October 15, 1924. It was the 660-foot ZR-III, the 126th airship built by the Zeppelin Company, and was obtained by the United States as part of Germany's war reparation. It was delivered by Dr. Eckener after a flight of 81 hours from Germany. This airship was later christened *Los Angeles* and used by the United States Navy.

Of 125 Zeppelins built by Germany between 1914 and 1918, only six survived the war.

Italy also entered the dirigible field. Umberto Nobile flew the Amundsen-Ellsworth Expedition over the North Pole in the *Norge* on May 12, 1926.

Airship LZ-127, the Graf Zeppelin, crossed over the North Pole on one of its record-making flights. Previously, on the first leg of its round-the-world cruise in August, 1929, the Graf Zeppelin set an unofficial endurance mark of 6,980 miles indirectly from Germany to New Jersey. Eckener was in command, and the Graf Zeppelin carried 25 passengers for the cost of $2,200 each. It also flew from Germany to Rio de Janeiro, Brazil. The Graf Zeppelin crossed the ocean 144 times and flew more than one million miles with no accidents.

The Goodyear Company became the United States builder of airships. They built the *Akron* for the United States Navy in 1932, and the *Macon* in 1933. The *Akron* was designed to be a flying aircraft carrier; it had an internal hangar bay that carried five Sparrow Hawk biplanes. Both of these airships crashed in storms within a year or two after being built. It was believed that the large rigid type of airship was not successful because its structure was inflexible in rough, unstable air. This ended the United States attempts with rigid airships. The *Akron* crashed off the New Jersey coast on April 3, 1933, when a storm forced its tail into the water. Of the 76 men aboard, only three survived.

The United States Navy dirigible airship *Macon* was 785 feet long and 144 feet wide. It contained 7.5 million cubic feet of helium that provided a gross lift of 410,000 pounds. It was commissioned in April, 1933 and built for the Navy by the Goodyear Corporation. It was powered by eight 12 cylinder 560 horsepower engines capable of producing a top speed of 85 miles per hour; cruise speed was fifty-five. The *Macon* was a flying aircraft carrier holding five Curtiss F9C-2 Sparrow Hawk fighters that could be launched and recovered on board while the airship was in flight.

The *Macon* was lost during bad weather on the evening of February 12, 1935, while returning from maneuvers to its base at Sunnyside Naval Air Station, now Moffett Field. It had been cruising at 1,250 feet when it was hit by a gust of wind that snapped its vertical tail fin and punctured three helium cells in the aft section. The crew fought for thirty-three minutes to regain control but it settled onto the ocean about fifteen miles offshore. Eighty-one of those on board were saved; the two who died could have been saved also had they not panicked and jumped overboard before the airship settled into the water. Commander Wiley, a survivor of the *Akron* disaster, guided the airship to the surface of the ocean. The *Macon* had logged 54 flights and 1,798 hours of flight time before it was lost.

Wreckage of the *Macon* was found June 24, 1990, under 1,500 feet of water in the Pacific Ocean near Monterey, California. Parts of the aircraft may be salvageable.

The English experimented with launching fighter aircraft from the R-33 and R-34 dirigibles while in flight. The United States also was successful in aerial launchings of aircraft from the *Akron* and *Macon*. This idea was known as "sky-hook." The purpose was for the airplanes to scout ahead and return to the airship. They were also to serve as defenders to ward off attacking hostile aircraft.

The United States thereafter used lighter-than-airships only of the smaller blimp-type to be used for observation by the Navy.

The Germans made one final supreme effort with the dirigible. This was in the middle 1930s when the world's largest and greatest airship, the *Hindenburg*, was built.

The United States had the monopoly on helium, the only practical nonflammable gas. They refused to provide it to Germany because of fear that the new Nazi government would use it for military use. The *Hindenburg* flew successfully carrying passengers across the North Atlantic to Lakehurst, New Jersey. It was unique in that it provided staterooms, restaurant facilities, smoking lounges, and other extreme comforts. It

The United States had the monopoly on helium, the only practical non flammable gas.

was the largest airship ever built, being over 800 feet long and 135 feet in diameter. Unfortunately, this giant of the skies exploded and burned while attempting a landing at the Lakehurst, New Jersey, Naval Air Station on the evening of May 6, 1937. Of the 97 people aboard when it crashed, 62 miraculously escaped death. Films of the *Hindenburg* crash were not shown in Germany until after World War II.

Eckener built a sister ship of the *Hindenburg*. It was completed in Germany and made forty flights in 1938–1939, however, it was later melted down to make fighter aircraft needed for World War II. However, he would not put it into operation unless he could secure helium. Hitler and the Nazi government did not think the airship had a place in their war machine, so Eckener and his new dirigible lay useless during World War II. Finally the airship was melted down for the use of its metal.

During World War II the United States Navy used nonrigid airships in convoy operations. Some 170 of them escorted 89,000 ships without loss of a single vessel. It has been estimated that they logged 500,000 hours' flying time and in 1944 were patrolling three million square miles of ocean.

After World War II nonrigid airships were used as units of the Early Warning Squadron. They were the ZPG-2-type built for the United States Navy by Goodyear. One of them, the *Snow Bird*, broke the Graf Zeppelin endurance record when it landed at Key West, Florida, after logging 9,070 miles. Commercial blimps are seen today over sporting events carrying television cameras and at night displaying lighted advertisements.

The Future of Zeppelins

A serious approach to the redevelopment and use of the Zeppelin has recently taken place. The company in Germany named Le Bourget Deutsche Zeppelinreederei, built and became the first operator for their designated Zeppelin NT. The company plans to initially conduct revenue flights using the airships for sightseeing tours over Lake Constance where Ferdinand von Zeppelin built and flew his first airship in 1900. The first prototype of the Zeppelin NT (NT for new technology) took part in daily flights at the Paris Air Show in June 2001.

Zeppelinreederei was established by Zeppelin Luftschifftechnik, which developed and built what is now the largest Zeppelin and the only one with rigid internal structure. The manufacturer chose to operate the first three airships itself. Luftschifftechnik president said he wants to prove the three airships operational feasibility and confirm calculated data. He said, "We know we have the best performing Zeppelin but we want to demonstrate that in reality."

Public interest in Germany is overwhelming. Initially the company received more than 2,000 reservations for flights in the Zeppelin NT that seats up to twelve people. A one-hour flight over Lake Constance cost about $300 per seat.

Potential buyers have been placed on a waiting list. The company will decide whether to focus more on building and selling the Zeppelin, or expanding its role as a commercial operator. A final decision depends on the market and how successful operations will result with the first three airships. The company's market studies forecast a demand for 100 to 1000 airships for use in advertising. There is uncertainty as to use of the airship for passenger transportation.

Commercial blimps are seen today over sporting events carrying television cameras and at night displaying lighted advertisements.

Luftschifftechnik plans to reach a break-even point at 20 to 25 air ships. Its present facility has the capacity to build two and one-half airships per year with an assembly time of eight-and-a-half months per Zeppelin. If successful the company plans to construct a larger version of the Zeppelin NT. The company sees a demand for a thirty-seat airship having a payload of five to six tons. Their present Zeppelin NT has a two-ton payload limitation with a range of 559 miles. It is 246 feet long and 64 feet in diameter and reaches a speed of 125 KPH.

Airships have served a purpose in the effort to fly and they may be used successfully in the future. The idea still persists.

CHAPTER 2

Development of the Airplane

During the years of lighter-than-air ship development, other inventors were experimenting with heavier-than-air craft known as the airplane.

If the studies of da Vinci had not been ignored for so long, then the dream of flight might have been realized centuries ago. From his death in 1519, until the first half of the 1800s, aeronautical progress was attempted by dedicated experimenters. In 1670 the Jesuit priest, **Francesco de Lana**, made an approximate determination of the weight of air at sea level, but he never succeeded in raising his craft using only four light copper spheres to create a vacuum. The Brazilian Jesuit **Laurenco de Gusman** firmly believed in his Passarola but he never succeeded in building anything except a fragile glider. Although Bacqueville, Letur, and De Groof were personally able to test their ornithopters, dozens of others never went beyond the drafting of plans for their fantastic projects. When people finally understood that muscle power alone was not enough to lift a man off the ground and sustain him and the machine in flight, no matter how light it was, they designed plans for the use of the steam engine. Neither the steam engine nor the gunpowder motor, planned by **Pomès** and **De la Pauze**, could offer the correct solution for heavier-than-air flight because the weight of these engines was too great in proportion to the power created.

Humankind owes much to the scientists and dreamers of the last four centuries for having continued to believe, in spite of their failures, that someday people would be able to fly.

Humankind owes much to the scientists and dreamers of the last four centuries for having continued to believe, in spite of their failures, that someday people would be able to fly.

No major effort was made in the development of airplanes from 1505, when da Vinci wrote his scientific paper on aircraft, until 1810. At that time an English scientist, **Sir George Cayley**, (See Figure 2-1) published a series of articles proposing that a surface could be made to

FIGURE 2-1 Sir George Cayley.

support weight. Cayley is considered the father of British aviation. This was the first scientific thought of what we know today as the airfoil. All successful airplanes have been constructed according to Cayley's principles on the surface of a wing.

George Cayley

In 1842, Cayley designed the Aerial Carriage, (See Figure 2-2) which had rotors, propellers, and a canvas covered fuselage to hold its steam engine. The Smithsonian Institution built a model of this aircraft according to Cayley's plans and found it workable.

Cayley is remembered for writing a collection of "how and how not" into a sensible mono-graph concerning flying. This approach to flight was based on the principle of mechanics. Not content with telling how not to attempt flight, he pointed the way toward eventual success. Cayley even foresaw the future development of the engine. He called his engine "an explosion machine," for want of a better name. Cayley was about 55 years ahead of his time, because it was not until 1865 that Lenoir completed the first practical gasoline engine.

Sir George Cayley theorized on the problems of aerodynamics and addressed velocity, wing loads, bending movements, and the necessity for combining lightness and strength. He foresaw and wrote about the need for streamlining, and he realized that a cambered or curved wing was necessary for flight, and that a flat wing could never do the job. Cayley was the first to think about the construction of a multiwing airplane.

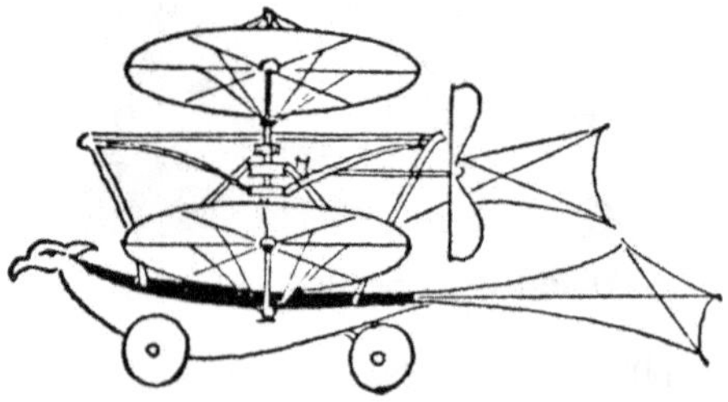

FIGURE 2-2 Cayley's Aerial Carriage.

Cayley said that wings to lift and sustain great weights in the air ought not to be made in one plane but in parallel planes one above the other at a convenient distance. Such an arrangement would reduce the likelihood of damage to a large single wing without sacrificing the required wing surface. He went on to propose wings in the form of a three-decker, each deck being eight or ten feet from the other to give free room for the passage of the air between them. In time, this multiwing idea suggested by Cayley would be adopted by many other aviation pioneers, Wilbur and Orville Wright among them.

His idea for a biplane was used in building a glider which he fitted with horizontal and vertical rudders; these were the beginning of the controls of today's airplane. History does not show that Cayley himself ever attempted a flight, but he wrote of his glider's success.

William Henson and John Stringfellow

The idea of the airplane was given further thought by two other Englishmen. **John Stringfellow** was born in 1799 and became very prosperous. His greatest challenge was to build a lightweight steam engine to power an aircraft. **William Samuel Henson** born in 1812, was an engineer in England's lace-making industry before he embarked on designing aircraft. He also had worked on developing a breech loading cannon, an ice making machine, and had patents for a safety razor and an improved steam engine.

Henson did not follow Cayley's advice for a multiwing but designed a single-wing airplane for his proposed aerial steam carriage. Lacking funds to build a prototype, Henson and Stringfellow proceeded with a grand scheme for launching an international airline known as the Aerial Steam Transit Company in 1843. They sold stock valued at £2,000. The intended purpose of this company was to build and operate flying machines. They designed a giant of an airplane. It had a 150-foot wingspan and two sixblade propellers 20 feet across; it was powered by a 30-horsepower steam engine. They attempted to raise funds with massive publicity. Descriptions and illustrations of the proposed craft appeared in such popular journals as *The Illustrated London News* and *L'Illustration of Paris*. Fanciful scenes were reproduced not only in magazines but on souvenir items such as decorative cloths and handkerchiefs, showing the aerial carriage steaming over London, the English Channel, the exotic Plains of Hindustan, the coast of France, the Pyramids, and China.

Claims for the aircraft were so great and the promotion so lavish that interest turned to skepticism and then to ridicule. Potential backers withdrew their support in the face of jeers and laughter. The Aerial Transit Company soon ceased.

Not discouraged, Henson and Stringfellow set out to build and test a scale model of their flying machine, hoping for results impressive enough to attract funds for a full-size aircraft. The experiments seemed promising but support was not forthcoming, and in September, 1846, William Henson appealed to Cayley for help. He wrote, "We therefore apply to you as the 'father of aerial navigation' to ascertain whether you would like to have anything to do in the matter or not."

Cayley's reply was gracious and he must have relished being addressed as the "father of aerial navigation." Cayley said that, while he approved of Henson's zeal, he could spare no money to support his experiments. Of course he would be most gratified if Henson could show some proof of progress in achieving mechanical flight, in which case, Cayley said he "perhaps might be able to aid you in some manner by my experience in connection with other mechanical persons."

In 1847 Henson and Stringfellow conducted final tests with a miniature aircraft that had a 20-foot wingspan and two steam-powered propellers. The model rolled down its inclined launching ramp and executed a brief descending glide, but that was all. It failed to achieve sustained flight.

The term "aviation" was originated by the Frenchman, *La Landelle*, about 1863 when he helped found the French "Society of Aviation."

Because both the model of their airplane and the stock sale were hopeless failures, disheartened Henson gave up the whole business, got married, and emigrated to the United States. Stringfellow continued his work on the problems of flight. He built a steam engine and a new model airplane with a wingspan of ten feet. Some historians call this engine the first practical model capable of successful performance. The glider on which the engine was mounted actually became airborne in a test flight in 1848. It flew a distance of 120 feet and proved to be the first successful power-driven airplane to be flown in free flight; however, no one was aboard the airplane. He was never able to construct a full-size airplane that could be successfully flown. Stringfellow tested an improved Henson model in 1848, but it was no more successful than its predecessor.

The contribution of the two men was more significant than their failures suggest. Henson's design was logically conceived, if only on paper, and his aerial carriage was a plausible expression of what an airplane should be. The showy publicity brought ridicule to the Aerial Steam Transit Company, but it also called worldwide attention to the pursuit of mechanical flight and fixed the vision of an elegant flying machine upon the public mind. Though the tests with the models were unsatisfactory, they were, nonetheless, a serious attempt to fly a propeller-driven, engine-powered airplane. The concept of the Aerial Steam Carriage was a lasting stimulus to other inquiring minds.

George Cayley, whose enthusiasm for aviation did not diminish by his advancing age, was influenced by the machine. Following his own advice to Henson, Cayley made a major departure from his earlier work and began experimenting with a three-winged glider, which he optimistically dubbed a "flyer." As Cayley described it in his note-books, the aircraft consisted of a boat-shaped wheeled car with a pilot-operated elevator and rudder; three tiers of wings were affixed atop the car at a dihedral, or shallow V-shaped angle.

By the spring of 1849 this triplane glider had been put through some brief tests, one of them similar to the flight of a kite. Cayley wrote:

> A boy of about ten years of age was floated off the ground for several yards on descending a hill, and also for about the same space by some persons pulling the apparatus against a very slight breeze by a rope.

Then came a 1853 ride of a reluctant coachman on board a glider Cayley called the "new flyer." It sailed across a valley in true non-powered flight. It is reasonably certain that the vehicle used was an improved version of the boy-carrying glider. Cayley then spent his time with his very ill wife. After she died he went back to his work. His strength was failing and he died on December 15, 1857, twelve days after his 84th birthday. His only surviving son had the historic flying machine dragged to a barn where it served as a roost for chickens.

Jean-Marie Le Bris

Jean-Marie Le Bris, a French sea captain, who conducted aerial experiments in Brittany from 1856 to 1868, was a man not given to understatement. Le Bris did practically nothing to advance aviation, but the mixture of fact and fantasy regarding his experiments have become firmly fixed in the history of aeronautics. Le Bris gained knowledge during his voyages around Cape Horn and the Cape of Good Hope. He watched the graceful flight of large albatrosses and to him this was an insight into the secret of flying. He captured and killed a bird to study the shape and structure of its wings by holding it aloft in the wind. Obviously the air blowing over the wings of the bird caused it to rise slightly due to its shape. Captain Le Bris exclaimed:

> I took the wing of the albatross and exposed it to the breeze, and lo in spite of me it drew forward into the wind; notwithstanding my resistance, it tended to rise. Thus I had discovered the secret of the bird! I comprehended the whole mystery of flight.

Abandoning long sea voyages for short runs along the coast of France, Le Bris devoted his time to the construction of a full-sized glider inspired by the albatross. The body of the craft, built to support a pilot, was shaped like a canoe; the narrow arching wings, each 23 feet long, were moved by pulleys and cords, provided a lifting surface of 215 square feet. In 1857 Le Bris tested his device by having it pulled down an incline on a horse-drawn cart and then releasing it against a brisk wind. He succeeded in making a short glide on his first try, but a crash on the second attempt left him with a smashed glider and a broken leg. By 1868 Le Bris had developed a larger version of his glider which made several successful unmanned test flights before it crashed. Nothing more was reported of his aerial activities.

Francis Wenham

In 1866, at the first meeting of the Aeronautical Society of Great Britain, an address was given by **Francis Herbert Wenham**, a marine engineer interested in internal-combustion engines. Entitled "Aerial Locomotion," his report was to become an aviation classic like Cayley's articles of 1810. It was based upon his years of carefully studying the flight of birds. He is remembered for having contributed a great deal to the understanding of technical problems which finally led to successful flight.

Wenham had conducted extensive tests with airfoils and airplane models. He advised his fellow enthusiasts that a cambered, or curved, wing was more effective for lift than a flat wing and that the leading edge of the wing, or the front portion, provided most of the lift required for sustained flight. It was obvious, he said, that a long narrow wing would, therefore, exert more lift than a short stubby one. Wenham further concluded, as Cayley had earlier, that several such wings, one above another, would provide the greatest lifting area with the least strain. He also advised that prospective fliers should master the control of unpowered gliders in the air before attempting to leave the ground in powered aircraft.

Wenham had little practical success with his own gliders, which he built with five wings. He advanced aeronautics significantly by building the first wind tunnel for simulating the movement of variously shaped wing models through the air. His apparatus consisted of a long wooden box with a steam-driven fan at one end. And while Wenham himself conceded that his tests were somewhat crude and incomplete, later and more sophisticated versions of his equipment became essential tools in aviation research.

Wenham's work would prove useful to both Octave Chanute and the Wright brothers in the United States. It had a more immediate influence on John Stringfellow, the Englishman who had been absent from the aeronautical scene since the failure of the single-winged Aerial Steam Carriage model he had built with William Henson in the 1840s. Spurred by Wenham's report to the Aeronautical Society and recalling that Cayley also had recommended multiple wing surfaces for heavy aircraft, Stringfellow resurrected the old aerial carriage as a triplane. Stringfellow showed his new model at the world's first exhibition of flying machines held in the Crystal Palace in London in 1868, sponsored by the Aeronautical Society.

Alphonse Penaud

In France, the brilliant but tragic **Alphonse Penaud** was one of the most outstanding airplane experimenters preceding the Wright brothers. He dominated the field of practical aeronautics in the 1870s. Prevented by the disabling hip disease from following his admiral father into the Navy, Penaud applied his training as a marine engineer to aeronautical matters. He built many models and in 1870, when he was only twenty, he discovered the twisted rubber band as a power source for model airplanes. He used his invention in a workable helicopter model which spurred toy makers to produce thousands of rubber band driven helicopter toys. Then he used his creation in his own more sophisticated airplane models.

In 1871 Penaud built what he called the "planophore," a monoplane twenty inches long with tapered dihedral wings, an adjustable tail assembly with dihedral tips and a rear-mounted pusher propeller. The wing and tail design made it much more stable than earlier airplane models. When Penaud demonstrated his planophore before a group of aeronautical enthusiasts in Paris, it flew 131 feet in 11 seconds; the first flight of an inherently stable aircraft.

Penaud's next major achievement came in 1876, when he patented his design for a full-sized amphibious monoplane. It included such forward-looking features as a glass-domed two-seated cockpit, a single control column to operate the elevators and rudder, and a retractable undercarriage with shock absorbers. It weighed over a ton and he calculated that it should fly at a speed of sixty miles an hour providing he could find a powerful enough engine. Penaud was too far ahead of his time. Lack of a lightweight engine stymied his progress and he was unable to raise funds to build his machine or pursue his research. Frustrated, discouraged, criticized for his work, and in failing health he committed suicide at the age of thirty. Later experimenters including **Wilbur** and **Orville Wright** would rank Penaud close to Cayley as one of the most significant 19th century aeronautical thinkers.

Penaud was one of a growing number of scientists and other researchers approaching aviation as an engineering problem that could be solved by good scientific methods.

H. F. Phillips

Another experimenter who contributed to the advancement of flying was **H. F. Phillips**. In 1884 he patented a series of curved wings for airplanes. In constructing these he copied the concave underside and the convex upper side of bird wings. He experimented wing models in one of the first wind tunnels.

Clement Ader

Not every experimenter was captivated by the idea of multiple wings, nor did all aeronautical researchers agree with Francis Wenham's conclusion that unpowered gliding would be a useful preparation for attempts at engine-driven flight. There was the Frenchman **Clement Ader** and the American-born Englishman Sir Hiram Maxim who also believed that successful flying could be achieved simply by attaching a sufficiently powerful engine to a pair of wings.

Clement Ader, born near Toulouse, France, in 1841 became one of the most controversial figures in aviation history. He spent 25 years and $120,000 in an effort to produce a successful airplane. Many of his early efforts were failures. Ader was a distinguished electrical engineer and inventor who was among the pioneers in the development of the telephone; he also studied the flight of birds and bats. In 1873 he built a bird-shaped aircraft with goose-feather wings. Ader did not try to fly this semi-glider; instead, he tethered his artificial bird to the ground, nosed it into the wind, and lay prone upon it as stiff breezes raised it to the limit of its anchor ropes.

His curiosity about gliding flight now satisfied, Ader began planning for a powered takeoff. By 1882 he had started the laborious construction of his first aircraft and the exceptionally light steam engine that would be its most remarkable feature. In 1886 he built a flying machine with bat-like wings powered by a 40 horsepower steam engine turning a four bladed propeller. Its wingspan was 46 feet and the entire structure weighed 440 pounds. The *Eole* crashed on its first test after going 150 feet.

Ader built another airplane in 1891 with a 54 foot wingspan. The *Eole*, or "god of the winds," was finally ready for testing. Resembling a huge bat, it was a monoplane with deeply arched wings that could be moved in four different directions. Its propeller was driven by a 20 horsepower steam engine that brought the craft to a total weight of only 653 pounds including the pilot. In several critical respects the design left much to be desired. The cockpit was located behind the steam boiler which provided almost no forward visibility, and Ader had given little thought to how he might handle the machine if he managed to get it off the ground. The mechanism for controlling the wings was complicated and difficult to operate. There was no elevator or rudder. Ader was brimming with confidence on October 9, when he transported the *Eole* to an estate at Armainvilliers for testing. Only a few assistants witnessed what happened after Ader wheeled his bat-like aircraft to level ground and slipped into the cockpit. By the inventor's own account "he ran up a full head of steam, sped along the ground and rose to height of about eight inches before touching down and rolling to a halt." The *Eole* had traveled through the air a distance of 165 feet. If Ader's account is true, and there is no reason to doubt it, his aircraft was the first manned airplane to take off from level ground under its own power. Aerodynamically unsound and too heavy for its engine, once airborne the *Eole* was uncontrollable and could not sustain flight. There is some doubt if it flew but it did cause the French government to advance money to Ader.

The French Ministry of War commissioned Ader to build an improved version of the *Eole,* which he named the *Avion III* with two engines. On October 14, 1897, with military observers watching on a gusty afternoon, he tried to get airborne. The aircraft was caught in a strong wind before its wheels had a chance to leave the ground. Ader shut off the engines but his airplane almost over-turned. Too damaged to do more testing and not having gotten the airplane off the ground, the War Ministry stopped their financial support.

Whether Ader had actually flown or not is beside the point. His efforts did spur others onward to further experimentation. He is remembered for successfully mating engines to airframes. His engines were strong enough to give sufficient power for flight.

Hiram Maxim

The enormous and highly publicized winged vehicle that **Hiram Maxim** built in the early 1890s was equally incapable of sustained and controlled flight. Maxim's primary concern was merely to build a flying machine that would lift itself from the ground. His experiments were marginally successful, though Maxim himself conceded that problems of aerial steering and control remained to be solved before practical aviation could be achieved. Maxim had been interested in aeronautics since his boyhood in Maine, where he was born in 1840.

Maxim's youthful interest in aviation resurfaced in the late 1880s when he began to study wing forms and propellers. After proving to his own satisfaction that heavier-than-air flight was possible, he turned his considerable talents to the development of a powerful but lightweight steam engine. In Kent he rented ample room to construct work sheds and a track for testing his proposed aircraft.

In June, 1892, Maxim described the kind of aircraft he had in mind. He said:

> It is neither necessary nor practical to imitate the bird too closely; because screw propellers have been found to be very efficient when operated at sufficient speed to produce thrust. Without doubt, the motor is the chief thing to be considered.

Maxim's flying machine reflected his preoccupation with powerful engines. It was an enormous biplane equipped with two 180 horsepower steam engines, each driving a pusher propeller nearly 18 feet in diameter. The platform for the engines, boiler, and a three-man crew was 40 feet long and eight feet wide; the craft measured about 200 feet from tip to tail and had a 107-foot wingspan. To provide stability, Maxim placed an elevator fore and aft and set the outer wing panels at a dihedral angle. The aircraft had a total lifting surface of 4,000 square feet and weighed an incredible 8,000 pounds, including crew.

Maxim built an ingenious two-level track for testing his machine. The lower track, built of heavy steel rails, was the runway, engaged by the four cast-iron wheels of the undercarriage. The upper track, made of 3 × 9-inch Georgia pine, was intended to keep the ponderous biplane from escaping its test track and lofting into uncontrollable flight. If the machine rose more than a few inches from its runway, a second set of smaller wheels would engage the wooden guardrails and hold the vehicle down.

Delayed by mishaps and breakdowns during test runs in 1892 and early 1893, the Maxim machine in motion was an awesome sight. A journalist, visiting Maxim in 1893, took a ride on the huge biplane. He wrote later about his experience saying:

> When full steam was up and the propellers spinning so fast that they seemed to become whirling disks, Maxim shouted, 'let go.' A rope was pulled and the machine shot forward like a railway train with the big propellers whirling, the steam hissing, and the waste pipes puffing and gurgling, it flew over the 1,800 feet of track in much less time than it takes to tell it.

The machine did not actually fly—then or ever. On its final test run on July 31, 1894, Maxim guided the aircraft at high boiler pressure along the track. After covering 600 feet and gaining a speed of 42 miles per hour, the machine lifted from its steel runway and rose until the secondary wheels reached the wooden guardrails. Suddenly, one of the rails snapped. The machine was liberated and briefly floated in the air. When a piece of broken guardrail smashed into a propeller Maxim immediately shut off the steam. His clumsy and uncontrollable flying machine settled to the ground, never to rise again. Maxim had demonstrated that a powerful engine could lift a heavy winged object from the ground. He was too busy with his other affairs to continue his aviation experiments.

Even though he contributed little to aeronautical progress, Maxim's reputation and his experiments helped convince the public that the age of heavier-than-air flight was nearly at hand. He said, "Even under the most unfavorable circumstances, aerial navigation will be an accomplished fact inside of ten years." How prophetic he was; the Wright brothers succeeded in 1903.

Otto Lilienthal

The man we call the father of glider experiments was a German engineer, **Otto Lilienthal**. As a young boy, Otto and his brother, Gustav, played in the meadows near their home where they chased storks they found there. Otto noticed that when the birds were frightened they would take off toward the boys even though this meant flying toward danger. Not until they were off the ground did the birds turn and fly away. Gustav later wrote that "rising against the wind might be easier than with the wind."

From those early days in 1860 until his death in 1896, Otto Lilienthal continued to search for the answer to the problem of flying. He and his brother, when in their teens, constructed model airplanes with 6 to 9 feet wingspreads. At first they attempted to fly in calm air, but remembering the storks, they were able to get their small gliders off the ground by turning them into the wind when it blew.

The Franco-Prussian War of 1870 interrupted the Lilienthals' work, but it was quickly resumed when the war ended. From 1871 to 1891 Otto concerned himself with the principles of aerodynamics. He derived many of his principles from the study of birds' flight. He recognized the superiority of a curved wing to the flat type. Otto recognized that nature utilized the advantage of the curved wing, not only in the bird kingdom, but also in the vegetable kingdom, by giving seeds of certain plants little curved wings which enable them to be picked up by the wind to fly along to a fertile spot.

Lilienthal believed that a cambered wing was necessary to obtain successful flight. In his workshop he built test equipment to measure the lift that would be produced by various wing shapes. His experiments demonstrated the superior lifting quality of the curved wing. In 1889 he published his findings in a book entitled *Birdflight as the Basis of Aviation* that became a classic of aeronautical literature. He described his studies of wing structure and the dynamics of bird flight. He included a description of how birds propel themselves by the twisting action of their outer primary feathers. He showed the lift provided by wings at various degrees of camber. Lilienthal's belief in gliding as preliminary to powered flight led him to experimenting with several unpowered craft that were fixed-wing monoplanes (See Figure 2-3). He later experimented with biplane designs. In a report published by the Smithsonian Institution in 1893, he stated that he approached flying one step at a time. In 1894, he constructed an artificial hill not far from his home. Running downhill from the summit and leaping into the wind, he repeatedly achieved glides of 150 feet and more.

The man we call the father of glider experiments was a German engineer, *Otto Lilienthal.*

Lilienthal then chose a hill near Berlin from where he achieved glides of up to 1,150 feet. Such performances were evidence that man could fly through the air. Lilienthal said that "he came to know the ocean of air; no one can realize how substantial the air is until he feels its supporting power beneath him. It inspires confidence at once." In an article for the Smithsonian Institution's Annual Report for 1893, he cautioned, "that the wind was a treacherous fellow; the man who would fly must proceed cautiously, achieving stability through the design of his flying machine and by his flight-control system. A flying machine's balance is constantly shifting and stable flight can be maintained only by a constant and arbitrary correction of the position of the center of gravity."

Lilienthal balanced his own craft by throwing his weight around in a series of acrobatic body movements that kept the glider flying on an even keel. A typical flight was described by a reporter for the *Boston Evening Transcript,* who accompanied Lilienthal to the hills of Rhinower one Sunday afternoon in 1896.

> Lilienthal was dressed in his usual flying costume of flannel shirt, heavy shoes, snug fitting cap, and knickerbockers with the knees thickly padded in case of a rough landing. With an assistant's help he carried the machine to the top of a hill and took his place in a rectangular frame beneath and between the wings. Lifting the craft from the ground, he grasped a crossbar and rested his forearms on a pair of small cushions that were attached to the main frame. The reporter, armed with a camera, wrote later that he took up a position downhill and waited for Lilienthal, who "faced the wind and stood like an athlete waiting for the starting pistol. Presently the breeze freshened; he took three rapid steps forward and was instantly lifted from the ground, sailing off nearly horizontally from the summit. He went over my head at a terrific pace, at an elevation of about 50 feet, the wind playing wild

tunes on the tense cordage of the machine, and was past me before I had time to train the camera on him. Suddenly the apparatus tipped side-ways as a gust got under the left wing. For a moment I could see the top of the aeroplane, and then, with a powerful throw of his legs he brought the machine once more on an even keel and sailed away below me across the fields at the bottom, kicking at the tops of the haycocks as he passed over them. When within a foot off the ground he threw his legs forward, and notwithstanding its great velocity the machine stopped instantly, its front turning up, allowing the wind to strike under the wings, and he dropped lightly to the earth."

For those occasions when his descent was not quite so gentle, Lilienthal had incorporated a shock-absorbing device into a glider he built in 1894. Called the prellbugel, or rebound bow, it was a flexible willow hoop fitted in front of the pilot's position to act as a bumper guard upon landing. At least once it saved Lilienthal's life. He wrote of the near-fatal occasion:

During a gliding flight taken from a great height the center of gravity lay too much to the back; at the same time I was unable, owing to fatigue, to draw the upper part of my body again toward the front.

FIGURE 2-3 Flights by Lilienthal

Otto Lilienthal about to test his first hang type glider in 1891. It had a wing area of 197 square feet and weighed 40 pounds.

Lilienthal made a flight in 1895; he traveled a quarter of a mile and obtained an altitude of 75 feet.

Lilienthal flying his biplane glider near Berlin in 1895.

Overloaded in the rear, the machine shot skyward, stalled and then nose-dived toward the earth from a height of more than 60 feet. The prellbugel splintered as it buried itself in the ground, but it absorbed the shock of impact and Lilienthal escaped with minor injuries. His crash landing did not dampen his enthusiasm for future glides.

For all his practical experience, Lilienthal had one misconception. He believed that flapping wings were the answer to aerial locomotion. First in 1893, and again in 1895, he built powered devices with ornithopter features. Each machine was fitted with a lightweight carbonic acid gas engine, producing about two horsepower, that was to make the wing flap and thus drive the craft ahead. The first model was so ineffective that he did not even attempt to fly it; the second malfunctioned during trial runs and did not achieve horizontal flight.

Lilienthal was never able to give these curiously old fashioned machines another test. On August 9, 1896, he took off from a hill in one of his gliders. A sudden gust tossed him upward at a sharp angle. Lilienthal immediately threw his weight forward and tried to bring the nose down. The craft stalled, its left wing dipped sharply and the machine plunged to the ground. This time there was no prellbugel to cushion the impact. Otto Lilienthal who frequently said that many sacrifices would have to be made on the road to manned flight, died the next day of a broken spine.

Otto Lilienthal's flight research had probably taken him about as far as he could go. His concept of wing-flapping propulsion was a blind alley, as was his approach to the problems of stability and control. He crashed because his machine was unstable and controllable only by human strength and agility. These qualities were not adequate for successful sustained flight, nor could they readily be incorporated in a practical flying machine.

Despite his limitations, Lilienthal's influence on aviation was incalculable. He was a fundamental airman, the personification of man in flight. His writings were translated and read throughout the world. Newspaper reporters and magazine writers who had made tentative test flights with his gliders wrote glowing accounts of the wonderful sensation of soaring. Widely distributed photographs of the German flying master in action showed a fascinated public that man could indeed sustain himself in the air on artificial wings. Other aviation enthusiasts studied his work, sought his advice and copied his machines. Perhaps Otto Lilienthal's greatest achievement was to arouse interest in others who continued the advancement of man's effort to develop the airplane.

During his years of experimenting Lilienthal performed a great deal of research into the problems of flying; he also realized that only by actual flying were the real answers to be found. Otto Lilienthal is regarded as one of the greatest pioneers of aviation. He proved the superiority of curved wings over flat wings. He brought the art of gliding to an everyday accepted fact.

Percy Pilcher

Among those inspired by Lilienthal's example was the Scotsman **Percy Sinclair Pilcher** who was an assistant lecturer in naval architecture and marine engineering at the University of Glasgow when he built his first glider which he named the *Bat* in 1895. The *Bat* was much like the standard Lilienthal monoplane hang glider pictured in the newspapers except that it lacked a horizontal stabilizer. With commendable prudence, Pilcher decided to visit Lilienthal and practice in proven machines before flying his own craft.

Lilienthal generously allowed Pilcher to gain gliding experience with him and made a number of suggestions for improving the *Bat*. Pilcher followed the advice and went on to build other gliders named the *Beetle*, the *Gull* and the *Hawk;* all of increasingly effective design. The *Hawk*, completed in 1896, the year of Lilienthal's death, was a particularly successful hang glider, with cambered wings, a tail unit that hinged upward and a wheeled undercarriage fitted with springs. Using a rope-and-pulley system to launch his craft, Pilcher was able to glide for distances of up to 750 feet.

Pilcher next intended to build a powered version of the *Hawk*, its four-foot propeller driven by a small gasoline engine of about four horsepower. Pilcher planned to take off in this machine by running down a hill; once craft was in the air the propeller would be started, and the machine would fly on or about 30 miles per hour. Pilcher completed his engine in 1899, but experiments with a new triplane glider and attempts to form a company to promote his machines kept him from completing and testing his powered craft. On September 30 he staged a gliding demonstration for a group of English aeronautical enthusiasts who had assembled on the estate of Lord Braye at Stanford Hall in Leicestershire. The day was damp and dismal and for some reason Pilcher's gliders had been left outdoors for several hours. All were sodden and heavy by flight time, but Pilcher decided to go on with the show. His first launch in the well-tested *Hawk* ended in a premature but gentle landing. On the second try, the waterlogged craft soared to an altitude of about 30 feet; then a soggy bamboo rod in the tail assembly gave way with a snap. The whole tail unit collapsed and the *Hawk* plummeted to the ground. The 33-year-old Percy Pilcher died two days later without regaining consciousness. Two years before his fatal crash, Pilcher had written prophetically that his gliders were mere practice vehicles to be employed only until the advent of engines that could drive an aircraft forward. He said, "Then a person who is used to sailing down a hill with a simple soaring machine will be able to fly with comparative safety." Pilcher's efforts proved that controlled powered airplane flight was possible and soon to be a fact.

That great leap from soaring to powered flight was not destined to be made in Europe. Following the deaths of Lilienthal and Pilcher, the scattered community of European aviation enthusiasts made little progress. The quest for aeronautical progress had already passed to the United States, which had its own source of inspiration and information in the somewhat unlikely person of a sexagenarian bridge builder named **Octave Chanute**.

John J. Montgomery

An American at the turn-of-the-century who did a great deal for the advancement of controllable powered flight, was **John J. Mongomery**, a science professor at Santa Clara College in California. He experimented with gliders at his family ranch near San Diego, California. He named his first craft the *Gull* which he tested in 1884. His feats in the realm of aviation have been called everything from great to doubtful. It is known that in 1893 Montgomery attended the International Conference on Aerial Navigation held in Chicago. His comments at this meeting demonstrated awareness of airflow and pressure on wings. He understood the importance of curved wing surfaces for lift. This subject was not widely known or understood. As early as 1883 Montgomery built a birdlike flying machine with flapping wings (ornithopter). He next constructed three gliders which demonstrated the need for curved wings. He continued his work and research on the problems of balance and control. Montgomery built several man-carrying gliders which were launched from balloons that carried the gliders into the sky and then released them. On one flight on April 29, 1905, his tandem glider called *Santa Clara* was flown by professional parachute jumper Daniel Maloney. A balloon carried the glider upward to 4,000 feet and released it. Maloney flew a controlled flight going eight miles in 20 minutes. He performed spirals and dives and landed at a pre-arranged location. Unfortunately Maloney died in a crash three months later.

Montgomery is remembered for his acrobatic stunts in public. He had a team of three pilots who, beginning in 1905, did some of the earliest aerial stunting. One of their tricks included a "corkscrew descent and side somersaults." The great San Francisco earthquake of 1906 interrupted Montgomery's experiments for five years. In 1911, while Montgomery was making a flight, a little whirlwind caught his craft and smashed it to the ground. Professor Montgomery landed on his head and, although he did not appear to be hurt, he soon after complained of severe pains and died.

Octave Chanute

Throughout the development of the airplane many names appeared, several of whom made important contributions to furthering the reality of the airplane. One such great name immediately preceding the Wright brothers was **Octave Chanute**. He was a learned, successful engineer, and his approach to the problems of flight was carried through on a scientific level. Chanute contributed much in the progression of successful flying.

Octave Chanute was born in Paris in 1832 and came to the United States with his parents at the age of six. He was a civil engineer who achieved notoriety and fortune as well as a well-deserved reputation. In 1891 he was chosen to be president of the American Society of Civil Engineers. He had designed the first bridge to span the Missouri River at Kansas City. His name is well remembered for designing and constructing the vast Union Stock Yards in Chicago.

Chanute probably became intrigued with the problems of flight after reading an article written by Alphonse Penaud in his report to the Aeronautical Society of Great Britain. Aeronautics was simply a hobby for Chanute until he reached his middle sixties. Prior to this time his aviation activities were confined to gathering and writing information on the subject. He was chairman of the International Conference on Aerial Navigation in Chicago in 1893. The following year he published a book called *Progress in Flying Machines,* which was a large volume and discussed the history of aviation up to that time. Past the age of 60, after becoming successful and having retired, Chanute seriously began the study of flying. He once said that he was influenced by the work of others. After Lilienthal was successful in man-carrying gliders, Chanute turned his major interest to flying. Whereby Lilienthal tried to control his gliders by shifting his weight, Chanute proposed a mechanism that would control the glider by moving the wings.

In 1896 and 1897 Chanute made over 2,000 gliding flights along the shores of Lake Michigan. For many of these strenuous experiments the older man had to rely upon the services of younger enthusiastic men. Chanute's earliest airplanes consisted of five and six wings, but after experimentation he eliminated some wings. Chanute introduced to airplane building the civil engineering concept of strut and diagonal wire bracing. The best flight by one of his pilots was 359 feet in 14 seconds. Chanute was convinced that further experiments with gliders would lead to stable controllable aircraft that could be driven by an engine.

Chanute constructed five man-carrying gliders in all and flew them hundreds of feet. He published his findings in the *Journal of Western Society of Engineers* and invited other experiments to improve upon his results. No one accepted the invitation until 1900 when Wilbur Wright wrote to Chanute inquiring about his construction, materials, and experiments on airplanes. Wright, in his letter, said that he had no ideas of his own but could think of no better way to spend his vacation. Chanute was happy to comply with Wilbur's request. This was the beginning of the long road to the Wright brothers' eventual success.

Octave Chanute and the Wright brothers became fast friends and had several interchanges of letters. When the Wright brothers were experimenting with their gliders at Kitty Hawk, North Carolina, in 1903, Chanute was present. However, due to mechanical delays and cold weather, he was not present when the successful flight took place on December 17, 1903. Chanute felt as a proud father to the Wright brothers' experiments and remained interested in their efforts until his death in 1910. Octave Chanute had given the Wright brothers guidance and inspiration. They made true his greatest wish, the hope and belief that the goal of useful airplanes could be reached.

Samuel P. Langley

Another great scientist known for his aeronautical experimentation was **Samuel P. Langley** (See Figure 2-4), who was a mathematician, astronomer, and physicist. Like Octave Chanute,

FIGURE 2-4 Manly and Langley.

he enjoyed a wide reputation and notoriety by the time he began experimenting with aircraft in his later years.

After the Civil War he became an assistant at the Harvard Observatory, and in 1866 he became a teacher of mathematics at the United States Naval Academy. For the next 20 years his major concern was astronomy, and by the time he was 52 years old he was recognized for his contributions to that science.

In 1889, Dr. Samuel Langley went to the Smithsonian Institution as its secretary. It was there that he began publishing his aerodynamic findings and was recognized as the country's leading authority on aviation. Wilbur Wright said that Langley's knowledge and position as the head of such a scientific institution was an influence which further led them to investigate the possibility of flying. Langley had built some 40 rubber-powered airplane models. He directed experiments using all kinds of engines such as hot air, compressed air, gas, electricity, carbonic acid, and gunpowder. He analyzed the problems of airplane design and construction and studied propellers. He also directed his attention toward airplane stability and control. Because of his duties and responsibilities as head of the Smithsonian Institution he had to rely upon assistants for much of the actual work.

Before Langley could launch a successful airplane he needed a lightweight, high-powered engine. He decided steam had the greatest possibilities and directed a program to perfect a steam engine. Trials for the launching of his aircraft were conducted from a houseboat moored on the Potomac River near Quantico, Virginia. A catapult mechanism was designed to launch the airplane from the top of this houseboat. During 1893 and 1894, because of weather conditions and mechanical problems, not much was accomplished. However, the fifth model was successfully launched, flew 35 feet, and fell into the water after a three-second flight.

The next model Langley built was observed in May 1896 by Dr. Alexander Graham Bell. The model weighed 26 pounds, and flew 3,200 feet attaining a speed of 25 miles per hour. To Langley the flights of these models meant success, and he wrote:

> For the first time in the history of the world a device produced by man had actually flown through the air and had preserved its equilibrium without the aid of a guiding human intelligence.

Langley was in his middle sixties and his administrative responsibilities at the Smithsonian Institution grew heavier. During the Spanish-American War, President McKinley asked Dr. Langley to construct a flying machine that could be used as a weapon of war. Langley accepted this mandate on December 12, 1898. He estimated the cost of this work to be $50,000 and stipulated that his work was to be done privately. Langley enlisted the aid of a Cornell University graduate, **Charles Manly** (See Figure 2-4), to take charge of the technical aspects of construction. Eventually models were constructed and several successfully flew, but there was always some kind of trouble. Despite several troubles a larger houseboat was made ready for the eventful day when a full-sized airplane would be flown. The houseboat measured 40 × 60 feet and provided living quarters and workshops. Atop the houseboat was a steam catapult for launching the airplane.

The engine to power this new aircraft was not available. The New York manufacturer who had agreed a year earlier to design and build a lightweight engine according to Langley's specifications, was unable to produce the final product. Langley and Manly spent six weeks in Europe searching for an engine manufacturer who could produce an engine of their desired high power and light weight. Wherever they went they were told that it was impossible to construct such an engine—12-horsepower and weighing less than 225 pounds. If it were possible, the European manufacturers would have already constructed one.

Not dismayed by their findings, Langley and Manly returned to America where Manly designed and built a revolutionary five cylinder radial engine. It weighed 125 pounds, developed 54 horsepower at 850 RPMs and became the forerunner of airplane radial engines. With this new engine the Langley airplane was ready for a flight attempt in the summer of 1903. This airplane, which was designed to carry the first man in controlled flight, had a wingspan of 48 feet and a body 52 feet long which was constructed of steel tubing. The airplane had two pusher propellers powered by Manly's engine. The wings were covered with cotton cloth and the total weight of the airplane was 850 pounds including the pilot. Every effort was used to give the airplane control.

On July 4, 1903, with ground testing complete, the airplane was placed upon the houseboat and moved down the Potomac. Langley had hoped to launch his airplane in private, according to his agreement with the Army. However, the Potomac soon became a haven for boats carrying newspapermen wanting to report the story. For three months the press moved up and down the river waiting for the eventful day. Everything was ready for the historic flight of the big airplane. Manly, who had designed and built the engine was to be the pilot. He wore, in addition to his work clothes, a life preserver, goggles, and tennis shoes on the first attempted flight.

For almost a month bad weather prevented his first effort and finally on October 7, 1903, the trial did come off. Ironically, Langley was not present. He was detained in Washington, D.C. on official business. Shortly after noon the airplane was lifted to the launching platform where Manly climbed aboard and started the engine. George Brown, a newspaperman, wrote for the *Washington Post:*

> Manly looked down and smiled. His face hardened as he braced himself for the flight which might mean for him fame or death. Propellers were whirling a thousand times a minute a foot from his head. A man forward fired two skyrockets. A mechanic stooped and cut the cable holding the catapult; there was a roaring, grinding noise and the Langley airplane went off into the Potomac River and disappeared.

Except for a good wetting Manly was unhurt; he blamed the failure on incorrect balancing of the aircraft. However, it was later determined that a part of the launching mechanism of the catapult had fouled and pulled the airplane down to disaster. Manly's engine was undamaged and the airplane was repairable. (See Figure 2-5.)

Langley had spent the original $50,000 appropriation from the Army as well as $20,000 provided by the Smithsonian Institution. The airplane was rebuilt and on December 8th, in a last desperate attempt to fly it, something again went wrong. The wings collapsed, and the aircraft again crashed into the river. This time Manly had to fight his way out of the submerged wreckage, and as he did, he bumped his head on a block of ice which was floating in the river—but he was finally picked up and carried to shore.

During all these trials the Army had been patient, but had hoped for quick success. The Army was criticized, and Langley became a point of ridicule. In 1906 Dr. Samuel Langley died. It was reported that he was the victim of a broken heart. Langley's accomplishments in the human effort to fly were notable, even though his man-carrying airplane failed. In June 1914, the Langley Aerodome with modifications and fitted with floats was flown from Lake Keuka near

FIGURE 2-5 Launching the aerodrome

October 7, 1903, Langley's "aerodrome" on launching catapult.

The ill-fated take-off.

Boats moved to pick up Manly out of the Potomac River.

Hammondsport, New York by Glenn Curtiss. By some it was considered proof that except for malfunction of the launching gear Langley's airplane would have flown in 1903. With equal conviction others said it was a fake operation. This led to controversy which lasted until 1942.

The Wright Brothers

It was on December 17, 1903, at the sand dunes of Kitty Hawk, North Carolina with the wind blowing 27 miles an hour that Orville Wright climbed aboard the flying machine and started along the launching tract. After traveling 40 feet the Flyer lifted into the air where it remained for 12 seconds before coming back to earth a total distance of 120 feet from where it started.

This episode marked the end of the long hardships of trials and errors of several men for the preceding centuries. Success had finally been achieved in an airplane that was flown in controllable powered flight.

Although this marked the point of success in the long progression of inventors and experimenters, it was not the end, but really the beginning. From that time on the airplane developed upon a sound scientific basis to achieve the success that it enjoys today.

How did the Wright brothers (See Figure 2-6) succeed where so many others failed? Earlier that month of December, 1903, Samuel Langley's airplane had failed for the second and final time. This ended 17 years of Langley's effort and more than $70,000 in funds.

Before Langley's failing efforts others had attempted and also failed. Maxim and Ader, who were amply financed, came to grief. Although an engineer and a scientist had attacked the problem of flight, they fell short of the ultimate goal achieved by the Wright brothers. Lilienthal and Pilcher were killed while experimenting with their inventions, and Penaud had committed suicide due to his frustrations and inability to accomplish what the Wright brothers achieved.

The Wright brothers did not invent the airplane; this is a misconception.

It was more than blind luck that led the two young men from Dayton, Ohio, to succeed where others had failed. The Wright brothers were bicycle manufacturers in their own little shop and were not scientists or engineers. In fact, neither of the boys had graduated from high school. This was due to their lack of interest in diplomas rather than a lack of intelligence. They were the sons of a bishop and they developed the habit of reading. It was through their intense interest in reading that they learned of the work of Lilienthal and Chanute and which led them to begin a correspondence with Chanute.

Actually, it was a series of events which led the Wright brothers on the path to succeeding where all others had failed. When they were young boys their father brought home a rubber band-powered flying toy. Being inquisitive mechanics, this toy started them thinking along the avenue which produced the eventual results. They read of Lilienthal's experiences and decided that he was correct in his approach to flying, but that he had not gone far enough when he was killed. They figured that in five years, Lilienthal spent no more than a total of five hours actually gliding in the air.

The Wright brothers decided that if people were ever to fly successfully they would have to learn how to design and build wings, how to devise power for flight, and, most important of all, how to control the machine once it got into the air. Of these three problems the first two had been reasonably worked out by the efforts of others. Although these two were very important problems to overcome, they would be for nothing if the airplane could not be controlled once it became airborne. The Wright brothers devoted their attention to this third problem. The Wright brothers did not invent the airplane; this is a misconception. What they are given credit for is producing a man-carrying airplane which was capable of controlled powered flight.

FIGURE 2-6 The Wright Brothers (top). The Wright Flyer (middle). December 17, 1903; the flight the world remembers (bottom).

As boys, the Wright brothers spent their early years building models which they flew as kites. They continually made larger kite models, and each one showed improvement over the previous kites. They eventually produced a large glider whose wing tips could be twisted by the pilot to direct the wind and change the direction of the airplane. These wings were controlled by a wire arrangement down to the pilot's position.

Wilbur wrote a letter to the Smithsonian Institution on May 30, 1899, inquiring about publications on the subject of flight. In part he said:

> I believe that simple flight at least is possible to man and that the experiments and investigations of a large number of independent workers will result in the accumulation of information, knowledge and skill which will finally lead to accomplish flight.

Since the area around Dayton, Ohio was not satisfactory for their flying experiments, they looked for a suitable place. They required an area which had moderate wind currents, hills from which they could launch their gliders, and, most of all, privacy from onlookers who doubtless would ridicule them. After much seeking they chose the sand dunes of unpopulated Kitty Hawk, North Carolina. They went there first in the summer of 1900. Their glider rose into the air where the wind was steady for good gliding, but it did not have much lifting ability. The next year they returned to Kitty Hawk with a new glider which had larger wings, but this one did not operate as they had planned. They decided that, in order to save time and money, they would make changes in their airplane by reproducing these changes on a model. They put the model in a long, open-end box wind tunnel where they were able to see the effect of their design changes. The wind tunnel idea which the Wright brothers conceived is still in use today. In fact, the engineering aerodynamics of jet aircraft and rocket-powered missiles are extensively tested in huge wind tunnels before they are put into full scale-sized production. From their little wind tunnel test the Wright brothers succeeded in building an airplane which they considered to be successful. From the test results of the glider they decided that the time had come to design and begin construction of an aircraft equipped with an engine.

On December 14, 1903, the Wrights were still at Kitty Hawk. They had stayed late in the year hoping for eventual success. They agreed to toss a coin to see who would be the one to make the first test. Wilbur won the toss of the coin and made the first try. It was a cold day but the wind was very mild. The Wright Flyer barely got off the ground for two and one-half seconds. Minor damage resulted, and the boys agreed that this flight didn't count as a real flight. Three days later they had completed the minor damage repairs, and it was then Orville's turn to try. This was to be it! Twelve seconds and 120 feet later history had been made!

Three additional flights were made that same day with the brothers taking turns at the controls. The best and the last of these flights was made by Wilbur—just under one minute, for a distance of 852 feet.

Octave Chanute had been at the Wright's camp in 1901 and 1902 as well as 1903. Unfortunately he was not there on December 17, 1903, to see the first successful flight delayed by trouble which developed in the propeller shafts and had to be corrected.

This successful Wright Flyer had been constructed at their bicycle shop in Dayton, Ohio, where they had a flourishing repair business going. They had hired a mechanic, Charles Taylor, to work for them in their bicycle business. It was Taylor who built the engine for their first airplane. The crankcase and block were aluminum; the cylinders and pistons were made of cast iron. The engine weighed only 170 pounds but could develop 12 horsepower. It was not a fancy product but it was dependable. The problem of constructing adequate propellers was a big one. They looked to ship builders for assistance, since boats used propellers. They discovered ship propellers were still being built on a trial and error basis. They correctly thought that an airplane propeller in the air would act the same as a ship propeller in the water. Also, they figured the propeller was actually a wing in itself and would behave as such. From these ideas they hand-built their own propellers. The efficiency of the propellers was remarkable and was one of the essential factors leading to the success of their flights.

The Wright brothers were craftsmen and excellent mechanics, and they knew how to fabricate with wood and metal. They learned how to conduct scientific research like true scientists. They could visualize results of their designs and fit all of this together into a workable result.

The Wright brothers continued their experimentation and by 1905 they had made more than a hundred flights totaling 160 miles. They were soon to become famous, not only in this country, but also in Europe. However, the acceptance of their successful flight did not take place immediately. As a result of their successful flights on December 17th, the Wright brothers sent a telegram to their family at Dayton. The message read:

> Success, four flights Thursday morning. All against 21 mile wind. Started from level with engine power alone. Average speed through air 31 miles. Longest 59 seconds. Inform press. Home Christmas.

Signed Orville Wright

The telegram did not reach the Wright brothers' father until that evening and when it arrived it had gathered errors in transmission. The brothers' sister, Katherine, sent a message to Chanute reporting their success. Their father prepared a brief message for the press. The reporter in the *Journal* office was annoyed over being expected to accept such a story, and nothing about the Wright brothers' success appeared in the news paper. The headlines were concerned with the stores being filled with Christmas shoppers. When the story was finally printed in the newspaper it was to the extent that the Wright brothers would be returning home for Christmas.

Orville Wright wrote his own account of the world's first powered, sustained, controlled flight.

The wind tunnel idea which the Wright brothers conceived is still in use today.

The article was published ten years after their first successful flight. It appeared in the December 1913 issue of the American Aviation Journal. It is entitled, "How We Made the First Flight." A copy may be obtained from the Smithsonian Institution. He stated:

> After running the motor a few minutes to heat it up, I released the wire that held the machine to the track, and the machine started forward in the wind. Wilbur ran at the side of the machine, holding the wing to balance it on the track. He was able to stay with it till it lifted from the track after a forty-foot run. The course of the flight up and down was exceedingly erratic, partly due to the irregularity of the air, partly to lack of experience in handling the machine. The control of the front rudder was difficult on account of its being balanced too near the center. This gave it a tendency to turn itself when started; so that it turned too far on one side and then too far on the other. As a result the machine would rise suddenly to about ten feet, and then as suddenly dart for the ground. A sudden dart when a little over a hundred feet from the end of the track, or a little over 120 feet from the point at which it rose into the air, ended the flight.
>
> (After the fourth flight) "While we were standing about discussing this last flight, a sudden strong gust of wind struck the machine and began to turn it over. Everybody made a rush for it. All our efforts were in vain. The machine rolled over and over. The ribs in the surface of the machine were broken, the motor injured and the chain guides badly bent, so that all possibility of further flights with it for that year were at an end.

A full decade after the Wrights, the noted Harvard astronomer, William Pickering, firmly denounced the idea that the airplane would someday become a major transportation vehicle. He wrote:

> The popular mind often pictures gigantic flying machines speeding across the Atlantic, carrying innumerable passengers. It seems safe to say that such ideas must be wholly visionary. Even if such a machine could get across with one or two passengers, it would be prohibitive to any but the capitalist who could own his own yacht.

Pickering was not as pessimistic as Simon Newcomb, a mathematician and scientist. In 1902 he was handed a model of the Wright Flyer (See Figure 2-7) by the two brothers who proposed to build it. They wanted his opinion and he gave it—handing the model back with a smile, he shook his head. "Man will never be able to fly a heavier-than-air machine," Newcomb decreed.

The most famous aviation expert in the United States, 10 years after Kitty Hawk, was asked about the chances of flying across the Atlantic. The expert was dubious.

> It is a bare possibility that a one-man machine without a float and favored by a wind of, say, 15 miles an hour, might succeed in getting across the Atlantic. But such an attempt would be the height of folly. When one comes to increase the size of the craft, the possibility rapidly fades away. This is because of the difficulties of carrying sufficient fuel. It will readily be seen, therefore, why the Atlantic flight is out of the question.

That statement was made by none other than Orville Wright.

Wilbur Wright became ill after eating fish at a Boston Hotel. He became feverish and went to bed. His father said he was suffering from typhoidal fever. He died quietly on May 29, 1912.

Orville died on January 30, 1948, after suffering a heart attack while fixing his doorbell. He was seventy-seven years of age and had lived to see the extent of his efforts.

It took decades, but on October 24, 1942, the Smithsonian finally apologized to Orville Wright and acknowledged that he and his brother Wilbur were the first to pilot a powered airplane. Now, Orville (Wilbur died in 1912) would summon his famed Flyer home from the Science Museum in London, where it had been on display for two decades, and donate it to the Smithsonian, thus ending one of the century's most bitter feuds. It all started back in 1903, nine days before the Wrights' successful Kitty Hawk flights, when an airplane designed by Smithsonian Secretary Samuel P. Langley crashed into the Potomac River. In 1914, his Smithsonian colleagues, in a case of misguided loyalty, altered Langley's Aerodrome to make it fly, then hung it in the Smithsonian as the "First Machine Capable of

FIGURE 2-7 The Wright Flyer

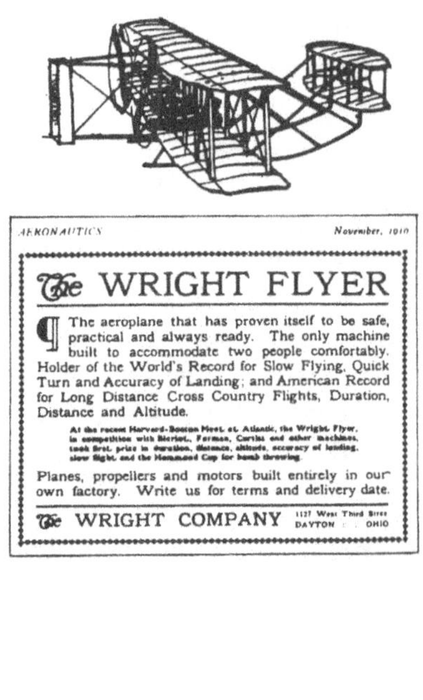

A Wright military aircraft in flight at Fort Myer in 1908.

Flight Carrying a Man," a claim that Langley, who died in 1906, had never made.

Indignant, Orville Wright told the *New York Times* on May 17, 1925, that he had held out hope for years that "objectionable conditions in the Smithsonian would change." On December 17, 1948, the Flyer went on display in the Smithsonian's Arts and Industries Building. Today, visitors can see the Wright Flyer's intricate controls up close in its new eye-level position at the National Air and Space Museum on the Mall.

> **"Man will never be able to fly a heavier-than-air machine," Newcomb decreed.**

Those who foresaw the time when mass, long-distance air transportation would become a reality were considered crackpots. For example, Rufus Porter preached the forthcoming glories of commercial flight from the prestigious pages of the *Scientific American* magazine of which he was the editor. In 1850, Porter actually tried to raise money for construction of an aerial ship capable of carrying 150 passengers which would easily carry them to California or London in three or four days.

Yet even as late as 1925, only two years before Charles Lindbergh's New York-Paris feat, Jose Navarro, an aeronautical engineer, proposed an airliner with a wingspan of 185 feet, a 38-man crew, nine engines, a bar on each of several decks, separate dining rooms, sleeping compartments, and electric kitchens. Navarro's goliath would carry at least 150 passengers and the suggested New York-London route, with refueling stops at Newfoundland and Iceland, would take only a day.

Lieutenant Thomas E. Selfridge (1882–1908) of the United States Army Signal Corps was the first person killed in an airplane crash. Selfridge had become interested in flying as a member of the Aerial Experiment Association headed by Alexander Graham Bell, the inventor of the telephone. He helped build some of the association's experimental airplanes. Selfridge was appointed one of the Army officers to judge the military value of the Wright brothers' airplane. In a flight with Orville Wright on September 17, 1908, a propeller broke 75 feet in the air. The plane crashed, killing Selfridge and severely injuring Orville. Selfridge Field, an Air National Guard Base near Mount Clemens, Michigan, was named in his honor in 1917.

Henry Ford was a friend and supporter of the Wright brothers pioneering efforts. He was also instrumental in developing the early aviation industry. In 1909, Ford provided a Model T engine for a primitive monoplane. His company manufactured engines for airplanes during World War I. He constructed the first modern airport. He invested in the Stout Metal Airplane Company that built the first United States commercial all-metal airplane and then the popular Ford Tri-Motor. He established the Ford Air Transport Service, which is believed to be the first scheduled airline. This company was the first private operator to receive an air mail contract from the United States Post Office in 1925. In 1927, the Ford Motor Company established its corporate flight department with Charles Lindbergh as the chief pilot. For his contribution to aviation, Henry Ford was enshrined in the National Aviation Hall of Fame in 1984.

Controllable powered aircraft was now a fact; many noted airmen and aircraft developers appeared. After the success of the Wright brothers in 1903 in the United States and that of Santos-Dumont in 1906 in Europe, a revival of interests in heavier-than-air machines took place. History has not recorded all the names of the scientists, engineers, mechanics and geniuses who put all they had into their effort to build aircraft. From 1904 to 1912, hundreds of prototypes appeared. Among some of the most important of these flying machines was the Ellehammer, built by the Dane **Jacob Christian Hanson Ellehammer**, which rivaled the 14-BIS of Santos-Dumont for the title of the first airplane to fly in Europe. On September 12, 1906, his airplane hopped 140 feet. **Henri Fabre's** Hydravion deserves acknowledgment as the first seaplane in the world. On March 29, 1910, it flew almost four miles and made a perfect landing. Other aircraft which should be noted are the Dunne D-5, the first all-wing airplane built by **John William Dunne**; the

Muller and the Chiribiri, among the first to be designed and built in Italy; the Cody built by the American expatriate **Samuel Franklin Cody**, who lived in Great Britain, and was the first to fly in that country.

> **Lieutenant Thomas E. Selfridge (1882–1908) of the United States Army Signal Corps was the first person killed in an airplane crash.**

No concrete proof exists that the Givaudan, with drum-shaped wings, the Safety, with looped wings, the Edward, with rhomboidal wings or the Seddon, ever got off the ground. Nevertheless, they did testify to the dedicated commitment to research and stood as objects of study and inspiration for all who endeavored to solve the problems of flight.

The aircrafts that made brief flights were Captain Dorand's Aeroplane, destined for military use; the Koechlin-De Pischoff; the Cygnett II, designed by the American **Alexander Graham Bell** and built by the Aerial Experiment Association. It seems impossible that this strange aircraft, built like a honeycomb, was able to fly before crashing on frozen Lake Keuka, in New York State in 1909.

> **In 1906 Alexander Graham Bell said: "There are two critical points in every aerial flight; its beginning and its end."**

Horatio Frederick Phillips' Multiplane flew only a short distance, but the studies he made on wing-surface curvature and later tried to test became very important for the future. Nor did the young Rumanian engineer Henri Coanda's aircraft ever fly though he presented at the second Salon de L'Aeronautique in Paris in 1910. But its design anticipated the advent of the turbine engine. Twenty-five years would pass before others would succeed where Coanda had failed.

Glenn Curtiss (1878–1930) (See Figure 2-8) began racing bicycles as a teenager, which led him to the use of air-cooled engines and, therefore, motorcycles. In 1902 he began helping the balloonist, Tom Baldwin, produce a powered dirigible for the United States Army, and as a result, in 1905, the United States had its first military airship.

Curtiss joined Alexander Graham Bell in his Aerial Experiment Association in 1907, and in 1908 he produced the *June Bug*, which won the very first aviation prize to be offered in the United States. Curtiss started the first aircraft factory in the United States where he built and sold his first airplane for profit. This action caused the Wrights to file suit claiming Curtiss infringed upon their patents, thus beginning several years of controversy and court action which was finally won by the Wright brothers.

Curtiss built the first successful seaplane in the United States in 1911 and became the "father of naval aviation." A Curtiss pilot, Eugene Ely, made the first shipboard take-off doing so from the cruiser *Birmingham* on November 14, 1910. Two months later Ely made the first landing on a naval vessel.

Another member of Bell's group, **James A.D. McCurdy**, in August, 1910, received and sent the first radio messages from an airplane over Sheepshead Bay, New York.

FIGURE 2-8 Glenn Curtiss in his *June Bug*.

In the meantime, flying made great progress in France. **Alberto Santos-Dumont** (1873–1932) (See Figure 2-9), a wealthy Brazilian, made the first airplane flight in France on August 22, 1906. **Louis Bleriot** (1872–1936) made the first international flight in an airplane. He flew across the English Channel from France to England in thirty-seven minutes on July 25, 1909. Bleriot became a prominent airplane manufacturer in France. He was the first to build monoplanes with the engine in the front, a stabilizing tail at the rear, and controls attached to a stick for the pilot. **Henri C. Farman** (1874–1934) made many daring flights, including the first night flight, in 1910. The first altitude flight of over 3,000 feet was made in an Antionette airplane in 1910 by **Hubert Latham** (1833–1912).

On November 7, 1910, Max Morehouse became the first to use an aircraft for business purposes. Owner of the Morehouse-Martens department store in Columbus, Ohio, he hired the Wright exhibition company to fly $800 worth of silk from Huffman Prairie Flying Field, in Dayton, to the Driving Park Race Track in Columbus. This flight was the first air cargo flight. Morehouse was a good businessman. He paid to charter the Wrights' Model B airplane, but he also sold tickets to three thousand people who came to see the airplane arrive. He charged them three dollars to park their vehicles at the track. He also sold swatches of the silk cargo, on a commemorative postcard, for five cents or by the yard at $1.35 per yard.

In 1910 the Curtiss Exhibition Company was putting on airplane shows and gave the Wright team competition.

FIGURE 2-9 Alberto Santos-Dumont, 14 BIS, France 1906. Wings and engine were at the rear of the plane.

On August 2, 1911, **Harriet Quimby** (See Figures 2-10 and 2-11) became the first woman pilot in the United States to obtain a Federation Aeronautique International certificate; it was number 37 and given to her at Mineola, New York. On April 16, 1912, she was the first woman to fly across the English Channel. She was killed later that year when she and a passenger fell out of her airplane while flying 2,000 feet above Boston Harbor.

September 23 through 30, 1911, the first officially approved public service airmail in the United States was flown during the International Aviation Tournament at Garden City, New York. Captain Paul Peck of the United States Army piloted an airplane with Postmaster General Frank Hitchcock carrying the first mail bag. Earle Ovington piloted later flights during the week of airmail service. This effort proved that airmail service was feasible.

Grover Loening was born in Germany when his father was a United States counsel. He earned his Bachelor of Science degree at Columbia University in 'aerial locomotion' and a Masters degree in aeronautics in 1910. His thesis on aircraft design was well received and a book was published from it entitled "Monoplanes and Biplanes." In 1912 he used an old float and attached a fuselage from a Bleriot airplane and wings. This was the world's first flying boat. In July 1913, he received a letter from Orville Wright offering him a job as general manager of their factory in Dayton, Ohio, at a high salary of $1000 a year which he accepted. With Orville Wright he designed the Wright model G Aeroboat. This airplane had a control wheel for pitch and roll. This was the first of its kind. Every pilot flying today has Loening to thank for that invention. In 1914, he became the chief aeronautical engineer for the United States Army Air Service located in San Diego. He built a fighter airplane but it came too late to be used as the war ended. He later began his own company where he designed and built the Air Yacht. This was used by the military in world flights and for many commercial purposes. Admiral Byrd used one in Antarctica in 1925. One is in the National Air and Space Museum in Washington, D.C. He held many

FIGURE 2-10 Harriet Quimby.

FIGURE 2-11 Wilbur Wright and the first female pilot in America, Harriet Quimby.

airplane patents and has received more than twenty awards and honors for his contributions to U.S. aviation. Until his death in 1976, Loening remained active in aviation as designer, author, consultant, and lecturer in the true spirit of the aviation pioneer.

Fearless pioneer fliers learned that airplanes could do aerobatics. **Adolphe Pegoud**, a French pilot, proved in 1913 that it was possible to fly upside down in an airplane. The next year a Curtiss exhibition pilot, **Lincoln Beachey** (1887–1915), "looped the loop" more than a thousand times. Beachey demonstrated the tailspin, flew inside a building, and was the first of the great daredevil stunt pilots.

In 1915 Lieutenant **Max Immelmann** (1890–1916) of Germany invented the Immelmann turn, which was a combination short climb and half-roll.

In 1914, at the beginning of World War I, all of the countries involved in the war had aircraft but none were suitable for reconnaissance purposes. Observing enemy movements from the air required an aircraft that would allow an observer, in addition to the pilot, to look over the countryside. A camera was necessary but a machine-gun was considered useless. Opposing pilots and observers eventually began to throw things at each other. Bricks were the favorite missiles. Some pilots were known to throw a chain hoping to snare the enemy's propeller and cause the aircraft to crash. Pistols were inaccurate and rifles were little better. The observer had to be the gunner while the pilot would maneuver to put the airplane in firing position. Unfortunately many observer-gunners shot their own aircraft structures and propeller blades, thus causing more damage than the enemy.

The next step was to carry a machine gun aloft. A British pilot did this by 'borrowing' a gun from the Company's storeroom. It was a very heavy Lewis machine gun. It was so heavy that the aircraft could fly no higher than 3,500 feet, while the German aircraft could easily fly at 5,000 feet. Eventually more powerful airplanes were developed to lift heavier guns as well as fly faster, higher and be more maneuverable. There was the problem of where to locate the machine gun without hitting their own aircraft. Some guns were placed on the top wing above the propeller. The pilot had to stand up to fire and reloading was next to impossible.

Roland Garros was the first flying "Ace."

The best innovator, **Roland Garros**, was the first real combat pilot. He was a music student who had gone to Paris to study the piano and became a concert pianist. While there he saw his first airplane and very quickly the airplane replaced the piano as his major interest. Garros met and convinced the famous pilot, Alberto Santos-Dumont, to teach him to fly. Garros was an excellent student and soon became one of the best pilots. He participated in air races and exhibition flights in Europe as well as the United States. In 1913 he was the first person to fly across the Mediterranean Sea from southern France to Tunisia. The 453 mile distance took less than eight hours.

When World War I was declared Garros was in Germany where he had been giving exhibition flights. As a French citizen he was an enemy alien. Although he was closely guarded he was able to take off and fly back to France. He accomplished this because it was not expected for anyone to fly at night. Upon returning to France, he joined the military and soon was flying above Germany as an observer. While flying, he thought about using a gun on the airplane and determined that a pilot should be able to fire in a forward direction and not to the side as the rear seat observers were doing. Garros heard about a French flyer, Eugene Gilbert, who had tried wrapping propeller blades with steel-wire tape to deflect any bullets that struck them. While Gilbert was trying out this idea, ricocheting bullets killed two men who were helping him.

Garros calculated that about seven percent of the bullets fired would actually strike the propeller. He designed triangular metal shields to place on the back of the propeller blades. They were angled so as to deflect any bullets away from the pilot and the propeller. Garros started his work in February 1915, and by April 1st was ready to try out his new invention. Shortly after takeoff he saw four German Albatross observation airplanes heading toward the French lines. He soon caught up with them and turned into the nearest aircraft and fired. The last thing the enemy pilot expected was a burst of bullets. The Albatross went down. Garros headed for the next airplane and with another burst of fire it blew up in midair. Seeing this, the two remaining enemy aircraft turned and headed home to report what had happened.

In the next two weeks, using his forward firing machine-gun, he shot down five more German aircraft and became the first allied ace. The word "ace" had been applied to anyone who had accomplished something outstanding, such as a long distance bicycle race, or a soldier for some special act of bravery. The local newspapers called him an ace among pilots and the word became legendary thereafter.

The German pilots, upon spotting Garros' aircraft, were very cautious and did not wish to be anywhere near him. However, his luck ran out when engine trouble forced Garros to make an emergency landing behind German lines. Upon landing he attempted to burn his aircraft but was not successful because the airplane was wet due to a local shower. He was taken prisoner and his airplane went to Berlin where the Germans asked Anthony Fokker, a Dutch airplane designer working for the Germans, to examine the aircraft. He was instructed to make something like that for German aircraft. Fokker, a brilliant engineer, decided this was not efficient because sooner or later the propeller would splinter, become out of balance and cause engine failure. Fokker determined that the two blades of the propeller passed the gun 2400 times a minute and the gun fired 600 times a minute. He reasoned that it would be possible to figure out a way to fire the gun only when the propeller blade would not be in the way. Fokker then designed a mechanism that would prevent the gun firing if the propeller was in the way of the bullet. Within forty-eight hours he had perfected an interrupter device which worked on ground tests. In less than a month, German aircraft equipped with the new device were shooting down French and British aircraft. Aerial combat had arrived.

During World War I there was a tremendous improvement in airplanes, engines, and manufacturing. Glenn L. Martin, who had started building airplanes in 1909, produced bombers of advanced design in 1918, using two powerful new

400-horsepower Liberty engines. Donald Douglas was chief engineer for Martin during the war, and in 1920 organized his own airplane manufacturing company. In Seattle, Washington, William Boeing began building planes during the war. Anthony Fokker of the Netherlands built many outstanding military planes for Germany and came to the United States after the war to build civilian and military airplanes. Willi Messerschmitt, a German, built airplanes for Germany, not only in World War I but also in World War II. In Russia, Igor I. Sikorsky built the first multiengine airplane and later developed the first successful helicopter. He came to the United States after the war. These manufacturers and others gained experience through the war and began making airplanes that were safer and more useful than had been previously possible.

After the war **Dornier**, in Germany, was the first to apply Duralumnin to aircraft. In 1922 **Fokker** offered his welded steel fuselage.

Between World Wars I and II many aviators, like the pioneer fliers of the earlier period, risked their lives to set new records. (See Table 2-1.) Nonstop transcontinental flights, transoceanic flights, and polar flights kept aviation news on the front pages of newspapers. Speed, altitude, endurance, and distance records were made and broken time after time. Newspapers, aviation organizations, and individuals offered large prizes to aviators to perform certain designated flights.

Tuskegee Airmen

On July 19, 1941, the Army Air Force began a program in Alabama to train black Americans as military pilots. Primary flight training was conducted by the Division of Aeronautics of the Tuskegee Institute, the famed school of learning founded by Booker T. Washington in 1881. Upon completing primary training at Moton Field cadets were sent to the Tuskegee Army Air Field for completion of training and transition to combat type aircraft. The first graduates were trained to be fighter pilots for the famous 99th Fighter Squadron for combat duty in North Africa. Subsequent pilots were assigned to the 322 Fighter Group, which flew combat along with the 99th Fighter Squadron from bases in Italy.

By the end of the war, 992 men had graduated from pilot training. Of these 450 were sent overseas for combat duty. During the war about 150 men died while in either training or combat flights.

The first great flight after World War I was the transatlantic crossing by the United States Navy seaplane NC-4. The plane started from Rockaway, Long Island, New York, on May 8, 1919, and arrived at Plymouth, England, on May 31. The route included stops at the Azores, Lisbon, Portugal, and Marengo, Spain. Three NC planes started the flight, but two were unable to complete the trip. Lieutenant Commander Albert C. Read was in command of the NC-4 and the crew of five other Navy and Coast Guard airmen.

The early transatlantic flights were made from west to east so that the fliers could take advantage of the prevailing west-to-east winds to increase their speed.

Sir John Alcock (1892–1919) and **Sir Arthur Whitten-Brown** won a $50,000 prize offered by the *London Daily Mail* when they made the first nonstop airplane flight across the Atlantic. They flew from Newfoundland to Ireland on June 15, 1919, in a Vickers "Vimy" twin-engine biplane.

The most noted flier of the 1920s was **Charles A. Lindbergh** who on May 10, 1927, flew a record solo flight from San Diego to New York in 21 hours and 20 minutes. Ten days later, on May 20, he took off from New York alone and headed across the Atlantic Ocean for Paris. His airplane was only slightly larger than the Wright brothers' first airplane. When he landed safely at LeBourget Field near Paris, he won over to aviation many persons who before his flight had regarded the airplane as impractical.

The most disastrous air race of the period was the Dole Race of 2,400 miles over the Pacific Ocean from San Francisco to Honolulu in 1927. Before and during the race ten fliers were killed. The $25,000 prize was won by Arthus C. Goebel.

TABLE 2-1 Aviation Records from 1906 to 1914

Altitude

Date	Place	Name	Airplane	Engine	Height Feet
1908 13 XI	Issy (F)	Henri Farman	Voisin	40 hp Vivinus	82
1908 13 XI	Auvours (F)	Wilbur Wright	Wright	24 hp Wright	82
1908 18 XII	Auvours (F)	Wilbur Wright	Wright	24 hp Wright	360
1909 18 VII	Douai (F)	Louis Paulhan	Voisin	50 hp Gnome	492
1909 29 VIII	Reims (F)	Hubert Latham	Antoinette	50 hp Antoinette	508
1909 20 IX	Brescia (I)	Rougier	Voisin	50 hp E.N.V.	633
1909 18 X	Juvisy (F)	De Lambert	Wright	24 hp Wright	984
1909 1 XII	Chalons (F)	Hubert Latham	Antoinette	50 hp Antoinette	1436
1910 7 I	Chalons (F)	Hubert Latham	Antoinette	50 hp Antoinette	3444
1910 12 I	Los Angeles (USA)	Louis Paulhan	H. Farman	50 hp Gnome	4110
1910 14 VI	Indianapolis (USA)	Walter Brookins	Wright	40 hp Wright	4379
1910 7 VII	Reims (F)	Hubert Latham	Antoinette	50 hp Antoinette	4539
1910 10 VII	Atlantic City (USA)	Walter Brookins	Wright	40 hp Wright	6237
1910 11 VIII	Lanark (USA)	Armstrong Drexel	Blériot	50 hp Gnome	6603
1910 29 VIII	Le Havre (F)	Léon Morane	Blériot	50 hp Gnome	7042
1910 3 IX	Deauville (F)	Léon Morane	Blériot	50 hp Gnome	8469
1910 8 IX	Issy (F)	Geo Chavez	Blériot	50 hp Gnome	8484
1910 1 X	Mourmelon (F)	Jan Wijnmalen	H. Farman	50 hp Gnome	9118
1910 31 X	Belmont Park (USA)	Ralph Johnstone	Wright	60 hp Wright	9600
1910 9 XII	Pau (F)	Geo Legagneux	Blériot	50 hp Gnome	10,168
1911 9 VII	Buc (F)	Loridan	H. Farman	70 hp Gnome	10,496
1911 5 VIII	Étampes (F)	Cap. Félix	Blériot	70 hp Gnome	10,988
	Chicago (USA)	Lincoln Beachey	Curtiss	60 hp Curtiss	11,578
1911 4 IX	St-Malo (F)	Roland Garros	Blériot	70 hp Gnome	12,824
1911 6 IX	Dinard (F)	Roland Garros	Blériot	70 hp Gnome	16,269
1912 17 IX	Issy, Villalonblay (F)	Geo Legagneux	Morane	80 hp Gnome	18,050
1912 11 XII	Tunisi (TN)	Roland Garros	Morane	80 hp Gnome	18,400
1913 11 III	Buc (F)	Édouard Perreyon	Blériot	80 hp Gnome	19,290
1913 29 XII	St-Raphaël (F)	Geo Legagneux	Nieuport	60 hp Le Rhône	20,060
1914 9 VII	Johannisthal (D)	Gino Linnekogel	Rumpler	100 hp Mercedes	21,653
1914 14 VII	Lipsia (D)	Harry Oelerich	D.F.W.	100 hp Mercedes	25,725

Speed

Date	Place	Name	Airplane	Engine	Speed per Hour
1908 21 IX	Auvours (F)	Wilbur Wright	Wright	24 hp Wright	27.2 mi
1909 31 V	Juvisy (F)	Léon Delagrange	Voisin	45 hp Antoinette	27.9
1909 3 IX	Juvisy (F)	Cap. Ferber	Voisin	45 hp Antoinette	29.7
1909 28 VIII	Reims (F)	Blériot	Blériot	60 hp E.N.V.	47.7
1910 29 X	Belmont Park (USA)	Alfred Leblanc	Blériot	100 hp Gnome	67.5
1911 1 VIII	Eastchurch (GB)	C. Weymann	Nieuport	100 hp Gnome	70.5
1912 9 IX	Chicago (USA)	Jules Védrines	Deperdussin	100 hp Gnome	105
1913 29 IX	Reims (F)	Marcel Prévost	Deperdussin	160 hp Le Rhône	124.5

(continued)

TABLE 2-1 Aviation Records from 1906 to 1914 (continued)

Duration

Date	Place	Name	Airplane	Engine	Time (h m s)
1906 12 XI	Bagatelle (F)	Santos-Dumont	Santos-Dumont	50 hp Antoinette	0 0 21
1907 26 X	Issy (F)	Henri Farman	Voisin	40 hp Vivinus	0 0 52
1908 13 I	Issy (F)	Henri Farman	Voisin	50 hp Antoinette	0 1 28
1908 21 III	Issy (F)	Henri Farman	Voisin	50 hp Antoinette	0 3 39
1908 11 IV	Issy (F)	Léon Delagrange	Voisin	40 hp Vivinus	0 6 39
1908 30 V	Roma (I)	Léon Delagrange	Voisin	50 hp E.N.V.	0 15 26
1908 6 VII	Issy (F)	Henri Farman	Voisin	50 hp Antoinette	0 20 19
1908 6 IX	Issy (F)	Léon Delagrange	Voisin	40 hp Vivinus	0 29 53
1908 21 IX	Auvours (F)	Wilbur Wright	Wright	24 hp Wright	1 31 25
1908 18 XII	Auvours (F)	Wilbur Wright	Wright	24 hp Wright	1 54 53
1908 31 XII	Auvours (F)	Wilbur Wright	Wright	24 hp Wright	2 20 23
1909 27 VIII	Béthény (F)	Louis Paulhan	Voisin	50 hp Gnome	2 43 24
1909 27 VIII	Béthény (F)	Henri Farman	H. Farman	50 hp Gnome	3 4 56
1909 3 XI	Mourmelon (F)	Henri Farman	H. Farman	50 hp Gnome	4 17 53
1910 9 VII	Reims (F)	Labouchére	Antoinette	50 hp Antoinette	4 19 0
1910 10 VII	Reims (F)	Jan Olieslaegers	Blériot	50 hp Gnome	5 3 5
1910 28 X	Étampes (F)	Maurice Tabuteau	M. Farman	70 hp Renault	6 0 0
1910 18 XII	Étampes (F)	Henri Farman	H. Farman	50 hp Gnome	8 12 23
1911 1 IX	Buc (F)	Fourny	M. Farman	70 hp Renault	11 1 20
1912 11 IX	Buc (F)	Fourny	M. Farman	70 hp Renault	13 17 57
1914 4 II	Johannisthal (D)	Langer	L.F.G. Roland	100 hp Mercedes	14 7 0
1914 24 IV	Étampes (F)	Poulet	Caudron	50 hp Gnome	16 28 56
1914 24 VI	Johannisthal (D)	Basser	Rumpler	100 hp Mercedes	18 10 0
1914 28 VI	Johannisthal (D)	Landmann	Albatros	100 hp Mercedes	21 50 0
1914 10 VII	Johannisthal (D)	Boehm	Albatros	100 hp Mercedes	24 12 0

Distance

Date	Place	Name	Airplane	Engine	Distance
1906 14 IX	Bagatelle (F)	Santos-Dumont	Santos-Dumont	50 hp Antoinette	8.6 yd
1906 12 XI	Bagatelle (F)	Santos-Dumont	Santos-Dumont	50 hp Antoinette	244.4 yd
1907 26 X	Issy (F)	Henri Farman	Voisin	40 hp Vivinus	855.5 yd
1908 13 I	Issy (F)	Henri Farman	Voisin	50 hp Antoinette	0.625 mi
1908 21 III	Issy (F)	Henri Farman	Voisin	50 hp Antoinette	1.25
1908 11 IV	Issy (F)	Léon Delagrange	Voisin	40 hp Vivinus	2.50
1908 30 V	Roma (I)	Léon Delagrange	Voisin	50 hp E.N.V.	7.7
1908 6 IX	Issy (F)	Léon Delagrange	Voisin	40 hp Vivinus	15.3
1908 17 IX	Issy (F)	Léon Delagrange	Voisin	40 hp Vivinus	41.5
1908 21 IX	Auvours (F)	Wilbur Wright	Voisin	24 hp Wright	60.9
1908 18 XII	Auvours (F)	Wilbur Wright	Wright	24 hp Wright	62
1908 31 XII	Auvours (F)	Wilbur Wright	Wright	24 hp Wright	77.5
1909 26 VIII	Reims (F)	Henri Farman	H. Farman	50 hp Gnome	112
1909 3 XI	Mourmelon (F)	Henri Farman	H. Farman	50 hp Gnome	150

TABLE 2-1 Aviation Records from 1906 to 1914 (continued)

Distance

Date	Place	Name	Airplane	Engine	Distance
1910 10 VII	Reims (F)	Jan Olieslaegers	Blériot	50 hp Gnome	139.5
	Reims (F)	Jan Olieslaegers	Blériot	50 hp Gnome	245
1910 28 X	Étampes (F)	Maurice Tabuteau	M. Farman	70 hp Renault	290
1910 30 XII	Étampes (F)	Maurice Tabuteau	M. Farman	70 hp Renault	362.7
1911 16 VII	Kiewitt (D)	Jan Olieslaegers	Blériot	50 hp Gnome	393.7
1911 1 IX	Buc (F)	Fourny	M. Farman	70 hp Renault	448.3
1911 24 XII	Pau (F)	Gobé	Nieuport	70 hp Gnome	460
1912 11 IX	Étampes (F)	Fourny	M. Farman	70 hp Renault	633
1914 28 VI	Johannisthal (D)	Landmann	Albatros	100 hp Mercedes	1178

He flew the distance in 26 hours, 17 minutes and 33 seconds.

Many engineers, designers, and manufacturers were responsible for the improvements in the power, speed, and safety of airplanes. **William B. Stout** produced a successful all-metal airplane in 1924 which later led to the Ford Trimotor passenger transport.

Commercial airlines developed during this period and began carrying airmail and then passengers. This, however, is the story of the growth and development of air transportation. Throughout the early years of its development, until the end of World War I, the airplane had yet to be used commercially. The first commercial use of the airplane took place in 1918 when the United States Post Office began scheduled airmail flights. Throughout the history of airplane development human beings achieved their goal of flight and have now embarked on the next step—space flight and reaching for the stars.

Female Aviators

Many women became famous as fliers. **Amelia Earhart** (1899–1937) was the first woman to fly across the Atlantic in 1928 and to solo across the Atlantic in 1932. She disappeared during an attempted round-the-world flight in 1937. **Ruth Nichols**, who learned to fly in 1922, was the first woman ever to hold three international records at one time for altitude, speed, and distance. In 1932, she became the first woman airline pilot. **Jacqueline Cochran** won the Bendix Transcontinental Race in 1938 in a competition with nine male fliers. She also set a number of national and international records. In 1953, Cochran became the first woman to fly faster than the speed of sound. **Bessie Coleman** became the first African-American woman in the world to earn a pilot's license.

Claude Dornier

Claud Dornier was an industrialist who was born at Kempten, Germany, May 14, 1884. His early career as an engineer in metal construction and super structures provided him an excellent background for his later involvement in aircraft construction. After graduation from the Munich Institute of Technology in 1910, he served briefly with the Graf Zeppelin Company where he was introduced to aviation. Unable to persuade Zeppelin officials to diversify into airplanes, he formed his own company with his brother and enjoyed immediate success with flying boats and some of the world's earliest, large, all metal aircraft. His DO-X was the largest passenger airplane in 1929. In the face of widespread skepticism concerning its practicality, the giant 12-engined craft flew from Germany to New York in 1931. In 1932, Dornier turned his talents to military aircraft, developing two distinctive thinfuselaged, twin-engined bombers, the DO-17 and the

DO-217, which became mainstays of the Luftwaffe's bomber and reconnaissance fleet in World War II. As a consequence of World War II, Dornier lost his business. In 1947, he moved to Zug, Switzerland where he consulted in aeronautical engineering until his death in 1969. Claude Dornier's vision coupled with his technological skills permitted him to lead the world in the advancement of metal aircraft. In 1987 he was installed into the International Aerospace Hall of Fame.

The Treaty of Versailles of 1919, which banned construction of engine-powered aircraft for fifteen years in Germany, forced Claude Dornier to look elsewhere in order to continue his aircraft business. Most of the Dornier flying boats included the two that were purchased by Roald Amundsen for his 1925 Arctic expedition. It is interesting to note that the huge Savigliano hangar that was used for major structural repairs on the DO-X, still exists in Marina di Pisa. Dornier moved to Spain and then to Switzerland due to the aircraft production prohibition. Dornier's com-pany was banned from producing aircraft in Germany from 1945–1955 under a general government ban on aircraft production. The company continued to exist and was owned by Claude Dornier until his death in 1969. His sons managed it after his death. The company eventually became a part of the Fairchild Aircraft Company.

DO-X

Ten years before the building of the first Boeing Flying Boat Clippers, Dr. Claude Dornier in Germany created world interest with the DO-X. It was the largest, heaviest and most powerful aircraft in the world at the time. He anticipated using it for transatlantic passenger service. The German Transport Ministry financed it. It was a colossal flying boat powered by twelve engines. It had a wingspan of 157½ feet, a length of 131 feet 5 inches and a height of 33 feet. Takeoff weight was 110,250 pounds. It had a metal hull and wings of metal framework covered with the fabric. It had seating capacity for 66 to 100 passengers depending on the length of the flight. On October 21, 1929, with 169 persons aboard consisting of 150 passengers, 10 crew members and nine stowaways, it weighed 48 tons when it took off from Lake Constance, the water boundary between Germany and Switzerland. It took fifty seconds to lift off the water surface. It slowly ascended to a maximum of 650 feet; it flew forty minutes at a top speed of 105 mph before successfully landing on the lake. Never before had that many people been carried aloft at one time in a single aircraft. The famous Graf Zeppelin was limited to seventy passengers. The test was successful and a flight to the United States was planned.

The twelve engines were Siemens-Halske Jupiters placed on top of the wing in six tandem tiers, delivering 6000 horsepower. They could lift fifty-two tons of weight off the water in fifty seconds and attain a top speed of 134 mph. The aircraft could remain airborne on eight engines. The triple-decked duralumin hull was equal to that of a ship; its main passenger deck was 64 feet long. On board was a corridor with doors opening into sleeping compartments leading to a luxurious lounge with soft carpets, expensive chairs, deep plush cushions, and a phonograph. The cabin also contained a recreation room, bathroom, kitchen, and dinette.

The aircraft proved to be vastly under powered. To correct this, Dornier substituted Curtiss Conqueror engines for the European engines—thus increasing horsepower to 7200.

At last the planned flight started from Lake Constance on November 5, 1930 with a crew of nineteen under command of Captain Friedrich Christiansen. On board were 180,000 pieces of mail. The route was along the coast of Africa then across the Atlantic Ocean to South America. The DO-X reached New York on August 27, 1931, by following the route flown by Pan American Airways. The crew was given a ticker-tape parade and received medals from Jimmy Walker, the mayor. The DO-X created public interest wherever it went. However, the grand tour was less than triumphal because of a series of unpredictable events. It took ten months to reach New York, a distance of 12,000 miles. After visits to Holland and England, the leviathan was forced down at sea by fog and had to taxi sixty miles to a French harbor for repairs. Repairs took more than a month. A wing was damaged due to a fuel tank fire. After continuing the tour, the hull was badly damaged in Las Palmas, Canary Islands, by waves during takeoff. This repair took three months.

The twelve engines consumed 400 gallons of fuel an hour. To permit a maximum fuel load, the crew was reduced to twelve for the next long flight and the airplane was stripped of all expendable weight, including spare clothing. However, it still could not gain altitude and flew the 1400 miles to Fernando de Noronha at an altitude up to thirty feet.

At New York it was placed in dry-dock for the winter for overhaul. It left for Germany on May 19, 1932 with a woman added to the crew. She was a German actress and a pilot named Antonie Strassman. She became the second woman to cross the Atlantic in a heavier-than-air-craft.

Two more DO-X airplanes were constructed and designated the X2 and X3. They were sold to Italy where they were used primarily by the military for prestige flights. In 1933, Mussolini ordered them to be destroyed.

Little more was heard of the DO-X after that. It was eventually placed in the Berlin Air Museum in 1934 but was destroyed in 1943, during World War II, by allied aerial bombardment.

While the DO-X was not a commercial success, it was an important experiment. It was extraordinary because of its physical strength and size and earned a place in seaplane history.

Hercules HK-1 SPRUCE GOOSE

This aircraft is probably the largest aircraft ever flown. There was critical need to fly over enemy submarines that were sinking Allied ships during World War II, especially in the Atlantic Ocean. It was originally conceived by Henry Kaiser, famous for the production of liberty ships during World War II. He contacted Howard Hughes who had acquired fame as a designer, builder and pilot of fast aircraft. Originally designated the HK-1 for Howard Hughes and Henry Kaiser, it later became the H-4 Hercules when Henry Kaiser withdrew from the project. This huge flying boat became known as the "Spruce Goose."

The project was a remarkable story of sacrifice, determination, and technological development. The HK-1 was far ahead of its time in the early 1940s. It revolutionized jumbo aircraft having large lift capability.

The aircraft was a cargo-type flying boat designed to transport men and materials over long distances. It had a single hull with fixed wing-tip floats as large as small yachts, cantilever wing and tail surfaces. The entire airframe and surface structures were composed of laminated wood, primarily birch not spruce. All primary control surfaces except the flaps were fabric covered. The hull was divided into two areas: a flight deck for the operating crew and below it a large cargo deck. Access between the two decks was provided by a circular stairway. Below the cargo decks were fuel bays divided by watertight bulkheads. Each engine had 28 cylinders, with two spark plugs each, totaling 56 spark plugs per engine, or 448 spark plugs per aircraft. For in-flight inspection and repair, the engines were accessible through the spar of the 13-foot thick wing. A sight station located behind each engine allowed a mechanic/engineer to observe the engine.

While being developed it encountered and dealt with tremendous design and engineering problems. These included the testing of new concepts for largescale hulls and flying control surfaces; the incorporation of complex power boost systems that gave the pilot the power of a hundred men in controlling the Hercules. Eight of the most powerful engines available were installed. A fuel storage and supply system was created to allow for the long, over-water flights. Hughes and his engineers built the aircraft with non-essential materials, which some industry persons considered impossible. All of this was accomplished under the constrictions of building materials during World War II.

The Hughes flying boat was to be the biggest airplane ever built and probably the most prodigious aviation project of all time. Only the courage and solitary dedication of Howard Hughes, and his small development group, caused this project to advance. He was subject to much criticism and doubt about the aircraft's ability to fly. During a senate investigation committee, a disgruntled United States Senator dubbed it the "flying lumberyard." This resulted in Hughes' announcement that he was going to begin taxi tests in Long Beach Harbor, near where it was constructed. On November 2, 1947, Howard Hughes and a small crew taxied the airplane for taxi tests and thrilled thousands of people assembled along the shore. This was as a result of criticism by the congressional committee as well as many other skeptics. The purpose was to show that it could operate on the water and not sink. After making a few long powered taxi runs, Howard Hughes, at the controls, lifted the flying boat seventy feet off the water. It flew one mile in less than a minute at a speed of eighty miles an hour before gently settling on the water for a perfect landing. The flight was ample proof that Hughes was correct. The event is recognized as a great moment in aviation history. Hughes, feeling vindicated, took the aircraft back to its dry dock. It never flew again. Many years later it was put on display in Long Beach Harbor next to the Queen Mary. After being there for many years, where it was seen by millions of visitors, it was disassembled and floated to Seattle, Oregon, where a museum is being built to house it.

The Spruce Goose is now regarded as a true American icon and deserves to be preserved so future generations can see this true marvel.

Because of its uniqueness, specifications for this flying boat are listed below for comparisons to other aircraft now known or planned in the future.

CHAPTER 3

World War I

Country	Aircraft	Trained Pilots
France	260	171
Russia	100	28
Germany	46	52
Great Britain	29	88
Italy	26	89
Japan	14	8
United States	8	14

TABLE 3-1 Aircraft and Trained Pilots in 1914
Source: Holley, *Ideas and Weapons*, p. 29.

In 1913, the state of aeronautical advance in the United States was primarily represented by the accomplishments of Glenn Curtiss. Conversely, leaders in Europe had invested heavily in aircraft technology. Competitions were regularly sponsored to encourage advances in aircraft speed, range, and altitude. Europeans had also incorporated aircraft units in their armed forces prior to the war. In 1913, the United States had only six pilots in the entire U.S. Army.

As in most wars, technological advances in weapons and support were greatly accelerated between 1914 and 1918. The United States was late entering the conflict (1917) and did not participate in the major aircraft innovations that occurred during the war. European manufacturers and designers had jumped ahead in aircraft and engine design, partly out of necessity. On the Allied side, the French Nieuport, followed by the SPAD, manufactured by the French company Societie des Productions Armand Deperdussin (hence the acronym), and the English S. E. 5, Sopwith Pup and Sopwith Camel, provided the fighter aircraft. In 1918, close to the end of World War I, Glenn Martin was responsible for contributing the only American design for combatant aircraft in the war with his MB-1 bomber. On the Axis side, the German manufacturers Junkers and Albatros Werke Gmbh produced formidable fighter aircraft, but the Fokker designs proved to be the best, particularly the D.VII, which is widely regarded as the best fighter of the war. This aircraft was flown by Hermann Goering, who was to become the confidante of Adolf Hitler in the 1930s and the leader of the German Luftwaffe during World War II.

Although the Curtiss Aeroplane and Motor Company was the largest aircraft manufacturer in the world during the war, producing 10,000 planes by 1918, the JN-4 did not come close to matching European models in speed, power, and reliability. First produced in 1916, the Jenny mounted the OX-5 engine, 90 horsepower and water-cooled. When Curtiss improved his O model engine in 1913, he wanted to publicize its advances by designating it the "O Plus." But neither the "O Plus" nor the "O+" designation looked particularly good when printed, and the "+" could even be confused with the letter "T." Someone suggested rotating the Plus sign by 45 degrees, depicted as "OX," and the new series of engines became known as the OX-2.

The Curtiss models, although behind the Europeans, were higher performance machines than those built by the Wrights. The Wright Company had fallen by the wayside in aircraft design by World War I, its designs being almost entirely based on the outmoded Flyer models.

Manufacturing in the United States upon its entry into the war was made subject to the oversight and control of the Aircraft Production Board, which decreed that the United States should gear up to produce 22,000 aircraft for delivery to France within a year. Isolated between two great oceans, America was far removed from the vast destruction war had brought to Europe just since 1914, and the world was about to appreciate how valuable the heavy manufacturing reserve of the United States could be. But America was too far behind the design performance standards of aircraft already in use in Europe, so much of the American production effort was limited to manufacturing aircraft under license from European designers.

The DeHavilland DH-4, a single engine bomber/observation plane, was the primary military airplane built in the United States during the war. The Dayton-Wright Company built the most, 3,106 planes, followed by Fisher Body at 1,600, and Boeing, which produced 150.

The Wright Company concentrated on the production of engines under license, notably the 150 horsepower Hispano-Suiza aircraft engine, much in demand by the French. The French government in 1915 placed orders in the United States for 800 engines, which were required to be built in the United States due to the lack of capacity in Europe during the war. The Wright Company contracted to supply 450 of these. To facilitate filling the order, the Wright Company arranged a merger with Glenn L. Martin in 1916 to form the Wright-Martin Aircraft Company. By the end of the war, the company had produced over 10,000 of the engines, known as the Wright-Hispano. In 1919, the Martin combination was dissolved and the company became the Wright Aeronautical Corporation. Much was to be heard from this company for its contributions to aircraft engine development during the 1920s.

With the entry of the United States into the war, the Aircraft Production Board ordered the production of 44,000 American-built aircraft engines to be used in conjunction with the ambitious goal of manufacturing over 22,000 aircraft. The immediate problem was, however, that the United States did not possess an aircraft engine capable of providing sufficient horsepower or speed for military airplanes. Packard Motors happened to have in its design inventory an experimental, but tested, eight-cylinder automobile engine that was to prove to the basis for America's greatest contribution to the war effort. On May 29, 1917, automobile engineers at Packard began a redesign of the engine with the purpose of supplying the ordered military aircraft engine, and five days later, a revised design was presented for aircraft use. But it was still an automobile engine, having battery ignition instead of magnetos, for example. It was redesigned again, this time expanding its power to twelve cylinders like the British Rolls engine, and with magneto ignition. The new design, the water-cooled Liberty, weighed only 710 pounds and, producing 410 horsepower, it surpassed the performance of all other aircraft engines in the world. By war's end, some 17,935 Libertys had been produced, of which 5,827 had been delivered to Europe for use in aircraft there. The Liberty was installed in the DH-4, and by November 1918 deliveries to the Army numbered 3,431 airplanes. Of these, 1,213 arrived in Europe, but only 248 ever flew at the front.

When the Armistice was signed, so many airplanes and engines had been produced for war use that engine and airplane manufacturing literally stopped cold. Surplus equipment was everywhere, and it was cheap. Curtiss Jennys were so numerous that, for $500, a student pilot could receive his instruction and upon solo be awarded a Jenny in the bargain.

As peace settled once again over the world, as the railroads were returned to their owners by the government, as the automobile began hitting the open roads being built by the government, and as all of the planes appeared to be sitting on the ground, many wondered what would become of aviation in America.

The National Advisory Committee for Aeronautics (NACA)

Although the United States had lost the initiative in aviation even before the start of World War I, in 1915 the government acknowledged the importance of aviation to the country when it established the National Advisory Committee for Aeronautics (NACA). From the beginning, NACA has been on the cutting edge of aeronautical research, its mission being to "direct and conduct research and experimentation in aeronautics with a view to their practical solution."

It was apparent to those in the Wilson administration that the United States lagged far behind European countries in its aviation capabilities, and those in government and in aviation found the situation untenable, to say nothing of the loss of national pride. As we will see, this competition between Europe and America has continued into the twenty-first century. Europe was clearly ahead of the United States in individual and governmental accomplishments, organized research, and government funding for military aviation. In 1917, NACA established the Langley Memorial Aeronautical Laboratory in Virginia for the pursuit of advanced test and experimentation in aeronautics.

Among its initial activities, NACA advised the national government on such matters as the desirability of beginning the airmail service, including the feasibility of night flight for mail delivery purposes. NACA was instrumental in securing the Manufacturers Aircraft Association agreement that curbed patent litigation and resulted in the sharing of ideas and inventions during World War I.

As we progress through the history of aviation and the role of government in this book, we will continually come back to the role of NACA as appropriate to illustrate its specific role in the evolution of aviation and the law related to aviation in the United States.

CHAPTER 4

Commercial Aviation in the U.S. pre-WW II

It was not long after the Wright brothers were first successful in marketing their airplane to the French and to the U.S. Army, in 1908 and 1909, that the idea occurred to someone in the Post Office Department that the airplane could be useful in delivering the mail—and faster than the railroads. Federal funding for airmail delivery was not forthcoming in spite of a bill introduced in Congress in 1910 by Congressman Morris Sheppard for that purpose. Beginning in 1911 without specific government funding, limited experimentation with airplanes hauling mail (15 pounds a load) was initiated. Congress was not convinced that the entire process of flying mail to a point over a United States Post Office, and dropping it from various heights to the ground, was not too hare-brained to be dignified by appropriations. Only in 1916 did Congress finally approve limited funding ($50,000 from the "Steamship Fund") for the establishment of a trial airmail route, in large part because of the rapid improvement and reliability of aircraft. In 1918 specific funding was finally approved (the Sheppard bill had been hung up in Congressional debate for 8 years) with a $100,000 appropriation for the purchase, operation, and maintenance of airplanes for use by the Post Office Department.

A Rough Beginning

Operations began by using airplanes and pilots furnished by the Army Signal Corps. It soon was clear that the airmail experiment was, in reality, a training device and exercise for the Army, and that delivery of the mail often amounted to an afterthought. It also became clear that the lack of training and experience of Army pilots, particularly in cross-country flying and navigation, was going to be a problem. Otto Praeger was Second Assistant Postmaster General of the United States from 1915 to 1921. He believed that the carriage of mail by air would be a logical next step in mail service to the country, and he also believed that the carrying of mail would have the secondary benefit of proving the use of the airplane for commercial purposes. After World War I, it seemed that business interests in the United States could not figure out how to put the airplane to any beneficial or productive purpose. This was the age of barnstormers, daredevils, adventurers, and a sideshow mentality that overshadowed most other thinking on the subject of airplanes. Banner towing, the selling of rides, and the occasional charter hop from one municipality to another was about the extent of commercial benefit associated with aviation. Besides, flying was fraught with danger.

The aircraft available after World War I were numerous, but they were mostly JN-4s, the latest version of which was the H model. This airplane had an average speed of 50 miles per hour, 60 tops, and could carry some 150 pounds of mail. The route fixed as the first experimental airmail route was between New York and Washington, D.C., a distance of 218 miles, with an intermediate stop at

Courtesy of the National Postal Museum, Smithsonian Institution.

FIGURE 4-1 President Woodrow Wilson at the inauguration of airmail—May 15, 1918.

Courtesy of the National Postal Museum, Smithsonian Institution.

FIGURE 4-2 Major Reuben Fleet (on the left) briefs airmail pilot Lt. George Boyle before he begins his flight on May 15, 1918.

Philadelphia, and the date set for its inauguration was May 15, 1918. An airplane would depart both New York and Washington at the same time. In Washington, President Woodrow Wilson was in attendance, attesting to the magnitude and portent of the event, as was Otto Praeger and other Post Office dignitaries. (See Figure 4-1.)

The pilot selected for the Washington departure, Lt. George Boyle, was chosen more for his family contacts than for either his experience or skill. (See Figure 4-2.) As the President watched, Lt. Boyle called "contact" and the propeller was pulled through for start, but nothing happened. After several attempts, amid an embarrassing silence from the august assembly, someone thought to check the airplane's gas tank. It was empty. Upon being filled, the engine coughed to life and presently brand-new airmail pilot Boyle was finally airborne, and the airmail service had been launched, much to the relief of the Post Office and Army officials gathered there. (See Figure 4-3.) But there was yet another problem.

A pilot wishing to fly from Washington to Philadelphia is required to follow a generally northerly course, owing to the fact that Philadelphia is north of Washington. Lt. Boyle, however, turned to the south shortly after take off and landed in a pasture farther away from Philadelphia than when he started. The day was saved by the southbound mail, which arrived in Washington three hours and twenty minutes after it left New York. The second leg of the northbound route, from Philadelphia to New York, was salvaged when Boyle's difficulties became known, whereupon the second leg pilot loaded his airplane with Philadelphia mail and took off for New York.

Scheduled Airmail Service

The experimental airmail service continued for about three months, until August 10, 1918, with an impressive record of 88% completion of flights attempted. The experiment using Army personnel had come to an end, and since it was the intention of the Post Office to use civilian pilots to operate the new, permanent airmail system, six new pilots were hired and new planes were put in service. On August 12, 1918, the world's first regularly scheduled airmail service was begun between New York and Washington. On May 15, 1919, service was commenced between Cleveland and Chicago, the first segment of what was to ultimately become the transcontinental airmail route of the United States Post Office. Service on the segment from New York to Cleveland was deferred due to the adverse terrain, the Allegheny Mountains, which lay between those two cities.[1] Attempts to inaugurate that service in December 1918 had failed due to the

FIGURE 4-3 Lt. George Boyle takes off for Philadelphia.

Courtesy of the National Postal Museum, Smithsonian Institution.

fact that every airplane sent aloft had been forced down by weather. But by July 1, 1919, that service had been begun as well. The New York-Chicago route segment would come to be known as the "graveyard run," and it would claim the lives of eighteen airmail pilots.

The mail was mostly flown in Curtiss Jennys from the beginning of experimental service, but it was clear that more powerful and larger airplanes were needed. The Army had developed an appreciation for Glenn Martin's airplanes during World War I, and was about to order the improved MB-2 bomber when the Post Office took over airmail delivery from the Army. In 1919, the Post Office applied some of the Congressional airmail appropriation to order six Martin MPs (mail planes), specially designed with nose cargo compartments capable of holding up to 1,500 pounds of mail, which were put into service in 1919 and 1920. Pilots crashed four of the new Post Office MPs on the New York to Chicago route, and the Post Office finally transferred the other two to the Army Signal Corps.

The Jennys gave way to the De Havilland DH-4 with Liberty engines, also leftovers from the war, that had earned the name "flying coffins" because of their propensity to catch fire on crashing, a not uncommon occurrence. These planes generally had as instrumentation an airspeed indicator, an altimeter of sorts, and an oil compass. The planes had to be flown visually, by reference to horizon, sky, and land outside of the cockpit. Navigation was also by outside reference, referred to as "pilotage," "contact flying," or "dead reckoning."[2] Landmarks on the ground were the prime navigational reference for these early cross-country pilots, and when clouds or fog obscured these, finding one's way became problematical, indeed. The airmail service did not operate at night; the mailbags were delivered to the trains for continuation of the journey until the next day, when once again the mail flew.

These two incapacities severely hampered the fledgling airmail service from fulfilling its promise.

- First, the lack of instrumentation to fly "blind" or by instruments alone, was a problem that had to be addressed by the aircraft manufacturers, their vendors, and by the pilots themselves.
- Second, the lack of any navigational infrastructure by which airplanes might find their way at night or in adverse weather conditions was a problem too big for individuals or the fledgling aircraft community.

A navigational infrastructure was an undertaking for government.

The first airmail pilots like, Max Miller and Wesley Smith, began pushing the limits of "blind" flying, usually in order to extricate themselves from situations inadvertently encountered, like flying into clouds or fog. Smith is said to have taken a half empty bottle of whiskey aloft, which he placed on top of his instrument panel, to practice flying wings level with the whiskey level. Soon, he found that a curved tube filled with liquid and containing a ball, like a carpenter's level, was available, and this he fastened to his instrument panel. And so it went. It was found that turns made at a constant, steady rate could be timed and the airplane could be rather accurately rolled out on predetermined headings. Sperry introduced a two axis gyroscopic instrument that allowed a pilot to determine whether his airplane had inadvertently entered a turn. This was followed up with a three-axis instrument that disclosed changes in pitch attitude.

At the same time, experimentation was proceeding on various fronts, including with radio, not only as a means of voice communication from air to ground, but also as a means of navigation. By May 15, 1920, airmail service had been extended westward from Chicago to Omaha, Nebraska, establishing a through route all the way from New York. On August 16, 1920, a route was added southward from Chicago to St. Louis. On September 8, 1920, the transcontinental route was completed to San Francisco. Although the airmail service operated only during daylight hours, the railroad coast-to-coast mail time was bettered by 22 hours.

The promise was yet unfulfilled. Moreover, Otto Praeger was concerned that the entire airmail program might be cancelled if better results were not soon achieved. Night flying was the only way to free the airmail service from its earthbound dependence on the railroads. Flying at night had been experimented with, and successfully under certain conditions, like clear, moonlit nights, for short distances. But what about transcontinental distances on a regular schedule? Could it be done?

Transcontinental Airmail

On February 22, 1921, the first attempt at a through, continuous transcontinental airmail service was made. The plan called for a westbound plane to leave New York to fly the initial segment of the route to San Francisco, and an eastbound plane to leave San Francisco initiating the first segment to New York. (See Figure 4-4.) The trip each way would be sequentially flown by fresh airplanes and pilots, like the Pony Express, handing off the mail at predetermined points along the route. Two aircraft were assigned to begin at each end of the route.

The first airplane to leave New York discontinued shortly after take off. The second plane flew to Chicago but was grounded due to weather. The first plane out of San Francisco crashed in Nevada, but the second plane made it to Reno, and 12 hours after leaving San Francisco, the mail arrived in Cheyenne. Another plane took the mail on to North Platte, Nebraska, and there it was turned over to the next segment airmail pilot, Jack Knight.

A combination of ground personnel and volunteers built bonfires along Jack's route, which was to be traversed at night, and he made it to Omaha, his segment complete, by 1:00 A.M. There

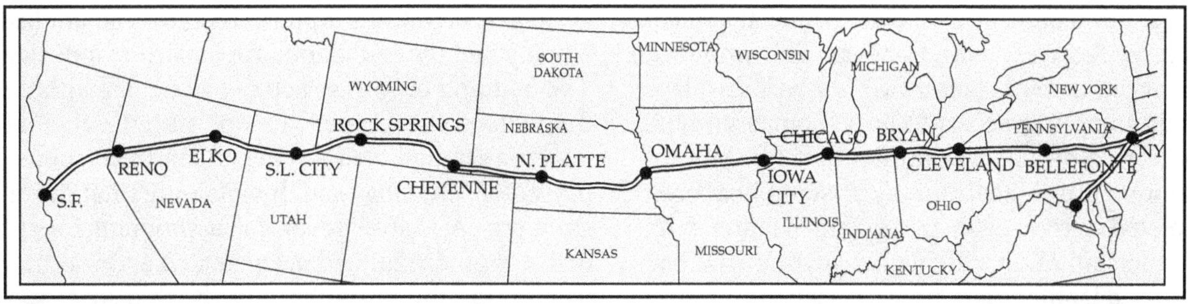

FIGURE 4-4 Original Transcontinental Airmail Route—compare to the Transcontinental Railroad, Figure 3-3.

he learned that the plane scheduled to meet him in Omaha had not left Chicago due to weather. He volunteered to continue, armed only with an automobile road map to guide him over unknown terrain, a landscape he had never flown. This part of the country is chilly in February, and this night was accompanied by cold, ice, and snow, along with the low clouds that produce snow. He was unable to land at Des Moines, Iowa, as planned. He continued to Iowa City, and arriving, searched for the airport that was unlighted because the ground crew had left for the evening, believing that no sane person would fly in the prevailing conditions. A lone employee at the airfield heard his engine, lit a flare and watched as Jack Knight glided in with an empty gas tank. After refueling and accepting a quick cup of coffee, Knight gamely flew on to Chicago, finally landing at Checkerboard Field at 8:40 A.M. From there, the mail relay was continued to New York, and when the results were announced, the mail had been successfully carried coast to coast in slightly more than 24 hours.

The best that the Post Office had been able to do up to that time using the railroads was a transcontinental transit of 3 days. The experimental policy of flying the mail during daylight hours and handing the mail off to the railroads at night had only marginally improved savings in time, and was generally considered to be not cost effective. But with the grand experiment of February 21, 1921, it was now clear that flying the mails for the entire route could be done.

The success of this first attempt caused Congress to favorably consider appropriations sought by the Post Office, granting a splendid sum for that time, $1,250,000, for airmail extensions. Paul Henderson, who became Second Assistant Postmaster General in 1922, was committed to the Otto Praeger principle that the mail could be flown. But it was clear that the mail had to be flown both night and day, and that bonfires as a means of nighttime navigation probably had only the most limited of possibilities.

The Lighted Airway

In 1923, Congress granted an appropriation to fund the construction of a system of sequential lighting on the transcontinental airmail route. The first segment completed was between Chicago and Cheyenne, a stretch of flat country most conducive to this original effort, and centrally located along the route. Airplanes launched at first light on either coast could reach the lighted segment by nightfall under most circumstances.

Beacons were placed on 50 foot steel towers constructed every ten miles along what had come to be known as the "airway." The beacons rotated so as to allow pilots to see their flash from 40 miles away in good weather. If the beacon were located at a landing field, the beacons showed a green course light, if not, then red. Morse code, which is still used as part of the FAA navigation scheme today, was introduced into the airway system at the time of this first construction. Each beacon flashed an identifier in Morse code that corresponded to the number of the beacon within the airway segment.

Regular scheduled airmail service was begun on July 1, 1924, just over six years from the inauspicious kick-off in Washington, D.C. with Lt. Boyle. Due to prevailing winds, eastbound mail crossed the nation in 29 hours, while that mail bound for the West took 34 hours. This provided a savings in time of at least two full business days in mail going from coast to coast compared with mail carried by rail. Within the first full year of operation, the lighted airway system had been completed coast to coast. American aviation, with the leadership of an enlightened government, had made a quantum leap into a technologically-advanced civilian navigation system. Nothing like it existed in the world.

The American scene was almost set for the beginnings of a viable commercial aviation transportation system, but not quite. Still to come were more reliable engines, sturdier airplane designs, flight by reference only to aircraft instruments, involvement of the banking and financial community, acceptance by the public of aviation as legitimate transportation and, lastly, those visionaries and adventurers, both physically and financially, who would make it all happen.

The Privatization of Airmail

The railroads had been taken over by the government during World War I. At the end of the war it was argued by some that the government should remain in control and operation of the railroads, as was the practice in Europe. But that was not to be in this country, and the railroads were again delivered to the owning railroad companies. The railroads had for decades delivered the intercity U.S. mail, and with the return of the railroads to their owners, they once again were the mainstay of the mail system. The airmail service operated by the Post Office was actually a relatively small part of the total intercity carriage of mail, but for several reasons the railroads were vigorously opposed to the government being in the long distance mail carriage business, as were all sectors of private enterprise.

It was never the plan for the government to remain in the airmail transport business, anymore than it was its plan to be in the railroad business. The time had come in 1925 to turn over the operation of the transportation of the airmail delivery system to private enterprise. The government had already laid the foundation for an airline industry. The building of infrastructure had begun with the funding of the lighted airway system, and the government had clearly undertaken the responsibility for creating and maintaining that infrastructure. A viable government program for the delivery of airmail had been created and was now in place as the basis for subsidy paid to private carriers. A small but growing aircraft and engine manufacturing sector had emerged and had demonstrated its engineering, production, and fiscal competence, and the banking and financial sector had been drawn into the picture and was becoming comfortable with financing the fledgling industry. The confidence of the public in the legitimacy and safety of aviation and the desirability of air travel, while still skittish, was beginning to grow.

From the Wright's first flight in 1903 to the middle of the 1920s, there had been only two known attempts to start a scheduled passenger flying service, or airline, in the United States. The first was the St. Petersburg-Tampa Airboat Line inaugurated on January 1, 1914 to serve the 18-mile over water route between the two Florida cities with a 26-foot Benoist XIV flying boat. Although the little air service carried 1,200 passengers over the twenty-three minute route during the next several months, at a fare of $5.00 each, business sagged with the departure of the northern tourists and their money in the spring, and the company folded.

Manhattan Cadillac dealer Inglis M. Uppercu was the force behind the other airline attempt. Uppercu ran an aerial sightseeing service in New York that he had started as an offshoot of having manufactured seaplanes (Aeromarine Corp) for the Navy during World War I. In 1920, he bought out a small Key West-to-Havana mail line and began to supplement the cargo with passengers. Prohibition (which outlawed the sale of alcoholic beverages) was in effect in the United States beginning in 1920, and Uppercu correctly figured that Cuba and the Bahamas, with their plentiful rum and sunshine, would make a fetching destination for thirsty and cold Americans.

He formed Aeromarine West Indies Airways and during his first year carried 6,814 passengers in seven flying boats, flying 95,000 miles. He noted that people who seemed to be terrified of flying at altitude over land appeared to have no fear of flying a few feet above water in a flying boat. The airline published schedules and met them. During its second year, the fleet was expanded to 15 aircraft, which carried 9,107 passengers on two thousand flights. After two widely publicized accidents, resulting in the deaths of several passengers, and receiving exceedingly bad press that emphasized the complete absence of any kind of government mandated safeguards for the flying public, the bloom was off the rose. Uppercu shut down his airline in 1923.

Henry Ford, who by the middle of the 1920s was quite successful as the manufacturer of automobiles, saw that he had a legitimate business use for airplanes in the middle 1920s. Ford Motor Company had automobile plants in various locales, including Detroit, Cleveland, Chicago, and Dearborn, and it was necessary to carry parts and machinery between them on a regular basis. Ford became acquainted with William E. Stout, an idea-man and former airplane designer, who had a passion to build an all-metal airplane. The craft would be built of duralumin, not quite as light as aluminum but twice as strong. Ford decided to back Stout who did, in fact, produce a single engine high wing monoplane constructed almost completely of metal, all as advertised. Its corrugated metal sides and thick wings looked remarkably like those produced in Germany by the Junkers Company, but no one said anything. It was powered by the Liberty water-cooled engine, carried 8 passengers and was dubbed "Maiden Detroit."

Ford not only bought the plane, he bought the plant as well. He started Ford Air Transport and began a regular service between his plants. He then set Stout on a course to develop a bigger all metal airplane, one that would go down in history as the Ford Trimotor.

In 1924, Stout's efforts resulted in the trimotor Ford 3-AT, a bulbous-nosed monstrosity configured with the pilot seated in an open cockpit above the high-winged fuselage. The design was so horrendous that Ford retired Stout and turned the design function over to his team of engineers, which included William McDonnell. McDonnell's name was destined to later lead the merged McDonnell Douglas Corporation in 1967.

The story goes that progress on converting the mongrel 3-AT to a more aesthetically pleasing and efficient design was slow, until one day in 1926 when a Fokker F-7 trimotor monoplane showed up in Dearborn, under the command of Admiral Richard E. Byrd. The airplane was gratuitously hangared for the night at Ford's field, and it is said that Ford's design team did not get much sleep that night. In due course, the Ford 4-AT trimotor emerged from the Ford team's plans and sketches, bearing a striking likeness to the Fokker F-7. The Ford Trimotor, affectionately dubbed the "Tin Goose," sported the same heavy cantilevered wing without wire bracing as did the F-7, and its dimensions and engine mountings were similar. The airplane conveyed a sense of sturdiness and stability to the fourteen passengers it could carry at 100 miles per hour over a distance of 250 miles. Refinements in this basic design were continued into the 1930s, with 199 Trimotors ultimately produced.

The Airmail Act of 1925 (The Kelly Act)

"Recently a man asked whether the business of flying ever could be regulated by rules and statutes. I doubt it. Not that flying men are lawless. No one realizes better than they the need for discipline. But they have learned discipline through constant contact with two of the oldest statutes in the universe—the law of gravity and the law of self-preservation. Ten feet off the ground these two laws supersede all others and there is little hope of their repeal."

Walter Hinson, 24 July 1926, *Liberty Magazine*

Congress passed the Kelly Act (so-called after the name of the bill's sponsor) on February 2, 1925. The act was appropriately titled "An Act to Encourage Commercial Aviation and to Authorize the Postmaster General to Contract for the Mail Service." The statute called for the Postmaster General to seek competitive bids to operate the airmail feeder routes to the transcontinental main airmail trunk line between New York and San Francisco.

Advertisement for bids was published in the middle of 1925, and bids were received from 10 companies. Although eight routes were to be awarded, financial responsibility concerns caused the Post Office to delay assigning some of them until later.

Seven contract airmail (CAM) routes were awarded at the beginning of 1926:

1. CAM 1-*Boston-New York*, awarded to a group including Juan Tripp, later to found and operate Pan American Airways. Colonial Air Transport operated the airmail service.
2. CAM 2-*Chicago-St. Louis*, awarded Robertson Aircraft Corp., a forerunner to American Airlines. Robertson hired Charles Lindbergh as Chief Pilot.
3. CAM 3-*Chicago-Dallas*, awarded to National Air Transport, a forerunner of United Airlines.
4. CAM 4-*Salt Lake City-Los Angeles*, awarded to Western Air Express, a forerunner of TWA.
5. CAM 5-*Elko, Nevada-Pasco, Washington,* awarded to Varney Air Lines, a forerunner of United Airlines.
6. CAM 6-*Detroit-Cleveland*, awarded to Ford Air Transport.
7. CAM 7-*Detroit-Chicago*, awarded to Ford Air Transport.

Ford was the first to begin service, on February 15, 1926, with the others following within four months. The last to begin service was Colonial Airways on CAM 1. Subsequent awards that year were:

8. CAM 8-*Los Angeles-Seattle*, awarded to Pacific Air Transport, a forerunner of United Airlines.
9. CAM 9-*Chicago-Minneapolis*, awarded to Charles Dickenson. Northwest Airlines began operating the route in 1926.
10. CAM 10-*Atlanta-Jacksonville*, awarded to Florida Airways Corp, a forerunner of Eastern Air Lines.
11. CAM 11-*Cleveland-Pittsburgh*, awarded to Clifford Ball, later absorbed by United Airlines.
12. CAM 12-*Pueblo, Colorado-Cheyenne, Wyoming*, awarded to Western Air Express.

The next year saw more awards of airmail routes. National Air Transport (NAT) added the lucrative Chicago to New York mail route to its Chicago-Dallas service. In addition to this, Western Air Express began flying passengers on a non-airmail route between Los Angeles and San Francisco. By 1928, over five thousand souls had flown the route with Western Air.

The aircraft available to serve the new airmail companies were limited. The Post Office had largely relied on the World War I DH-4, but its Liberty engines were pretty much used up. Varney had to begin airmail service with the underpowered Swallow biplane, and Western Air Express bought the Douglas M-2, all six of them. Ford had the first of the Ford Trimotors (see Figure 4-5), producing fourteen in 1926. Juan Trippe of Colonial Air Transport, impressed by Fokker's monoplane design and its absence of wires and struts, ordered the first three Fokker trimotors produced, but they would not be available until 1927. In 1928, Western Air added Fokkers for its Los Angeles-San Francisco passenger service.

The Post Office Department continued to make awards across the country for its feeder routes, as it phased itself completely out of the airmail carrying business. During the nine years that the Post Office Department carried airmail, 32 pilots—about one out of every six—were killed in the service.

One fall day in 1926, William E. Boeing, who had flown the first international mail route between Vancouver and Seattle in 1919 under contract with both Canada and the United States,

Courtesy of the Minnesota Historical Society.

FIGURE 4-5 Ford Trimotor.

learned that the main trunk transcontinental airmail route was soon to be offered for bid. The Chicago-San Francisco segment would be first. The most advanced airplane so far built by Boeing was the Model 40, but it had the old water-cooled Liberty engine of World War I fame. The route between Chicago and San Francisco was challenging viewed from any angle: the weather, the Rockies, the distance, and the fact that flying at night was a requirement. The B-40, with the Liberty engine, was a marginal airplane for this route, and both the businessman and the pilot in William Boeing knew it.

> "I've tried to make the men around me feel, as I do, that we embarked as pioneers upon a new science and industry in which our problems are so new and unusual that it behooves no one to dismiss any novel idea with the statement that "it can't be done!". Our job is to keep everlasting at research and experimentation, to adapt our laboratories to production as soon as possible, and to let no new improvement in flying and flying equipment pass us by."
>
> **William E. Boeing, founder, The Boeing Company, 1929**

As we have seen, Fred Rentschler, under the auspices of Pratt & Whitney Aircraft, had designed and produced the first air-cooled engine of any significant horsepower under contract for the Navy. The Wasp was in production primarily for the Navy at this time, but the Navy was not at war, and there were no rumors of war. An arrangement was made for Boeing to step ahead of the Navy for delivery of these engines, at the rate of five per month, and the Boeing 40A, with a single Wasp engine mounted up front, was born. Boeing low-balled the transcontinental airmail bid and won, knowing that carrying the mail in the B-40A was as much about selling airplanes as it was about carrying airmail. Varney Airlines, on the route between Elko and Pasco, was the first to order the Boeing with the Wasp.

Although the Kelly Act had been twice amended to increase the rate of compensation paid to airmail contractors, these airline pioneers needed and wanted to carry passengers along with the mail. It was generally acknowledged that the passengers who flew with the mail were, more than anything, adventurers of similar stripe to the pilots themselves. Most of the airplanes available in the middle 1920s were more suited to carrying cargo or other inanimate objects than human beings. The 40B-4, a revised version of the 40A, had design changes that allowed 4 passengers to be seated inside the fuselage forward of the open pilot cockpit, with 2 windows on each side. The B-4 layout provided relative comfort for the passenger, but flying was still a frightening thing for many.

The Founding of the Airlines

It is a curious historical fact the two men who arguably did the most for the advancement of commercial aviation in the 1920s were completely unacquainted while they each made their early contributions, but would wind up being related to each other by marriage. Dwight Morrow did not know of Charles A. Lindbergh when, in 1925, President Coolidge selected Morrow to head a board appointed by the President to investigate circumstances relevant to the fledgling aviation community and to establish a national aviation policy. Morrow was a Wall Street banker, intelligent, and well connected. He took his appointment seriously and conducted very effective hearings on the subject.

The Morrow Board

His board heard the testimony of Herbert Hoover, then Secretary of Commerce in the Coolidge administration, which said that the government was obliged to lend its support to commercial aviation, as it had always done in the maritime industry. Hoover pointed out that the government had for a century maintained aids to navigation in the coastal waters of the country, provided education and competency standards for ships' officers, required inspections of ships, and funded improvements in and about the navigable waters, including ports. He noted that the twenty-five years since the flight of the Wright brothers in 1903 had brought little advance in commercial aviation, and that America was lagging the Europeans in engaging the subject of transport by air.

The Morrow Board heard from another strong voice in support of governmental action. By the early 1920s, the National Advisory Committee for Aeronautics (NACA)[3] had become a loosely organized group of scientists and engineers who were developing into leaders in aeronautical research and experimentation. NACA conducted pure research in its Langley Laboratory unconstrained by bureaucratic influences. Independence from political pressures contributed greatly to NACA becoming the premier aeronautical research facility in the world beginning in the 1920s. By the time the Morrow Board was convened, NACA had even then gained a level of respect that caused the Board to heed its recommendations. NACA laid the foundation for initiating the examination and licensing of pilots and the imposition of airworthiness standards for aircraft, as well as for the creation of an Aeronautics Branch within the Department of Commerce to administer these activities.

Based on all of the testimony produced before his board, Morrow prepared a report that was to become the blueprint for the development of commercial aviation for years to come. The report concluded:

1. Aviation is vital to the national defense. The means of aircraft design and production must be supported in the national interest, and a military procurement program should be initiated.
2. Non-military aviation, comprising the largest potential for commercial development, serves a national purpose, and deserves the support of the government.
3. The government should enhance the safety and reliability of flying by establishing standards for pilots and aircraft. It should establish and maintain airways for navigation and enlarge its support for airmail contract carriers under contract with the Post Office. All this would have the collateral effect of bolstering both public and banking confidence in aviation.

The Morrow board was central to the second major federal statute affecting commercial aviation, the Air Commerce Act of 1926.

The Air Commerce Act of 1926

Prior to this enactment, there had been no official government statement identifying what role, if any, the federal government would play in the field of aviation. There had been no structure, no plan, no strictures, and no standards. In one fell swoop all of this uncertainty vanished, and in its place was laid a solid foundation for the building of a national commercial aviation industry.

The purpose of the act was to promote air commerce. It specifically charged the federal government with the obligation of creating and maintaining a national system of navigational aids and of adopting rules and regulations to promote safety of flight.

The Department of Commerce, in turn, was charged with the responsibility of promulgating and enforcing safety regulations, including the registration and licensing of aircraft, producing aeronautical charts, providing meteorological advice and reports, investigating accidents, and certification and medical examination of pilots. The Aeronautics Branch of the Department of Commerce was created to administer and carry out the requirements placed on the Department. This agency was renamed the Bureau of Air Commerce in 1934 and assumed all safety responsibilities. The Interstate Commerce Commission assumed all rate and fare authority.

The black letter law was on the books, the Commerce Department had its marching orders, the banking community had taken note, the manufacturing sector was enthused, and the entrepreneurs were emerging. Still, the hearts and minds of the public were with the railroads. Those in government and in aviation wondered how the public imagination could be captured.

Lindbergh

Charles Lindbergh had been hired by Robertson Aircraft, one of the original airmail contractors, following a short career in which he fully qualified as an all-around daredevil. He parachuted from a plane in 1922, even before he had soloed an airplane for the first time. He adopted an itinerate life first as a wing-walker and stunt man and then as a barnstormer pilot. With Robertson, he flew the mail between St. Louis and Chicago, a route known for its range of temperatures and volatile weather.

An offer of $25,000 prize money had been made in 1919 by a New York businessman, Raymond Orteig, to anyone who successfully completed a non-stop flight between New York and Paris. Although the Atlantic had been successfully crossed in 1919 in three separate efforts, including one non-stop flight from St. Johns, Newfoundland to Clifden, Ireland, no one had succeeded in claiming the prize by 1927. Several attempts had been made during the intervening years, including French World War I ace, Rene Fonck, in 1926. In early 1927, Fonck was rumored to be readying another attempt, and Admiral Richard E. Byrd was said to also be preparing to make the crossing in his Fokker trimotor. Advances in technology by 1927 made the chances of success increasingly likely, and the race was heating up with great publicity.

Lindbergh was backed by a group of St. Louis businessmen, but his budget was limited to $15,000. No airplane existed for that sum of money that had any chance of making the 3,600-mile flight successfully. He decided to fly solo, a controversial decision in an otherwise foolhardy endeavor, but a decision that lent itself to a smaller airplane, one that could possibly be built for a cost within his budget. The Ryan Airplane Company, a small aircraft manufacturer located in San Diego, California, agreed to build the airplane to his specifications for $6,000, plus the cost of the engine. He decided on the Wright Whirlwind engine, whose endurance had been proven earlier in 1927 when two pilots kept their Bellanca aloft with it for a period of 57 hours.

Lindbergh decamped to San Diego where he supervised the construction. Although the airplane type had never before been built (it was a custom job), it was completed in 77 days, and with the Wright Whirlwind installed, the total price was $10,580. To save weight, the Spirit of St. Louis, named in honor of his backers, had no brakes and no radio. Gasoline tanks occupied the forward portion of the cockpit where a windshield would normally be placed. To see forward he was required to use a small periscope. The airplane's range was 4,200 miles, just 600 miles over the flight-planned distance necessary to reach Paris.

Lindbergh had accumulated just over two thousand hours of flying time, but his airmail experience had given him exposure to practically all types of weather conditions. He felt that he was ready. He flew the Spirit of St. Louis from San Diego to New York on what was really a "shake down" flight, stopping in St. Louis to refuel, and in the process he set a coast-to-coast record of slightly less than 22 hours. The press coverage of the transcontinental flight only served to heighten the public attention that had been building.

The Spirit of St. Louis left Roosevelt Field on Long Island at 7:52 A.M. on May 20, 1927, with 450 gallons of gasoline, half the total weight of the airplane. Thirty-three hours and thirty minutes later, Parisians flooded the field at Le Bourget to welcome Lindbergh, and the entire world was entranced.[4]

"Science, freedom, beauty, adventure: what more could you ask of life? Aviation combined all the elements I loved. There was science in each curve of an airfoil, in each angle between strut and wire, in the gap of a spark plug or the color of the exhaust flame. There was freedom in the unlimited horizon, on the open fields where one landed. A pilot was surrounded by beauty of earth and sky. He brushed treetops with the birds, leapt valleys and rivers, explored the cloud canyons he had gazed at as a child. Adventure lay in each puff of wind.

I began to feel that I lived on a higher plane than the skeptics of the ground; one that was richer because of its very association with the element of danger they dreaded, because it was freer of the earth to which they were bound. In flying, I tasted a wine of the gods of which they could know nothing. Who valued life more highly, the aviators who spent it on the art they loved, or these misers who doled it out like pennies through their antlike days? I decided that if I could fly for ten years before I was killed in a crash, it would be a worthwhile trade for an ordinary life time."

Charles A. Lindbergh, *The Spirit of St. Louis*

If the flying feat itself were not enough to sufficiently impress the mind, then the proceedings that followed, conducted on the world scene, would certainly do the trick. He went on a triumphant tour of European capitals, and was given audiences with the kings of Belgium and England. President Coolidge sent a United States warship to fetch the young Lindbergh home, where he was met by the dirigible U.S.S. Los Angeles and a ticker tape parade. He was awarded the Congressional Medal of Honor and commissioned a colonel in the Army Reserve. He was also introduced to Dwight Morrow.

He went on a three-month tour, sponsored by the Guggenheim Fund, of all 48 states, parading in 82 cities, and flying over 22,000 miles in the process. He was a fine hero, conducting himself at all times in his trademark modest and dignified manner. (See Figure 4-6.) He was invited to Mexico by Dwight Morrow, who was then ambassador there, for a Mexican tour and then for a sojourn through Latin America. Ambassador Morrow's daughter, Anne Spencer, met Lindy on one of his visits to the ambassador's residence in Mexico and, mutually taken with each other, in due course they were married.

It would be difficult to overstate the effect that Lindbergh had on the nascent airline industry in the late 1920s. Dormant aviation stocks across the board ignited as money poured in from all quarters. In 1926, total passenger enplanements in the United States had numbered less than six thousand. By 1930, the flourishing airline industry carried over 400,000 adventurous souls. Production of aircraft soared.

Back in New York, what might have been the first of all celebrity endorsements occurred when Lindbergh joined the new airline, Transcontinental

Source: Library of Congress.

FIGURE 4-6 Charles Lindbergh pays a visit to Orville Wright at Wright Field, Dayton, OH, June 22, 1927.

Courtesy of www.panamair.org.

FIGURE 4-7 Juan Trippe and Charles Lindbergh.

Air Transport (TAT), lending his name to a commercial product in return for cash and stock. Juan Trippe, (see Figure 4-7) having been deposed from Colonial Air Transport, also signed him up as a technical adviser to his new airline venture, Pan American Airways. This was the beginning of a long-standing relationship between Lindbergh and Trippe[5] that would play a key role in the expansion of air commerce around the world, and with it, American influence.

The Daniel Guggenheim Fund for the Promotion of Aeronautics

In the middle of the 1920s, aviation in America was emerging from its long period of confusion and stagnation. But aviation in Europe had captured the imagination of the people and of industry almost immediately after the Wright brothers' tour of Europe in 1908–9, and it still led the way. The United States government had been significantly involved in promoting aviation at least since 1918 with subsidy and direct investment in infrastructure, and in 1926 it would begin to legally promote safety and standards in aviation in order to boost the public confidence. But the job was big. The list of individual citizens with ardent interests in aviation in the 1920s was long, and the record of their contributions was even then impressive. Corporate America, as well, had shared in the promotion and advancement of aviation. But the going was still slow.

One of the more beneficial byproducts of the American system of private enterprise is the philanthropic activity of its successful practitioners. The giving of one's time, interest, and assets to causes of one's choosing is a time-honored tradition in America. The Guggenheim family of New York made its money in the mining industry. In 1924, Daniel Guggenheim and his wife, Florence, established a foundation to promote a variety of charitable causes. One of their sons, Harry Guggenheim, was a pilot during World War I and became committed to the advancement of aviation. Father and son established, in 1926, a separate fund called the Daniel Guggenheim Fund for the Promotion of Aeronautics, and by 1930 the family had given almost $3 million to aviation-related projects.

The Guggenheims believed, like the government, that the public would embrace air travel if confidence could be established in its safety. A significant part of the safety of air travel depended on reliably designed and constructed aircraft, yet there was no such thing as an aeronautical engineer in 1925. The Guggenheims began to fund the establishment of schools at universities across the country, and by 1929 aeronautical engineering programs or research centers had been set up at the California Institute of Technology, Stanford, the Massachusetts Institute of Technology, Harvard, Syracuse, Northwestern, the University of Michigan, and others.

In 1927, the Guggenheims offered a prize of $100,000 for the construction of safe aircraft in a contest called the "Safe Aircraft Competition." The prize went to the Curtiss Tanager, which had incorporated into its design the first short takeoff and landing (STOL) characteristics ever demonstrated.

The Fund subsidized an operation in California by Western Air Express (WAE) in 1928 known as the "Model Airway," between Los Angeles and San Francisco. Airmail was not carried on this route since no award had been made by the Post Office to WAE. Instead, the airline carried only passengers along the corridor in an effort to show that commercial passenger service was feasible without airmail subsidies, as well as safe and reliable. The Fund provided a Fokker F-10 Super Trimotor to WAE for this scheduled service and had implemented a weather-reporting regimen along the route that utilized two-way radio. This was the first PIREPS (pilot weather reporting system) in history. The passenger operations ended in 1929, and although the experiment was not profitable, it did demonstrate that passenger only (non-airmail) service was feasible and popular with the public. At a time when one of the most frequent causes of airplane crashes was allocated to adverse weather conditions, not one weather-related incident was recorded. Upon the termination of the experiment in June 1929, the weather bureau assumed the reporting of aviation weather locally, and the practice ultimately spread all across the country.

When Charles Lindbergh in 1928 suggested that it would be helpful to navigation if the names of cities and towns could be painted on the roofs of large buildings, the Guggenheims funded the cost. The Postmasters of some 8,000 communities arranged for the painting on rooftops of their towns' names in large letters, with arrows pointing to the north and, if available, to the nearest landing field.

The Guggenheims made a very significant contribution to advances in instrument flying by funding research involving gyroscopic instrumentation invented by Elmer Sperry (directional gyrocompass and artificial horizon) and Paul Kollsman (precision altimeter). Guggenheim-funded engineers worked with the Aeronautics Branch of the Commerce Department and the Bureau of Standards to advance radio navigation, and with Jimmy Doolittle to test and implement instrument procedures in 1929 that made safe instrument flight routine within a decade.

Airways—from Lighted Beacons to Radio Navigation

By the end of 1927, the government had extended the lighted portion of the airway system from New York to Salt Lake City on the transcontinental route, and on portions of feeder and parallel segments, such as Los Angeles to Las Vegas, New York to Atlanta, Chicago to Dallas, and between Los Angeles and San Francisco. That year there were 4,121 miles of lighted airways operated by the Aeronautics Branch of the Department of Commerce. By 1933, there were 1,500 beacons in place, extending the lighted airway systems for a length of 18,000 miles. While the lighted airway was of significant aid in navigation, it had serious limitations in the context of an all-weather air carrier system. It was still a visual navigation system, dependent on reasonably good weather in order to operate.

The Bureau of Standards in the Department of Commerce began, in 1926, to work with radio as a means of communication and navigation. As government involvement in aviation began to kick in as a result of the Air Commerce Act, money and effort were applied to solve problems and to attempt to eliminate limitations on the commercial development of air commerce. In 1926, for instance, there was no two-way voice communication possible with aircraft in flight. This constituted serious limitations in safety, including a lack of pilot awareness of developing weather. By 1927, the first transmitter was established at Bellefonte, Pennsylvania, allowing communication with aircraft in a 150-mile radius.

In 1928, the Bureau of Standards developed a new radio beacon system of navigation, the first non-visual navigation system in the world. The Aeronautics Branch, which had authority over the lighted airway system, took over the installation and control of the new radio navigation system in 1929. The system was known as the "four course radio range," and it would provide the first step in allowing a true all-weather air carrier system to begin to develop, and it would remain the standard navigation system in use until World War II.

The four course radio range utilized low frequency radio waves (190 to 535 kHz radio band) transmitted from powerful 1,500-watt beacons spaced 200 miles apart on the airway. The beacons transmitted two Morse code signals, the letter "A" and the letter "N." In Morse code, these signals are opposite, "dot-dash" for A, and "dash-dot" for N. When the aircraft was centered "on the beam," these signals merged into a steady, monotonous tone. If the aircraft ventured to one side of

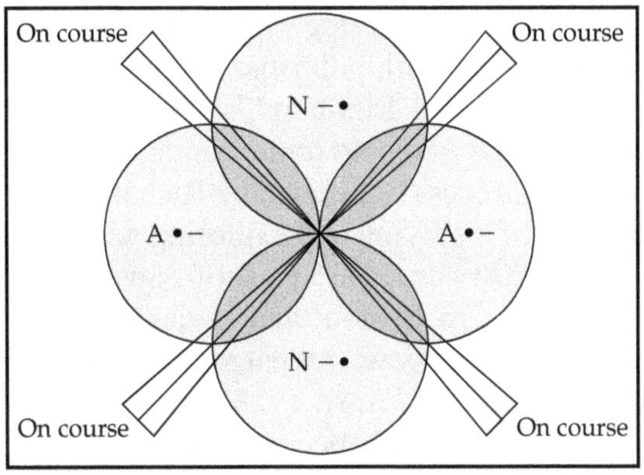

FIGURE 4-8 Schematic of the four-course radio range.

the airway, the signal heard was either the Morse A or N, depending on the aircraft's position from the beacon. (See Figure 4-8.)

Each beacon defined four airways, thus the name "four course radio range," and the beacon's identification was broadcast in Morse code twice each minute. The so-called beam width was 3 degrees, so that at the half-way point of 100 miles between beacons, the on-course deviation was about +/–2.6 miles. Station passage was marked by a "cone of silence," at which point the aural tone would disappear as the aircraft passed overhead. Distance from the station was later provided by marker beacons placed along the airway at intervals of 20 miles or so.

By today's standards, the four course radio range was primitive. Low frequency radio was subject to electrical static and other weather aberrations and distortions, but it constituted a quantum leap forward over the visual, lighted beacon system in use at the time. Pilots became very adept at flying the four course system, and as the airlines began establishing schedules on their new routes, all-weather navigation allowed adherence to schedules that theretofore would have been impossible due to cancellations for weather and low visibility conditions.

Amelia Earhart

Lindbergh was not to be the only aviation celebrity of the late 1920s. Building on the women pioneers before her,[6] Amelia Earhart was to emerge shortly after the Lindbergh flight as the foremost female aviator up to that time, and arguably, even to the present day. (See Figure 4-9.) She certainly captured the public imagination in much the same way that Lindbergh had, and thereby contributed to the enthusiasm that helped to create the airlines.

When Earhart soloed in 1922 at a small field in south Los Angeles, California, there were fewer than 100 female pilots in the United States. Her flight instructor, Neta Snook, was one of those women. The right to vote had been achieved by women only three years before, in 1919. Earhart purchased a yellow Kinner "Airster" prototype for $2,000 and began making a visible impression in the area, being featured in the ***Los Angeles Examiner*** declaring that she intended to fly across the continent. She immediately set a new women's altitude record (of 14,000 feet) and was advertised as one of only two female pilots in an air meet at the Glendale airport. She was awarded a flying certification in May 1923 from the Federation Aeronautique Internacionale, although flying licenses were not required, nor even issued by the U.S. government. The certification was necessary in order to have official recognition for any record breaking or record setting achievements.

Source: Library of Congress.

FIGURE 4-9 Amelia Earhart.

In 1924, Earhart moved with her mother to Boston, Massachusetts, arriving there by a "motor-car" which she drove all the way from California by way of Banff, Canada. This was a daring adventure at the time, as the U.S. Army had only performed the first sustained cross-country convoy in 1919. In Boston, her flying activities once again prompted curiosity from local interests, including ***The Boston Globe***, in which she was featured in an interview in June 1927, shortly after the Lindbergh flight. She was billed as "one of the best women pilots in the United States" and began to be mentioned increasingly in the local press. Earhart bore a remarkable resemblance to Charles lindbergh, and some say that was the main reason for the event that set her career skyward. She was to be the first woman to fly the Atlantic Ocean.

> "You haven't seen a tree until you've seen its shadow from the sky."
>
> **Amelia Earhart**

We have already seen what excitement Lindbergh's spectacular solo transatlantic flight had caused in 1927. By 1928, women in both Europe and America were making plans to be the first woman to make the crossing, but with a difference—they planned to have the flying actually performed by male pilots. (See Box 4-1.)

Amelia Earhart had no such plans. The flight that was to make her famous was entirely planned and paid for by others. She had nothing to do with the selection of pilots, the selection of aircraft, or with flight planning. She was, in effect, chosen. At the time, she was making a living as a social worker with immigrants in Boston.

The adventure began in early 1928 when another female patron, the wife of a wealth Londoner, Frederick E. Guest (she was formerly Amy Phipps of Pittsburgh), purchased a Fokker Trimotor from Commander Richard E. Byrd with plans to hire a pilot to fly her to England. Mrs. Guest was soon dissuaded from this venture by her family, but she stuck with the plan for the trip to be made by "an American girl with the right image,"[7] and a committee was formed to find her replacement.

On the committee was George Palmer Putnam, a New York publisher and writer. He had, in fact, published Lindbergh's ***We***, the firsthand account of the first solo transatlantic crossing, and was in the process of publishing Richard E. Byrd's chronicle of his flying and exploring adventures in the book, ***Skyward***. He heard of Amelia Earhart, then residing in Boston, and invited her to New York for an interview. It went well, and the agreement was sealed in April 1928.

Earhart would be a passenger on the "Friendship," a Fokker F-7 Trimotor fitted with floats that was already scheduled for a transatlantic attempt in June. The crew consisted of a mechanic, Bill Gordon, and the pilot, Lou Stultz, the latter of whom was proficient in multi-engine aircraft, float plane flying, and instrument flying. Amelia Earhart had none of these qualifications, nor for that matter, did many male aviators. Nevertheless, she was billed as the "Commander" of the flight, which left Newfoundland on June 17, 1928 and arrived in New South Wales the next day after a flight of 20 hours and 49 minutes.

Gordon and Stultz were soon forgotten, but the public embraced Amelia much as they had Lindbergh. This was a matter of some embarrassment to Earhart, who felt that she had done nothing to deserve such adulation and that the credit should go to the crew, and she had the courage to say so. But the public clamor continued. She received congratulations from many government quarters, including President Coolidge, and under the tutelage of George Putnam, she embarked on a lecture circuit during 1928 and 1929 that gave her worldwide recognition. She thereby became acquainted with aviation luminaries of the time like Admiral Byrd and Colonel Lindbergh. Because of the physical resemblance to Lindbergh and the Atlantic transit similarity, she soon garnered the moniker "Lady Lindy."

After her return from Europe in 1928, Earhart began to earn the celebrity that had been handed to her by fate. She crossed the continent solo from New York to Los Angeles in September 1928. She was swamped with offers to endorse products in advertising media, she began writing articles for national magazines, including

Box 4-1 Who Would Be the First Woman Across the Atlantic?

The first attempt was by Princess Anne Lowenstein-Wertheim, a German aristocrat who for some years had exhibited an affinity for aviation and who held several aviation records in her own right. She departed England on August 31, 1927, bound for Ottawa, Canada with two experienced Royal Air Force pilots, Leslie Hamilton and Fred Minchin, at the controls. Although the aircraft was spotted once enroute, it disappeared.

Another, Ruth Elder, was only a student pilot when she announced in August 1927 plans to fly the Atlantic. Her flight instructor was George Haldeman, and she acknowledged that he would do the flight planning and most of the flying. The route he chose was the longer southern route via the Azores. The aircraft chosen for the flight was a Stinson Detroiter, a single engine monoplane that Elder had named "American Girl." The project was a promotion for the manufacturer, Stinson, and was likened to a publicity stunt. The pair left Roosevelt Field, Long Island, New York, on October 11, 1927, for Paris and completed most of the planned flight route (over 2,600 miles) but the craft was forced down just 300 miles short of their destination by an overheating engine caused by a leaking oil line. A passing ship off the Azores rescued them. Their return to New York was celebrated almost as if the flight had succeeded, which, admittedly, it nearly did. Ruth Elder continued to fly and, in 1929, placed fifth in the first Women's Air Derby.

Frances Grayson was another female patron with money. She hired an all male crew for her attempt in a Sikorsky amphibian, which was readied for the flight at Curtiss Field on Long Island. Ruth Elder and George Haldeman were preparing for their departure at the same time from the same airfield. Francis Grayson flight planned the shorter northern route to Europe. This aircraft took off from Newfoundland on December 23, 1927, and was never seen again.

Cosmopolitan and *McCall's*, and she was hired by Transcontinental Air Transport (the Lindbergh Line) as Assistant to the General Traffic Manager. She acquired a Lockheed Vega and began entering air races around the country and set several speed records.

In 1929, she was largely responsible for inaugurating the Women's Air Derby, a grueling nine-day race from Santa Monica, California to Cleveland, Ohio, dubbed by Will Rogers as "the Powder Puff Derby," a name that has remained with the event. The race was limited to women who had been licensed and who had logged at least 100 hours of solo time. It was estimated at the time that only 30 women could qualify for the event. Twenty fliers started the race, fifteen finished, and there was one fatality. Earhart finished third.

Earhart improved her flying proficiency, particularly in instrument qualification. With her publicist (and now husband) George Putnam, she planned and advertised her intention of becoming the first woman to solo the Atlantic. On May 19, 1932, AE (as she had begun to sign her name) left Harbor Grace, Newfoundland for Paris. Mechanical difficulties enroute, including a leaking reserve fuel line, an inoperative altimeter, and a broken weld on an engine manifold, caused her to alter course for Ireland. Fifteen hours and eighteen minutes after leaving Harbor Grace, AE landed the Vega in a sloping field outside of Londonderry. Her acclaim rose higher and higher, and she had earned it.

Amelia Earhart was destined for even bigger accomplishments, and for tragedy, as the decade of the 1930s unfolded. We will jump ahead in our chronology of the development of aviation to briefly consider the rest of her story. So far, Earhart had only tried to duplicate what men had done. She had set no significant records for the first time in aviation on a genderless basis. But on January 11, 1935, she left Wheeler Field in Hawaii and successfully soloed her new Vega across 2,400 miles of Pacific Ocean to Oakland, California. No person, man or woman, had ever done that before. The Vega performed flawlessly and required only eighteen hours and fifteen minutes enroute. She followed this up with other aviation firsts, a nonstop flight from Burbank, California to Mexico City on April 20, 1935, and from there she flew nonstop across the Gulf of Mexico and on to Newark, N.J., in fourteen hours and nineteen minutes on May 8 that year.

In June 1935, AE took on a new role as visiting aeronautics advisor at Purdue University, which had begun an ambitious plan to develop one of the nation's first academic aviation curricula. As it turned out, this assignment would provide Earhart with the airplane with which she would make her final mark—her unsuccessful attempts to circumnavigate the globe. The Lockheed Electra 10E was purchased by Purdue to be used for research purposes in connection with AE's duties, but soon after taking possession of the Electra in July 1936, she flew it in the 1936 Bendix air race from New York to Los Angeles. Shortly thereafter, the airplane was fitted for long distance flight and the latest radio navigation equipment.

Earhart made two attempts to fly around the world in the Electra, both on a flight-planned route close to the equator of over 29,000 miles.[8] In both attempts she carried navigator Fred Noonan, who had flown the Pacific extensively with Pan American. The first attempt was planned from east to west, beginning in Oakland with the initial stop in Hawaii, and then on to tiny Howland Island. This effort was abandoned when Earhart ground looped the Electra on take off from Hawaii on March 19, 1937, with serious damage to the airplane. The plane was shipped back to the Lockheed plant where it was repaired.

For her next attempt, AE decided to reverse course and fly the route from Oakland to Miami, thence to South America, and on to Africa and points east. The Pacific itinerary was to be the last part of the flight, but it still included Howland Island, which is truly a relative speck of land in the vast Pacific Ocean. By late June 1937, the route had been successfully flown all the way to Lae, New Guinea, a distance of 22,000 miles. On July 2, Earhart and Noonan departed New Guinea for the long over water flight to Howland Island. In spite of all precautions, including the Coast Guard vessel ***Itasca*** standing off Howland to broadcast homing signals, and to plot her position, the Electra never made landfall at Howland. Although she was heard on several occasions attempting to make contact with Itasca, her location was never established, and no trace of her, Noonan, or the Electra has ever been found.

While Amelia Earhart had little direct effect on the establishment of commercial aviation in the United States, her efforts to overcome and transcend the boundaries encountered by the aviation pioneers of the 1920s and 1930s engendered public admiration and a greater acceptance of the new industry of flight. She became a model for both male and female aviators, and like many of them, her time was too short.

From Mail Carriers to Airlines

As the airmail contract carriers began to proliferate, and aviation groups not so lucky as to have garnered an airmail route simply started up, the field was beginning to get crowded. Before 1930, at least 44 different companies had jumped into the airline business. Some were better financed; others more experienced, and possessed of more natural talent and engineering expertise. All were eager.

During the late 1920s, the new aircraft industry came under the control of a few large companies that combined several different disciplines, including airmail carriers, airplane manufacturers, and engine manufacturers. Three groups, in particular, formed the core of these vertical holding companies that were to figure prominently in the future of the airline industry. These groups were

1. United Air Transport
2. North American Aviation (NAA)
3. Aviation Corporation of America (AVCO)

United Air Transport

Bill Boeing and William Rentschler had been personal friends ever since Boeing's airplanes had begun flying with Rentschler's engines. In 1929, amidst the great public enthusiasm generated by Lindbergh's flight, the two friends set up a holding company they named United Air Transport, combining the Boeing Aircraft and Transport Company and the Pratt & Whitney Aircraft works. Included in the new company were Chance Vought and the propeller manufacturer, Hamilton Aero Manufacturing Company. They added Sikorsky, Northrop, Stearman, Standard Steel Propeller, and Stout Airlines.

North American Aviation

North American Aviation was the brainchild of Clement Keys, a vice president of the Curtiss Aeroplane and Motor Company during World War I. He gained control of the company during the early 1920s and began acquiring an assortment of companies, including National Air Transport, which had the lucrative New York to Chicago airmail route. Next, he lined up and bought out several airmail routes operating between the Northeast and Florida (which would ultimately form the basis for Eastern Air Lines), while continuing to fly the mails on the midwestern routes. Before the great Wall Street crash of 1929, he merged the Curtiss holdings with Wright Aeronautical Corporation, creating Curtiss-Wright, completing the triad acquisition of companies capable of airplane building, engine manufacture and air carriage. General Motors would gain controlling interest in NAA for a time in the early 1930s. At this time, NAA's holdings included Eastern Air Transport, Transcontinental, and Western Air, and a substantial interest in Douglas Aircraft.

Aviation Corporation

Aviation Corporation was the work of Averell Harriman and Robert Lehman, New York financiers. Robert Lehman was of the Wall Street banking house of Lehman Brothers. Harriman was to later serve in key government posts, including lend-lease administrator, ambassador to the Soviet Union, Secretary of Commerce, and was to be a confidant of Presidents. Together they salvaged the Fairchild Aircraft Company in 1929 through a stock sale before the financial crash in October of that year. The new company proceeded to acquire five airmail carriers, themselves holders of eleven airmail routes; they bought all sorts of aviation properties, including aircraft and engine plants, and even some airports. The air carriers were combined to form American Airways.

Harriman and Lehman did not remain in the aviation business very long, losing out in a proxy fight to E. L. Cord in the early 1930s. Harriman then began his government service, while Lehman returned to Wall Street. But Lehman would be heard from again, this time with Yellow Cab magnate John Hertz at TWA.

Transcontinental Air Transport—A Novel Idea

Flying was not yet very appealing to travelers in the late 1920s. Forced landings, cancellations due to weather, airsickness from the heat, noise, and sometimes-violent motion of the primitive aircraft combined to make the experience more of an adventure than a reliable mode of transport. The passenger service that was available was short-haul, and there were parts of the country where there was no service at all, like between New York and Chicago, which was the heavily traveled train route through the Allegheny Mountains. The experience of the early airmail pilots over that route, as well as the subsequent trials of the contract airmail carriers, had convinced National Air Transport (Clement Keys), which flew the mails over the route, not to attempt passenger service. Still, Keys believed that airline service was destined to be a marketable and timesaving device for transcontinental passengers.

In 1928, Keys seized upon the idea of providing air transportation by day, then turning the passengers over to the trains for the standard luxurious Pullman service by night. He established a joint venture with the Pennsylvania Railroad in the East and the Santa Fe in the West to round out the full passage. The air carriage portion of the deal was carried out by the newly formed Transcontinental Air Transport (TAT), which had no mail routes, and it had to rely solely on passenger fares for its income. The westbound trip began in New York at Penn Station, where passengers boarded the Pennsylvania Railroad for the overnight run to Columbus, Ohio. Once beyond the Alleghenies, TAT took over at Columbus, flying passengers all day on the next leg to Waynoka, Oklahoma, where they again boarded a Santa Fe overnight train to Clovis, New Mexico. From there, they continued in Ford Trimotor discomfort to Los Angeles. The complete itinerary could be completed, at least on paper, in forty-eight hours, saving a full day over the fastest through train schedule then being operated. (See Figure 4-10.)

The operation, although highly touted by the railroads and accompanied with big-name fanfare (Charles Lindbergh himself was at the controls of the inaugural flight), was really only a novelty. The Ford Trimotors held ten to fourteen seats, but the passenger load was usually only six or seven passengers, and sometimes just three. With no airmail subsidy, the airline lost money, \$2.7 million in eighteen months. The Wall Street crash of 1929 impoverished many of the potential passengers, and by 1930, TAT was barely hanging on.

Progenitors of the Big Four

The premier airlines that actually did the flying were to become known as the "Big Four":

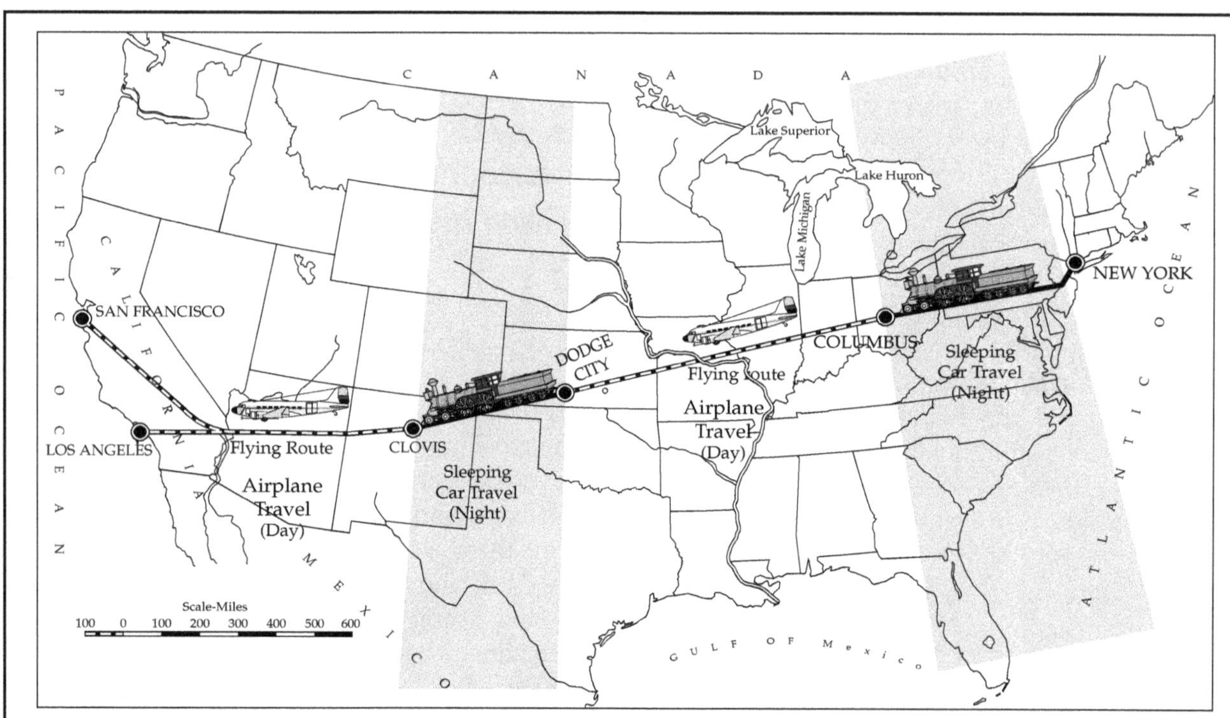

FIGURE 4-10 Route of Transcontinental Air Transport—depicting night and day portions of the route.

1. Boeing-Pratt & Whitney (United Airways) group,
2. the Robertson, Colonial, Texas Air Transport (American Airways) group,
3. the Western Air Express, Transcontinental Air Transport (TAT), Maddux Air Lines (TWA) group, and
4. Eastern Air Transport (Eastern Air Lines). There was also Panagra, the New York, Rio, and Buenos Aires Line (NYRBA), and Pan American Airways, which flew the routes outside of the continental United States. But Juan Trippe had decided that rather than compete within the boundaries of the United States for the available mail routes, Pan Am would simply carry the mail to and from the United States borders—all of it.

Other airlines, like Curtiss Flying Service and Gorst Air Transport, were substantial and proud, but fell outside of the area of main strength.

The Airmail Act of 1930 (McNary Watres Act)

Former Secretary of Commerce Herbert Hoover was elected President of the United States in November 1928, and took office in 1929. He appointed Walter Folger Brown as his new Postmaster General. Brown knew very little about aviation, and less still about the infant airline industry. But he was intrigued, and his appointment carried with it the mandate to "encourage commercial aviation." The Postmaster General was the only government office charged with awarding lucrative contracts to commercial air carriers, oftentimes the only thing that separated such carriers from extinction.

Brown set about to learn as much as possible about the fledgling airline business. He was well aware that, by law, government contracts had to be awarded competitively. He noticed that sometimes desperate, smaller companies foolishly submitted unrealistically low bids in order to get the business, and then played games with the weights of mail carried in order to increase their income. The Post Office set its postal rates, in part, to encourage the public to use the airmail system. The airmail contractors, however, were paid on the basis of weight. The difference between rate and weight amounted, in truth, to a subsidy paid to the carriers. It was not unknown for some operators to mail telephone books or steel machine parts to themselves, paying the Post Office postal rate on the front end, then subtracting that as an operating cost from the much larger air carrier rate charged to the Post Office.

Brown did not like this. It was about this time that a notorious proxy fight was being waged for control of National Air Transport (NAT), which held the New York-Chicago airmail route, between the Boeing-Rentschler group (United Air Transport), which had the San Francisco to Chicago mail route, and North American Aviation (NAA), the holding company to which NAT belonged. Rentschler's proposal to merge NAT with United in order to establish the nations first transcontinental carrier was rebuffed by Clement Keys of NAA, and what appeared to be a contest of personalities and wills had soon exploded. After a series of corporate maneuvers reminiscent of the Gould and Vanderbilt railroad era of the late nineteenth century, accompanied by contentious and unpleasant litigation between the route holders, the United group finally prevailed. United took over National Air Transport from NAA, and consolidated its route system coast to coast. Brown felt that this kind of personal and corporate self-serving activity was wasteful of both assets and energy, and not particularly in the national interest. After all, United now had a monopoly on transcontinental airmail service, which to Brown was not a good thing.

Within a relatively short time after taking office, Brown had formed well-researched and thought-out conclusions about what was wrong with the airmail service. Brown believed that the Post Office was paying too much for the carriage of mail, partly because the airline passenger business had not been developed. He believed that by developing passenger traffic, a new, untapped source of income would become available to the airlines. This new income would then be available

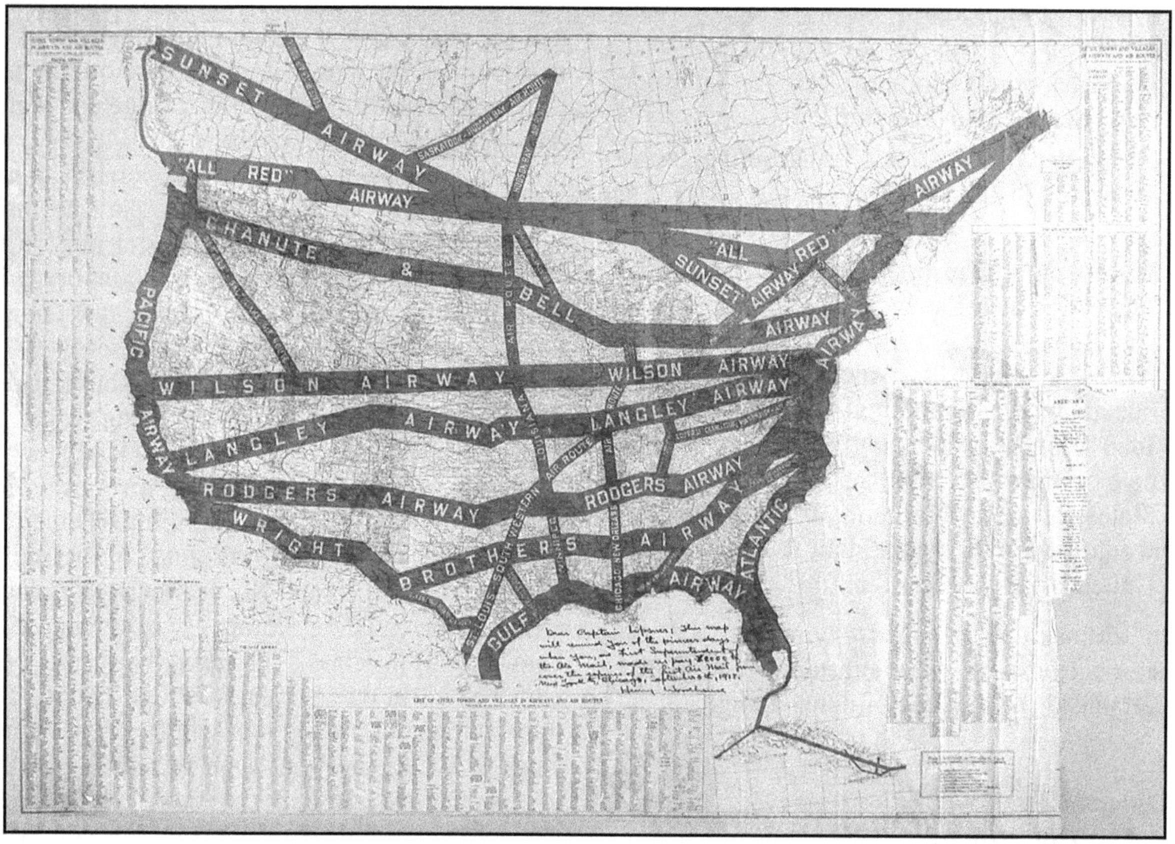

Courtesy of the National Postal Museum, Smithsonian Institution

FIGURE 4-11 Airmail airways.

to help offset the cost of airline operations. Brown thought that tying the mail contracts to the size of aircraft (bigger) and requiring on-board state of the art communication equipment (radios) and instrumentation would have the effect of making airline flying safer and more acceptable to the potential flying public.

He noted that a system of illogical short routes had been created through the process of competitive bidding, and that competitive bidding had resulted in a nonsensical pay schedule to the airmail contractors with rates of pay varying from 62 1/2 cents to the maximum of $3.00 per pound. He believed that competitive bidding on airmail contracts was counterproductive and that a system of appointing qualified, well-financed, and experienced operators rendered a more stable and efficient system. He felt that those operators who had 'pioneered' airmail routes in certain parts of the country, and who had expended effort and money in promoting the airmail system, developing good will, and encouraging aviation should be given preferred consideration in the award of airmail contracts. (See Figure 4-11.)

On December 9, 1929, Brown appeared before the Appropriations Committee of Congress and related his concerns and suggestions about the airmail system in place, and the lack of an efficient passenger service. Members of the Committee were receptive to Brown's ideas and requested that he draft a bill and send it to them as soon as possible.

On February 4, 1930, Brown sent to Congress a proposed bill that contained the majority of his ideas about how to fix the system. The bill expressly allowed the Postmaster General to award contracts based on negotiated rates rather than competitive bidding. Although the bill was reported favorably out of committee, a minority report was filed containing objections by two members of the committee to the proviso allowing awarding contracts on other that a competitive bid. One of those was Representative Kelly, the sponsor of the Airmail Act of 1925 (Kelly Act) and the 1928 Kelly amendments to the Act (reducing airmail postage and providing for ten-year airmail contracts).

The form of bill that was passed by Congress was substantially the same as the draft submitted by Brown, with the exception of the removal of the proviso allowing award of contracts based on negotiation and the removal of references to allowing consideration of 'equities' or 'pioneering rights' to carriers based on prior contributions. The bill became law when signed by President Hoover on April 29, 1930.

Accordingly, the main provisions of the act required that the carriers would no longer be paid on the basis of weight, but by the volume of the available space in the aircraft for the carriage of mail. This encouraged the carriers to invest in larger aircraft to earn more money. It further provided that bonuses would be paid if the aircraft were multi-engined and had certain navigational devices installed.

The act also contained certain provisions relating to the manner in which contracts were awarded. These provisions departed significantly from the traditional, straight competitive bid. The Postmaster was authorized, for instance, to circumvent the actual low bidder in favor of the "lowest responsible bidder." A "responsible bidder" was defined in the act as one that had flown daily scheduled service over a 250-mile route for a period of at least six months. Further, Brown went outside of the actual requirements of the act, and invoking what he considered his "discretionary" authority, subjectively required the applicant to have flown that route regularly not only during daylight hours but at night. In effect, the deck was being stacked in favor of the larger, more experienced airmail carriers.

Immediately after passage of the legislation, Brown summoned to Washington representatives from the major lines around the country, who assembled in the Postmaster General's office on May 19, 1930. It was his belief, he said, that the mail should be carried by substantial, established air carriers, the vast majority of which fell within the ownership of the three largest holding companies, United Air Transport, NAA, and AVCO. He explained his master plan to them, in conjunction with the expressed and unexpressed terms of the new law. He decreed that there could be no monopoly of transcontinental service, but that competition along that route would be required, to the dismay of United. Brown explained, in effect, that the country would be carved up among a few lines, with United flying the transcontinental route to San Francisco, another line flying the New York-Los Angeles route by way of Pittsburgh and St. Louis, and still another line proceeding from New York via Washington, Atlanta, and Dallas, thence on to Los Angeles.

Not surprisingly, the strong-willed leaders of the industrial and financial interests that controlled these carriers were unable to agree among themselves, as directed by the Postmaster General, as to how the country should be split up. Representatives of the carriers remained in Washington, attending meetings with each other until June 4, 1930. On that date, the carriers reported to the Postmaster General that they were unable to agree on allocating the five major routes in the country, including the two transcontinental routes, and submitted the issues back to Brown. The carriers advised the Postmaster General that they would agree to be bound by his decision as to the route awards.

During June and July 1930, negotiations and correspondence continued between the parties with a view toward an agreement that would be fair to all concerned and that would take care of smaller lines having some "equity" due to their "pioneering" efforts. Brown suggested that the central transcontinental route should go to the beleaguered TAT, the plane and train airline, which had been flying without airmail subsidy. TAT had no night flying experience, however, and was ineligible for consideration for the transcontinental route because Brown had added this experience requirement under his "discretionary" authority. Western Air Express, however, did have the requisite night flying experience. Brown, in effect, ordered the merger of TAT with Western Air Express. This merged airline was to be TWA, or Transcontinental and Western Air. As consolation for sacrificing its independence, Western Air Express was allowed to survive as an independent entity and retain its passenger service between San Diego and Los Angeles, and between Los Angeles and Salt Lake City.

United kept its New York-San Francisco route through Chicago, and was allowed to expand northwest. Eastern was assigned New York-Miami, along with Atlanta, New Orleans and Houston. TWA got New York-Los Angeles through St. Louis and Kansas City. American would fly New York-Los Angeles via Nashville, Dallas, and points in the southwest.

Although the smaller lines were not invited to the meetings in May, representatives of several operators showed up, including Southwest Air Fast Express (SAFE), owned by oilman Erie Halliburton, Pittsburgh Aviation Industries, U.S. Air Transport, Curtiss Flying Service, Delta Air Service, and Thompson Aeronautical Corporation. During the summer of 1930, prior to the request for bids being sent out by the Post Office, discussions and negotiations continued. The financial interests of some of the smaller lines were taken into consideration, like SAFE and Delta, and mergers and buyouts were agreed to between them and the larger carriers who would be serving the routes on which the smaller lines had "pioneered." The parties even agreed that Walter Folger Brown would be the arbiter of the value of the stock transactions made to complete the arrangements. Some of the smaller operators received "extensions" of the major routes as additional consideration for the overall agreement.

The airlines paid lip service to the requirements of the Watres Act by going through the motions of competitive bidding with all of the carriers duly submitting bids. The only thing was, none of the Big Four submitted competing bids on the routes that had been assigned by Brown to others. Lower bids on the assigned routes submitted by small carriers were rejected as "not responsible." In this way, modern commercial aviation was born.

Hindsight will not compel a uniform judgment of Brown's actions. It cannot be doubted that the struggling world of commercial aviation was given a mighty boost by the arrangements put in place, and that it evolved at a much accelerated pace over what would otherwise have been the case. At the end of Brown's tenure in 1933, passenger traffic was rising, and the airlines were competing on their transcontinental routes. The cost to the government was less than it had been 4 years before, down from an average of $1.10 per mile in 1929 to half that in 1933, $.54 per mile. The airlines were in good shape financially. It is clear, therefore, that the public interest was served. As we shall see in the next chapter, the 1932 election of Franklin D. Roosevelt as President of the United States (he assumed office in 1933) would have a profound effect on the new commercial aviation community. The allocation of airmail routes and the award of airmail contracts would be the subject of a Congressional investigation, and Walter Folger Brown, himself, would be the subject of intense scrutiny and criticism.

No evidence would be adduced that would even suggest any financial or material gain by Brown. His actions appear to have been the result of a sincere desire to promote aviation, and he did so with success. It is, however, beyond dispute that the procedures employed by Brown were outside of the requirements of the Watres Act. The Congress did not remove the requirement of competitive bidding in 1930, yet that requirement was not observed. The bill that was passed by Congress had removed consideration being given to "pioneering" efforts of some of the operators, yet such consideration was given. In structuring the airmail routes in 1930, Walter Folger Brown proceeded as if his views were superior to the mandate of Congress. With the aid of hindsight, it can be seriously argued that his vision for the future of aviation was far superior to any of his peers.

It should also be noted that Congress had established a benchmark in 1928 with the Foreign Airmail Act, which gave wide discretion to the Postmaster General in awarding overseas mail contracts.[9]

Although the results of Brown's actions would be undone at the beginning of the next administration, the reality is that the Big Four put in place by the Brown policy were still the Big Four for the ensuing 48 years, until deregulation in 1978, in fact. It was then, in 1978, that the country would finally have the chance to glimpse what might have happened during the 1930s had it not been for Walter Brown.

New Deal—The Roosevelt Administration

A New Broom Sweeps Clean

The Great Depression was getting seriously underway in 1933, at the time that the Republican Hoover Administration was vacating office and the Democratic Franklin Roosevelt Administration was sweeping in with reform on its mind. Big business had ruled during the roaring twenties. The stock market increasingly through that decade had reflected in price the explosion in commerce and development, much money had been made and people were happy. But in 1933, the bread lines were long and were filled with disillusioned and angry men. The majority of voters had voted, in effect, to "throw the bums out." And so it was that the Democrats arrived in town with an agenda, a mandate even, to begin to set things straight. In the process, a rare opportunity was seen to make a little political hay and find out who and what was to blame for the mess the country found itself in.

Hugo Black came to the U.S. Senate in 1926 as a Democrat from Alabama, where he had enhanced his political career by winning local judicial elections and with membership in the Ku Klux Klan. While a lawyer there, he mostly sued corporations representing personal injury claimants. He considered himself a populist, and was re-elected to the Senate in 1932. He was a supporter of Franklin Roosevelt's bid for the presidency, and after Roosevelt's election, he was an ardent supporter of New Deal (anti-corporate) initiatives.

Black was given the job of investigating government mail contracts. Soon, the original Brown meetings attended by the heads of the airlines in 1930 became known as the "Spoils Conferences." Small airline operators told of being excluded, and how on the subsequent submission of bids, their lower bids were rejected in favor of higher bids submitted by the larger lines.

To a populist in the 30s, this was rank favoritism, to say nothing of being blatantly against the law. Black used the subpoena power of his senate committee to seize correspondence and documents, using Interstate Commerce Commission agents with synchronized watches to swoop down on the aviation companies without warning. What he discovered led Black to conclude that the airlines had exploited the public through inflated contract rates charged to the government, that the airlines' manufacturing arms had made huge profits from military procurement contracts and that speculation in airline stocks had profited them all. And now the soup lines stood long.

Black was unsympathetic to evidence produced at the hearings that supported the Brown position and that what he had done was in the public interest. Nor was the undisputed fact that the airlines were stable and rendering to the nation the anticipated, though still formative, air transportation system of much consequence to him. Of what interest were airlines to a populace that did not have enough to eat? This was a pertinent, if shortsighted, question.

James Farley was appointed Postmaster General in the new Roosevelt administration. To him fell the duty to carry out the verdict of Black, sanctioned by the President by way of presidential order, that all of the existing airmail contracts held by the airlines then operating were to be cancelled forthwith. The airmail, decreed Farley, would be carried by the Army Air Corps.

Army pilots were basically untrained in cross-country flying and had neither knowledge of nor experience in flying the routes that the mails took across the country. Their airplanes were all open cockpits and contained few of the instruments that had become standard in just a few short years due to Brown's enticements to the airlines. By the end of the first week of flying, five pilots had been killed in accidents and six were critically injured. The Army pilots began flying only in daylight hours. Within five weeks, twelve Army pilots had died. It became clear that a major mistake had been made. New bids to reinstate private carriage of the airmail were called for, but under a revised set of rules.

A temporary arrangement had to be put in place immediately, pending adoption of legislation. Postmaster General Farley called a meeting of airline representatives, like Walter Folger Brown before him. But this time, none of the airlines involved in the "spoils conferences," nor any of the executives who attended them, could participate in the new round of bidding. No bidding airline could also be involved in the manufacture of aircraft designed to be used in the airline business, like the Boeing-Rentschler combination at United. Vertical holding companies that exercised control over the actual airlines were disallowed. Rectitude reigned supreme in public, but in private, practicality ruled the day. Cosmetic name changes by the Big Four were accepted as serious compliance with the new rules, changes like American Airways becoming American Airlines and Eastern Air Transport becoming Eastern Air Lines. In fact, after the bids were opened on April 20, 1934, the commercial airline industry looked very much the same as it did before Black started his quest for delayed justice the year before.

Two significant changes did occur. An upstart airline named Braniff Airways beat out United on the Dallas-Chicago route, and the crop dusting C.E. Woolman, operating as Delta Airlines, had secured the Dallas to Charleston, S.C. route. Each of these new airlines would ultimately make the most of their opportunity.

Walter Brown would continue unrepentant and dignified amidst the righteous alarums of Black and other politicians who sought to capitalize at the expense of his reputation. The airlines whose contracts had been terminated, although back in the airmail business within a short time, had lost a significant amount of money by continuing operations in the interim, but only the United group of airline companies resorted to litigation against the government because of the contract cancellations. That litigation would drag out for another ten years. And, when finally concluded, it would uphold the government's right to terminate the airmail contracts. ***Pacific Air Transport v. U.S., et. al.***[10] The court did, however, award money damages to the United group for contract payments earned prior to the cancellations. The opinion, which is 104 pages in length, is a detailed chronology of the events that transpired after passage of the Watres Act.

The Airmail Act of 1934 (The Black-McKellar Act)

The Black-McKellar Act, passed by Congress in June 1934, codified the arrangements for the award of airmail contracts made in April 1934, and repealed the powers and prerogative of the Postmaster General as established in the Watres Act. Competitive bidding was reinstated. The newly named airlines bid on the routes. Airline executives involved in the Brown meetings were prohibited from occupying positions of authority in the new airlines. The vertical structure of the airlines and manufacturing companies was prohibited. United Aircraft and Transport, for example, was dismantled and its operations split three ways:

- Boeing took over all operating properties in the West,
- All eastern U.S. functions were assumed by United Aircraft Corporation (today known as United Technologies), run by Rentschler,
- Finally the airline itself, United Airlines, became a separate and independent entity.

The air carrier industry was reorganized under the Act by separating oversight and regulatory authority among

1. The Post Office, which would continue to award contracts, designate routes, and establish schedules;
2. The Interstate Commerce Commission, which would establish reasonable rates through competitive bidding oversight;
3. The Department of Commerce, which through the Bureau of Air Commerce would attend to safety.

Ultimately, the effect of the Act would be to divest the other large airline operations from their holding companies. Aviation Corporation (AVCO) divested itself of American Airlines, now to be run by C.R. Smith as an independent corporation. North American Aviation (NAA), a complex and diverse entity, was the parent of both TWA and Eastern Air Transport, and had substantial manufacturing interests. NAA first sold off TWA, now run by Jack Frye, a pilot's pilot. In due course, Eddie Rickenbacker cobbled together Wall Street financing to the tune of some 3.5 million dollars and bought out the Eastern Air Lines interest. Thus, all of the Big Four were positioned independently for the advent of commercial air transportation and the first comprehensive federal regulation of it.

Pan American, meanwhile, had been unaffected by the so-called Brown scandal and still had its airmail contracts. The Brown philosophy that the international airmail business should not compete with the domestic airmail business, and vice versa, was intact. The international trade routes that had emerged from the Brown era were not in the least affected by the new law, nor by anything that Black had done, and Pan American was set to become the premier airline of all.

State of the Airlines before the Civil Aeronautics Act

The Big Four, having been established largely through the efforts of Walter Brown, and having survived the Black investigation and the resulting remedial legislation (Black-McKellar), were well positioned for the beginnings of the modern era of commercial air transportation. The airlines were hurting financially, however, due to the losses experienced during the stand down period when the Army had flown the mail after the cancellation of all CAM routes in February 1934, and because the new rates mandated by Black-McKellar were set to a maximum of 33.5 cents per mile, less than a third of the going rate in 1929.

But progress had been made. In 1929, the contract mail carriers (who were to become the country's major airlines) were still flying wood and wire airplanes, although a few had acquired the very latest technology in the Fokker or Ford trimotors. By the late 1930s, when the Civil Aeronautics Act was passed, great innovations in aircraft manufacture had occurred, largely due to a combination of commitments and risks undertaken by the airlines and by the aircraft manufacturers, and to government innovations achieved at the National Advisory Committee on Aeronautics.

The National Advisory Committee on Aeronautics (NACA)

NACA was created in 1915 during the administration of Woodrow Wilson as a governmental aeronautical research laboratory. It was recognized during World War I that the United States had fallen well behind the European countries in aviation, and the events in Europe during that conflict were pointing up that aviation was rapidly becoming an issue of national defense. The appropriation for the Committee ($5,000) was modest even for those times, but its mission was clear: "To direct and conduct research and experimentation in aeronautics with a view to their practical solution." It was also to establish a proud legacy during its existence from 1915 to 1958, when it was renamed the National Aeronautics and Space Administration (NASA).

In the early days, NACA advised the government in several pivotal issues, including the government's role in the early airmail service and the lighted airway system. It was largely on NACA's recommendations that the details of the Air Commerce Act of 1926 were fleshed out, including licensing and standardization of aircraft and pilots, inauguration of the aviation weather service, and the establishment of the Bureau of Aeronautics within the Department of Commerce. But one of its most notable early accomplishments was in the area of testing and experimentation in its large wind tunnel, which produced advanced technical information used by American aircraft manufacturers, particularly in connection with the design of engine cowlings, airfoils, and their drag characteristics.

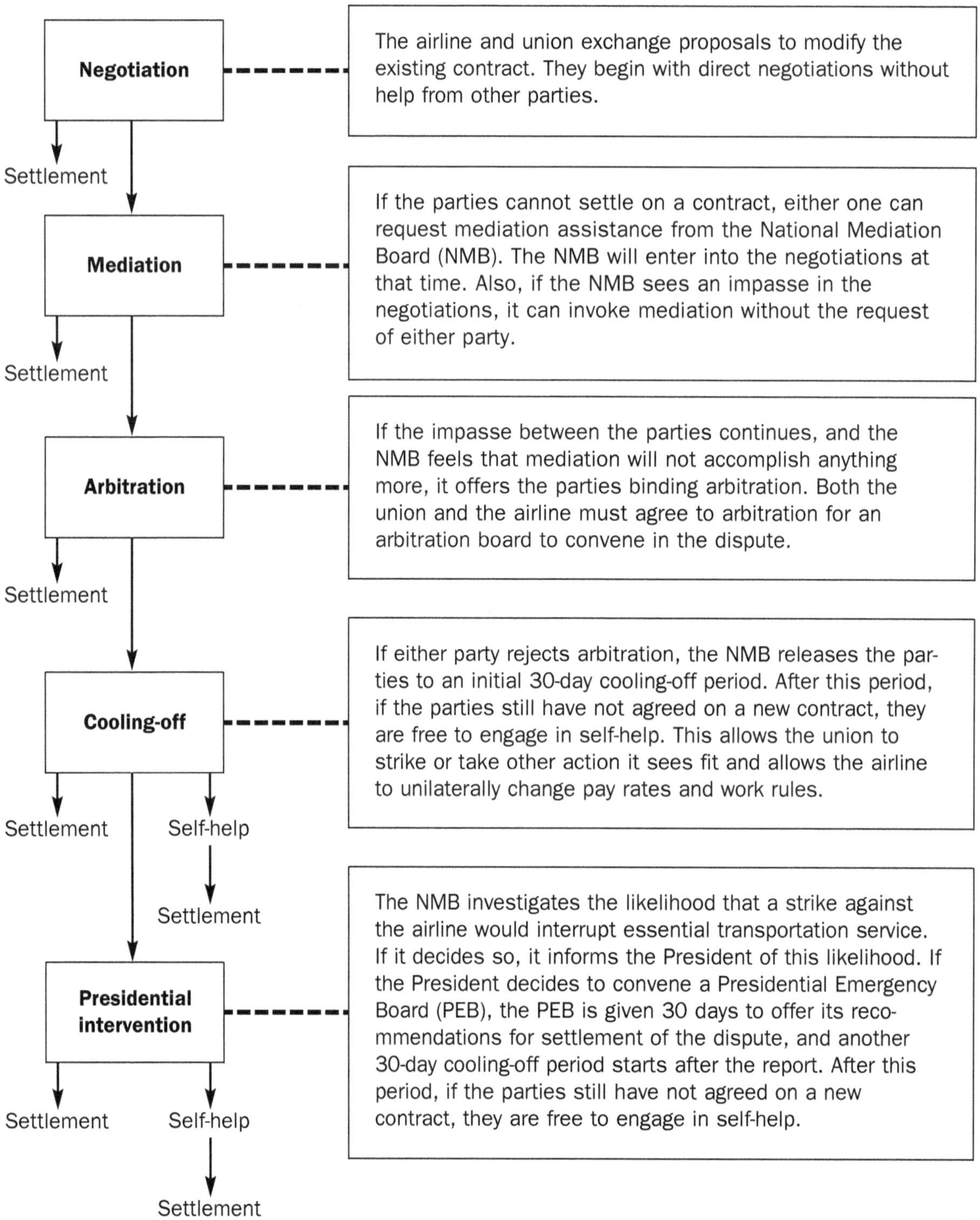

TABLE 4-1 Collective bargaining process under the Railway Labor Act

Source: GAO analysis of NMB data.

The Big Four

American Airlines

American Airlines emerged as the surviving corporate entity after Black-McKellar. Aviation Corporation (AVCO) had been formed in early 1929 as a holding company by a group of New York financiers, and it rapidly proceeded to acquire aviation-related companies left and right. Its holdings included airports, instrument manufacturers, and engine makers, and finally totaled more than 80 separate companies, some having absolutely nothing in common with the others. In early 1930, AVCO consolidated its aviation holdings into American Airways, the name under which the airmail contracts would be serviced in the early 1930s.

American Airways had been formed from eleven smaller lines. Some had their beginnings as early as 1925, notably Embry-Riddle Company (first to be acquired), and 1926, such as Colonial Air Transport, Southern Air Transport, Robertson Aircraft Corp, and Southern Air Fast Express (SAFE).

E.L. Cord, who had bested Harriman and Lehman for control of Aviation Corporation, made C.R. Smith, who had come from Southern Air Transport, president of the new American Airlines in 1934. Smith had a penchant for detail worthy of his training and experience as a bookkeeper and part-time bank examiner, and he ran the company with a hands-on approach that, by 1939, would make American Airlines the nation's foremost air carrier in passenger miles flown. Upon his election as president, and while American was flying a conglomeration of airplanes ranging from the trimotor monoplanes to the bi-wing Curtiss Condors, he immediately placed orders with Douglas Aircraft for the new DC-2. Then, in 1935, he ordered the DC-3, the airplane that was to revolutionize commercial air travel. The DC-3, in fact, was a direct result of collaboration between Smith and Douglas to make modifications that American wanted to the DC-2. The main change desired by Smith was a wider cabin, but in the process of this change, other significant changes in the DC-2 were made, including a new wing, a hydraulic system, insulation, and a much better landing gear and suspension system.

In 1937, American Airlines' route structure included New York to Los Angeles, the southern route, via Washington, D.C., Nashville, and Dallas. It also had New York to Boston and New York to Montreal, New York to Buffalo and thence to Chicago, St. Louis, and Oklahoma City.

Eastern Air Lines

Eastern Air Lines emerged in 1934 as the surviving entity following Black-McKellar. The predecessor company, Eastern Air Transport, was owned by the holding company, North American Aviation, which in turn was controlled by General Motors as of 1933. Eddie Rickenbacker, World War I hero and fighter ace, was hired by General Motors as a consultant and then was made general manager of Eastern Air Transport in 1934. Eastern Air Transport was successor to the original line, Pitcairn Aviation, and it later absorbed the Luddington Line and New York Airways before becoming Eastern Air Lines. When General Motors tired of the airline business in 1938, Rickenbacker purchased the company and steadily increased its business and its mileage.

In 1937, Eastern Air Lines had routes from New York to Miami and to Atlanta and points south and west, New Orleans, Houston, and San Antonio, all through Washington, D.C. It also flew the Chicago to Miami route through Indianapolis, Nashville, and Atlanta.

TWA

TWA was the designation taken by the airline combined at the behest of Walter Folger Brown. A combination of the former Transcontinental Air Transport (TAT) and Western Air Express, it flew the middle transcontinental route from New York to Los Angeles under the name Transcontinental and Western Air. After Black-McKellar, the airline simply added "Inc." after its name in order to comply with the prohibition of Postmaster General Farley that precluded those airlines which had participated in the Brown meetings from bidding on the new airmail contracts in 1934.

TWA had been a part of North American Aviation in the early 1930s, and General Motors controlled the holding company. After Brown-McKellar, General Motors sold its interests to John D. Hertz and Lehman Brothers, who then had effective control of TWA.

Jack Frye, at the age of twenty-six, was TWA's operational vice president in 1930. He had founded Standard Air Lines in the 1920s, after stints at flight instructing and stunt flying, and went with the company when it was purchased by Western Air Express. With the merger of Western and TAT, he suddenly found himself in charge of operations of a transcontinental airline. TWA, and most other airlines, relied heavily on the Trimotors in the early 30s. With the 1931 crash of the Fokker Trimotor in which Notre Dame football coach Knute Rockne was killed, government mandated inspections of that plane's wooden wing structure became cost-prohibitive, not to mention the fact that the flying public thereafter was not keen on stepping aboard that airplane. Frye needed new equipment.

In 1932, Frye had heard the buzz in the aviation community of a new prototype in the works at Boeing, the model 247. This airplane was to be a giant leap forward with its low mono-wing, and two engines instead of three that were mounted into the wings in nacelles (taking advantage of NACA research) that greatly reduced drag. The 247 used stressed all-metal skin, retractable landing gear (a first), insulated cabin walls, and hot water heating and double ventilation systems. This airplane would fly from one coast to the other in only nineteen and one-half hours, twelve hours less than with the trimotors. Fueling stops were reduced from fourteen to six. Frye decided that TWA had to have these airplanes.

When he inquired, he was advised that United Airlines (the sister company to the Boeing manufacturing arm) had already placed an order for sixty of the new planes, an order that it would take all of two years to fill, thus precluding any deliveries to other airlines. The 247 became operational in June 1933.

In the fall of 1932, Frye wrote to a number of aircraft manufacturers setting out airplane performance specifications for new equipment that TWA would be interested in purchasing. Although the specifications included that the airplane have three engines, the engineers at a small company located in California, known as Douglas Aircraft, believed that the performance specifications could be met with a twin engine design, including the requirement for a 10,000 foot minimum service ceiling on one engine (necessary to clear the Rockies).

A prototype was fielded in July 1933, the DC-1, designation for the Douglas Commercial Number 1. If this had been poker, the DC-1 would have called the B-247 and raised it. The DC-1 engine mountings and cowling were similar to the 247, incorporating the design developed by the National Advisory Committee on Aeronautics, but the landing gear of the DC-1 folded up into the engine nacelles. The engines, Wright Cyclones, had been engineered to produce 710 horsepower due to 87-octane gasoline having become commercially available during the period of the plane's construction. Although the constant speed propeller was still a few years off, the DC-1 did have a 2-speed propeller that could be set either for takeoff or for cruise (a first). Additional firsts included an automatic pilot and efficient wing flaps. Flight tests showed that Frye's performance specifications had been met. Only one DC-1 was built and that one was purchased by TWA. It was placed in limited service in 1933.

When Postmaster General Farley sent his notice dated February 9, 1934, canceling all airmail contracts effective February 19, 1934, Frye decided to make his own statement. With Eddie Rickenbacker of Eastern Air Transport as co-pilot, on February 18, 1934, Frye took off from Los Angeles in the DC-1 loaded with airmail and flew it to Newark, with fueling stops in Kansas City and Columbus, in thirteen hours and four minutes, setting the transcontinental speed record at the time.

The DC-2, with 14 seats, was brought to production in 1934, and 193 were built. The next year, in 1935, Douglas came out with the DC-3 (see Figure 4-12) (21 seats) with 900 horsepower Wright Cyclones (DC-3A with 1200 horsepower P&W engines), and Douglas would, before it was all over, build 455 of them for commercial use and 10,174 for the military. By 1936, the DC-3 had

Source: Florida State Archives.

FIGURE 4-12 DC-3—The plane that changed the world.

reduced the transcontinental flying time to about seventeen hours. The airplane was awarded the Collier Trophy in 1936 and became known as "The Plane That Changed the World." And, indeed, it was used all over the world—in World War II in Burma, this airplane which at normal configuration seated 21 passengers, set a load record of 72 refugees safely delivered, and 6 more stowaways were discovered on landing.

By late 1938, pressurized airplanes were on the drawing boards. Boeing designed a commercial transport, the 307 (see Figure 4-13), scheduled for delivery in 1939. It was based on the basic B-17 design with four 900 horsepower Wright R-1820 Cyclone engines. This airplane had a service ceiling of 26,200 feet and was the first commercial liner pressurized for high altitude flight. The airplane came to be known as the "Stratoliner." Jack Frye decided that TWA had to have them too, so he placed an order with Boeing for five of the new planes. But his board of directors, chaired by John D. Hertz of Lehman Brothers, did not agree. In December 1938, TWA's board voted to cancel Frye's order to Boeing for the B-307.

Jack Frye knew that this dispute represented an essential disagreement concerning his and the board's vision for the future of TWA. He also knew that this disagreement would likely mean his being removed if control of the company remained in the hands of the present directors. Jack Frye was acquainted with Howard Hughes, the eccentric multimillionaire and aviation pioneer in his own right, who was then living in Los Angeles and involved in the movie making business. Hughes had an abiding interest in aviation and had even worked for American Airlines, under an assumed name, as a co-pilot in 1932, flying between Los Angeles and Chicago. He listened to Frye, sided with his logic in the B-307 dispute with the board of directors, and agreed to buy the company. He began secretly buying up TWA stock. By April, it was public knowledge that Hughes was becoming a substantial stock-

Source: Florida State Archives.

FIGURE 4-13 Boeing 307.

holder in the airline, so much so that the significant interests represented by Lehman and Hertz decided to pull out of the company, the second time Lehman had departed the field. Control was effectively passed to Howard Hughes, the Boeing order for the 307 was reinstated, and the future of TWA remained firmly in the grip of Jack Frye, now with Howard Hughes. On July 8, 1940, the 307 was placed into service on the New York to Los Angeles route, reducing the transcontinental flying time to fourteen and one-half hours.

Hughes was a singular individual and unique in all known respects. He was born wealthy, son of the founder of the Hughes Tool Company of Houston, Texas. As soon as he could, he left Houston, began traveling the world, and wound up in Hollywood. He entered the film business and, in the process of directing his first film, *Hell's Angels,* a story of British pilots in World War I, he became fascinated with aviation and learned to fly.

Even as a young man, Hughes was obsessive, wanting to be the best, to know the most, and never to fail. With absolutely no concerns about money, he began the design and building of an airplane racer, the H-1, with which he would set a world's speed record of 352 miles per hour in 1935. He set a transcontinental speed record of seven hours and twenty-seven minutes with the H-1 in January 1936. He flew practically every commercial airplane in production over the next several years, gaining experience in long-distance navigation and planning, as well as execution at the controls, until he launched his most ambitious attempt yet: a round the world flight in the Lockheed Electra.

The record in 1938 stood from Wiley Post's solo circumnavigation in 1933 at seven days and eighteen hours. Hughes' route took him from New York to Paris in less than half the time it took Lindbergh, then across Europe into Russia and Siberia to Alaska. From Fairbanks he refueled in Minneapolis and returned to Floyd Bennett Field in New York triumphant in three and a half days, halving Post's record.

Hughes had some prior acquaintance with TWA; in fact, one of its vice presidents had been a stunt pilot for Hughes' movie, *Hell's Angels*. Hughes was also more pilot than businessman. As Jack Frye would later remark, "One thing about Hughes, he did have an understanding about the airplane." He fully understood the advantage of having an airplane that could top most of the weather, so he agreed with Frye's position on the Boeing 307.

United Airlines

United Airlines' name came through the Black and Brown affair unscathed. This was because each of the airmail carrying lines operated in their own names, for example, Boeing Air Transport, Pacific Air Transport, and United Air Lines Transport Corporation. The chief operating officers of the companies caught up in the Brown affair were banished. Thus, United's Phil Johnson left the stage and the presidency of United Airlines was assumed by Pat Patterson, a 34-year-old former banker who came up through the ranks from Pacific Air Transport.

Patterson is credited with initiating the in-flight passenger service staffed by young women, initially nurses, in 1930. The United group was the strongest of the airlines of the 1930s. It was United's lead that counted with the other airlines when Patterson decided to continue passenger and freight service in spite of the cancellation of the airmail contracts in 1934, as he said, "no matter what the losses." United maintained its schedules but at tremendous cost. Even with the return of airmail contracts, given the reduced rate then paid and the losses suffered during the cancellation period, United lost more than two million dollars in 1934, and continued to struggle financially over the last years of the 1930s, falling behind American Airlines with revenues less than half of American by 1938.

The United group, in their individual operating names (Pacific Air Transport, Boeing Air Transport, etc.) had been the lone airline group to sue the government over the airmail cancellation decision in 1934. That litigation would drag on and not be finally resolved until 1942, when the decision of the U.S. Court of Claims was handed down.[11] The court upheld the right of the government to cancel the contracts, but awarded damages to United for those sums representing United's airmail carriage up to the date of cancellation. The language of the opinion is generally considered to be favorable to United and not in keeping with the tone of the Black investigation and the negative airline press it generated.

It can be argued that United's difficulties beginning in 1934 were due to an unfriendly relationship between the Roosevelt administration (which would remain in office until 1945) and United due to the litigation, and exacerbated by a general anti-corporate attitude in certain government quarters. For example, the Interstate Commerce Commission had assumed responsibility for rates and mergers as a result of Black-McKellar. In 1936, the ICC refused to approve a merger between United and a moneymaking line serving New York-Washington. When the Civil Aeronautics Authority took over the ICC function in 1938, it denied purchase authority for United's bid for Western Air Express (the split off branch that did not merge with TAT to form TWA). United, with its 10 passenger Boeing Model 247s, was struggling to compete against the larger and faster DC-2s and DC-3s of the other lines.

In 1937, United still had its transcontinental route, but that was in heavy competition with the other routes awarded by Brown. It did not have the strength of the eastern lines that resulted from their consolidation of the shorter route structure between cities of the more populous eastern United States. Not until the 1960s would United be once again the airline industry leader.

The Lesser Lines

In 1938 it was the Big Four and then everybody else. Some of the smaller lines would fade away; others would prosper under the new law. When the Civil Aeronautics Act passed Congress, notables among the small lines were Delta, Northwest, Western Air Express, Braniff, National, and Continental. Much would be heard from them in the future. And then there was Pan American.

Pan American Airways

Pan American Airways was to occupy a singular place in the annals of American aviation and in the relationship of an airline company with the U.S. government. What Pan Am came to be was mostly a product of the efforts of Juan Trippe, a true visionary, an indefatigable worker and thinker, a man of exceptional personal and professional contacts in both the world of business and government, and a man who stayed at the helm of his company longer than any of his contemporaries.

Trippe was instrumental in the formation and early operation of Colonial Airlines, one of the original airmail contract flyers in 1926, that ultimately became part of American Airways. His vision for that airline was much too aggressive for its conservative directors and stockholders, and Trippe was soon out. He had actually formed a small airline in 1924, before the financial benefits of airmail carriage became available, but it had been unable to survive. After Colonial Airlines, he was soon underway with his concept of an international airline, lining up financing from his wealthy friends and his father's Wall Street contacts.

By 1927, Pan American was in the firm control of Juan Trippe and his friends. That year, wheeling, dealing, merging, and negotiating their way, the young men of Pan American had an airmail contract for the Key West, Florida to Havana, Cuba route. The contract stipulated that service must commence at the latest by October 19, 1927, and since other companies were waiting in the wings hoping Pan American would default, it became a matter of some importance to meet the deadline.

The Fokker Trimotors that Trippe had ordered to service the route had not shown up by that date, so the inaugural flight of Pan American Airways was hastily arranged on the dock at Key West on the drop-dead date. A transient floatplane pilot, bound for a job in Haiti, made a fortuitous fuel stop that day and unwittingly became a part of the grand history of Pan American, for a cash fee of $175.00.

Trippe's vision was fueled not only by his expansive imagination and unbridled determination, but also by the circumstances in which America found itself in the late 1920s. The 20s had been a decade of progress, experimentation, expansion, and success. World War I had caused Americans to look outward, mainly toward Europe, but now toward the untapped vast South American continent and the Latin American connection. The region was ideally suited to air transportation because of its island-hopping availability. South America was also largely undeveloped, ruled with mountains, smothered by jungles, and had largely skipped the era of railroad transportation. Transportation was about to go from pack mule and water skiff directly to air travel.

Europeans, mainly Germans seeking respite from the turmoil, frustration, and inflation of their defeated nation, had opened up aerial trading routes to South America in 1919. They were expanding their influence along its eastern coast and up into the Caribbean. The expatriates formed a company called Sociedad Colombo-Aleman de Transposes Aereos (SCADTA) under the laws of Colombia that had become, as one of the world's first airlines, an example of what aviation could do under extremely challenging conditions. In the process, SCADTA had become the pride of the people of Colombia.

The United States looked with some alarm at this development. The long-standing policy of the United States, as articulated in the "Monroe Doctrine,"[12] after all, essentially decreed the Americas for Americans, not Europeans, and certainly not the Germans. The governmental policy toward commercial aviation that was forming during the 20s held that, while competition among business interests within the United States was good for the public, competition between American businesses outside its borders could be harmful. To properly compete with foreign airlines that were strongly supported by their governments, American international aviation would have to have some form of American government support and should follow some kind of governmental policy.

Pan American was ideally positioned to take advantage of this political and economic situation, and Juan Trippe commanded the confidence of the right people to enhance Pan American's opportunities. The first Trippe ploy was to take advantage of a practice common in the domestic aviation market to 'extend' route authority by fiat of the Postmaster General. This he did by securing an extension authority to Miami from Key West. After Lindbergh's epic transatlantic flight and the ensuing public clamor and appeal that it engendered, Trippe signed Lindy up as a consultant, and Lindbergh became an integral part of the Pan Am strategy to extend its routes across the Caribbean and into Central America and then down into South America. In time, he would also figure prominently in Pan American's westward Pacific expansion.

The second significant development was the passage by Congress in 1928 of the "Foreign Airmail Act." This statute allowed the Postmaster General the discretion to grant routes to bidders that, in his opinion, were the "lowest reasonable bidders that can perform the service satisfactorily." The Act provided, in so many words, that only airlines capable of operating on a scale and in a manner that would project the dignity of the United States in Latin America would be granted the right to carry international mail.

The first three airplanes purchased by Pan American were land-based Fokker Trimotors. With these, the first passenger service between Key West and Havana was begun in January 1928. Given the lack of airports over the region and the fact that most of the flying was over water, Pan American made two significant decisions about its near-term future:

1. The line would employ flying boats to the exclusion of other types of aircraft and
2. The line would fly only multiengine planes. These decisions weighed favorably with the public and with the government.

It was also required that a form of navigation be developed that would allow flight over the trackless ocean. There were obviously no railroads to follow, no landmarks to navigate by, and no open fields to land in. Celestial navigation, long used in maritime transportation, was available, but it had serious limitations as the sole method of navigation for relatively fast moving airplanes. Voice radio was being experimented with on domestic air routes, but the equipment necessary to be placed on board approximated the size and weight of a small piano. Pan American had decided that radio was a near necessity from a safety standpoint, and it was searching for alternatives. An employee of RCA well versed in radio, Hugo Leuteritz, began experimenting with radiotelegraphy with devices that were installed on some of the airplanes. The equipment on board was very light, and the signals were clear and not beset by the static that made voice communication at these latitudes almost impossible. The procedure developed by Leuteritz utilized two land-based listening stations equipped with loop antennae that could pick up and then directionally locate the dots and dashes emitting from the enroute aircraft. When the two stations drew lines from their separate positions to that of the aircraft, and the two lines crossed, the latitude and longitude thus determined was transmitted by the shore station to the radio operator aboard the aircraft and its fix would be established. This method allowed pinpoint accuracy in making the desired landfall.

When the Sikorsky S-38 twin-engine flying boats arrived, Pan American's chief pilot, Captain Eddie Musick, began to make survey flights beyond Havana to anticipated destinations even before the Post Office advertised for bids. It seemed that Pan American had an uncanny knack for already knowing where the routes were going to be offered, and for sewing up the local political and logistical support, including landing rights, necessary to make the routes immediately feasible and successful.

The next two routes awarded to Pan American were (1) from Havana to the Mexican island of Cozumel, then down Central America to Panama, and (2) from Havana to San Juan, Puerto Rico which suddenly increased Pan American's annual airmail revenues from $160,000 to $2 million. Passenger service was then initiated on February 4, 1929, with Lindbergh at the controls flying the 100-mile per hour S-38. With the first flying boats, service was commenced directly between Miami and Panama. These two lucrative routes were soon followed by a

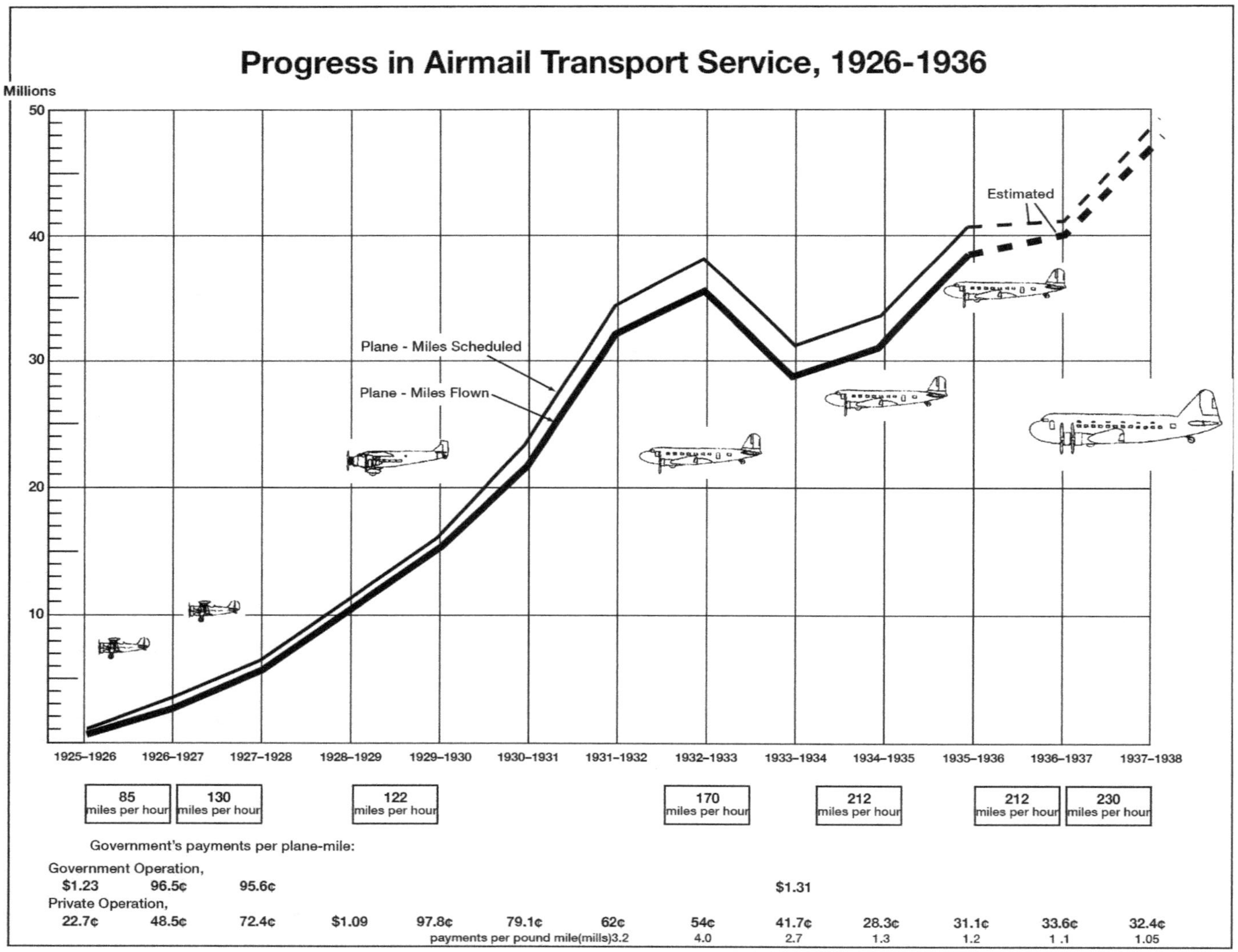

The public's use of airmail for business and social purposes has mounted steadily.

(The decline during the fiscal year 1934, and in the subsequent interval required for repairing the decline, was caused by the cancellation of the airmail contracts.)

As volume has mounted, the unit cost to the government has steadily decreased. (Note figures at extreme bottom of chart.)

Source: ATA Annual Report 1937.

FIGURE 4-14 Progress in airmail transport service, 1926–1936.

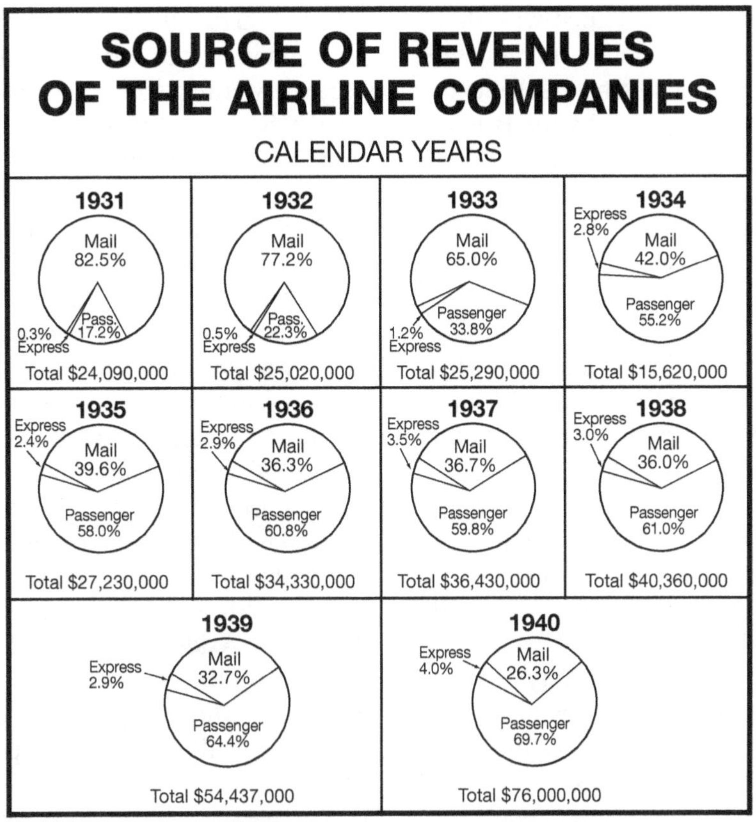

Source: *Chicago Tribune.*

FIGURE 4-15 Source of revenues of the airline companies.

third, from Miami to Mexico City, where linkups were made to the West Coast of the United States. Airmail revenues soon topped $3 million a year.

Airmail routes in the Caribbean, Central America and South America were consistently awarded only to Pan American in what was becoming the obvious policy of the United States government of allowing Pan American to be the 'Chosen Instrument' of U.S. foreign influence. This was despite the emergence of another American formed airline, the New York, Rio and Buenos Aires Airways (NYRBA), which began a head-on competition with Pan American in the region utilizing flying boats.

While Pan American went with the S-38, NYRBA ordered 14 of the Consolidated Commodore, an amphibian designed as a patrol boat for the United States Navy, but which was converted to commercial use by September 1929. The Commodore mounted two 575 horsepower Hornet engines beneath its high wing. It was put into service on the Miami to Santiago, Chile route down the west coast of South America, a 9000-mile route requiring seven days enroute. It was also used on the east coast route to Buenos Aires.

Big names associated with NYRBA, like James Rand (of Remington Rand), former Assistant Secretary of Commerce for Aeronautics William McCracken and William J. Donovan (credited with forming the Central Intelligence Agency), were unable by the middle of 1930 to secure even one foreign U.S. airmail contract. The airline was losing money on a then gargantuan scale ($50,000 a month) and, without help from the government, its backers saw no alternative to a sellout. On August 19, 1930, Pan American, with unofficial Post Office approval, bought out the NYRBA line. The next day, the post office department advertised the east coast of South America airmail route. Pan American, of course, was the only bidder and it bid the maximum allowable rate.

By 1930, Pan Am was flying 20,000 route miles to 20 different countries, and it was still within the Western Hemisphere. Trippe was obviously the American government's "fair haired child," but his efforts at establishing transatlantic service were continuously thwarted by the British. Although the British agreed in principle with the proposition of bilateral rights between America and England, the standing position was that they were not physically or financially ready, and until they were, no American rights would be granted. Europe was not considered a feasible destination without landing rights in Bermuda, and since that island was strictly English, no European schedules of any sort were considered possible. Trippe turned his attention to the Pacific.

The range, in miles, of available aircraft was the most severely limiting factor in attempting a traverse of the vast Pacific Ocean. Sikorsky was the first to complete an aircraft design that attempted to address this problem, the S-40 flying boat. This model boasted four engines, had a capacity of 44 passengers, and a range of 1,000 miles. The first S-40 was delivered to Pan American on October 10, 1931, and was christened by Mrs. Herbert Hoover at the Annapolis Naval Air Station. She broke a bottle of Caribbean seawater across the prow of the S-40, after which Juan Trippe dubbed the airplane a Pan American "Flagship." Thus was the appellation "Clipper" born.

Shortly thereafter, the S-42 (see Figures 4-19), with a range of 2,520 miles, came off the line. This was still a bit short for the 2,410 mile San Francisco-to-Honolulu run, if any reserve of fuel for weather or other contingencies were to be made. Trippe turned to Glenn Martin for help, while at the same time flying the S-42 configured with extra fuel tanks to assure another 500 miles.

With Lindbergh's help, it was Trippe's plan that the Pacific would be conquered by way of Alaska, Japan, China, and points south, the kind of Great Circle route Lindbergh had used in 1927 to Paris. No airmail contract had been awarded to Pan Am, but Trippe was proceeding anyway. He bagged a majority interest in an airline with operating rights in China called the China National Aviation Corporation, but then, in 1934, Japan was becoming militarily aggressive, and the U.S. State Department advised against the proposed route. To go straight across the Pacific would require a route including Honolulu, Midway, Wake Island, and Guam before reaching Manila, Philippines. Aside from the fact that the Sikorsky aircraft was limited in range, there were absolutely no facilities on Midway, Wake, or Guam.

In typical fashion, Trippe had a freighter loaded with the necessary equipment, supplies, workmen, and supervisors and dispatched it to each of the proposed landing sites to construct the necessary passenger and aircraft support facilities, including terminals and hotels. With this service archipelago in place, and with landing rights in Hong Kong, Pan American was poised to be the first transpacific airline with service from the American to the Chinese coast.

Source: Florida State Archives.

FIGURE 4-16 The S-42 had a range of 2,520 miles.

In October 1935, the first M-130 Martin flying boat was delivered (the first of three). This craft was larger than any other flying at the time. It had a range of 4,000 miles configured for mail and 3,200 miles with 12 passengers, a cruising speed of 163 miles per hour and redundant hydraulic and electrical systems. With the airmail contract secured, service was inaugurated for mail and cargo delivery on November 22, 1935, in a ceremony at the dock in San Francisco attended by Postmaster General Farley. In October 1936, with the support facilities now in place, passenger service across the Pacific Ocean began to Manila. A New Zealand route followed after Australia was blocked by the British, and then a second, southern transpacific route was initiated via Kingman Reef and Pago Pago. On April 21, 1937, the transpacific route was extended to Hong Kong, with connecting flights to destinations in China serviced by the Pan Am subsidiary, China National Aviation Corporation. Then, within a six-month period, December 1937 to the summer of 1938, Pan American suffered two highly publicized clipper accidents that brought unaccustomed criticism, both from the press and from government quarters. Chief Pilot Eddie Musick, who had surveyed the original Latin American routes ten years before, was at the controls of an S-40 off of Somoa when it exploded in midair. In July 1938, one of the three Martin 130 Clippers disappeared between the Philippines and Guam. The intense expansion of routes over the Pacific was taking a heavy toll and, while Pan Am banked over $1 million in profits from Latin American operations in 1938, it was losing large sums of money in the Pacific. Trippe turned his energies back to the Atlantic.

On February 22, 1937, the British Air Ministry issued Pan Am a permit to operate a regular air service between the United Kingdom and the United States via intermediate points in Canada, Bermuda, Ireland, and Portugal. The agreement by Pan Am to pool passengers and cargo with the British airline, Imperial Airways, had a lot to do with this breakthrough. Technological advances, however, followed shortly on the heels of diplomacy. On order from Pan American since 1936, Boeing in 1938 produced its B-314 clipper (see Figure 4-17), the largest aircraft to be used in scheduled service then or thereafter until the arrival of the jumbo jets of the late 1960s. This airplane was configured in two decks, had a speed of 193 miles per hour and a range of 3,500 miles, enough range to allow Pan American to fly right over Bermuda enroute to Europe. It carried 74 passengers seated or 40 passengers in the sleeping berth configuration. The Clipper went into the Pacific route service on February 22, 1939.

In the Atlantic, Pan American launched its passenger service between New York and Marseille, France, on June 28, 1939 with the Dixie Clipper, a Boeing 314A, followed on July 8, 1939, by Yankee Clipper service from New York to Southampton.

Summary of Airlines' Condition

The effects of the Great Depression were lessening by 1938. The economy was recovering, jobs were being restored, and manufacturing was picking up, including in the aircraft industry. Although the airlines' income from airmail carriage was down from what it had been before 1934, passenger revenues were up and exceeded airmail revenue for the first time. Airlines began to expand their passenger facilities and corporate infrastructure, and their traffic and sales departments. The

Source: Florida State Archives.

FIGURE 4-17 The B-314 Was the largest aircraft to be used in scheduled service until the arrival of jumbo jets in the late 1960s.

airline industry was beginning to have an impact on the public and on the economy. As the government lost more control over the airlines because of the lessening effects of airmail revenue, and as the airlines began to develop passenger traffic and revenue, it was time for the government to put into effect some kind of comprehensive control of the industry. The situation was not unlike that of the railroad industry with the Interstate Commerce Act of 1887, or the trucking and bus industry in the Motor Carrier Act of 1935, except for one thing: The airlines wanted regulation.

The Civil Aeronautics Act of 1938 (McCarran-Lea Act)

Since 1926, what little regulation the government had imposed on the aviation community had been administered by the Department of Commerce, specifically the Aeronautics Branch first and then the Bureau of Air Commerce beginning in 1934. In the meantime, flying had progressed from mail planes constructed of wood and wire with open cockpits to all metal stress skinned monoplanes flying in instrument conditions at speeds over three times that of early aircraft. In the middle 1930s, flying was still something of an adventure, for navigation facilities were primitive, instruments rudimentary, and weather prognostication an immature art form.

The TWA crash of the Fokker trimotor in 1931 that killed Knute Rockne was the most notorious domestic airline crash until the death of Senator Bronson Cutting on May 6, 1935, aboard another TWA airplane, a DC-2 on a transcontinental flight from Los Angeles to Newark. Cutting boarded the aircraft at Albuquerque, N.M., where it was reported that the plane's radio transmitter was faulty. The weather at Kansas City, which had been predicted to be good, had deteriorated to a ceiling of six hundred feet, one hundred feet below minimums, by the time of the flight's arrival in the area. Unable to communicate or to make the appropriate instrument approach, and with fuel low, the DC-2 crashed while attempting to fly visually at tree top level.

Senator Cutting was much beloved in the Senate. The congressional investigation of the crash centered on the Department of Commerce and its administration of aviation safety. It was determined that the Department had been lax in enforcing what few rules were in place. Then a controversy arose between the Department and TWA as to whether the 45-minute fuel reserve rule had even been properly published, or whether TWA was otherwise notified of the rule. There were questions of conflict of interest over the Department of Commerce investigating itself concerning the adequacy of existing rules and their enforcement. The Department was shown to have a propensity for laying blame on the pilot in command, a tendency, some may argue, that continues to this day. The press stirred the pot well, and the public reaction ranged from a loss of confidence in the system to outrage. The sense of the Congress was that the Department of Commerce had failed to keep pace with the ongoing progress of commercial aviation.

Then, on October 7, 1935, a United Airlines crash near Denver killed twelve people. On April 7, 1936, another TWA DC-2 crashed in Pennsylvania with twelve more fatalities. On August 6, 1936, a Chicago & Southern Lockheed went down in St. Louis with fatalities of all eight aboard, and on February 10, 1937, a DC-3 flown by United Airlines crashed in San Francisco and all eleven on board were lost. The winter of 1937, in fact, recorded five airline crashes with fatalities.

The airlines took it upon themselves to develop operating rules and regulations for the governance of their pilots, still a rather undisciplined lot, and who looked upon flying as another form of freedom, not restriction. The management of the airlines understood that, in order to win the confidence of the public and take their place as a legitimate form of public transportation that could compete with the railroads, order must be brought to the rather free-form society of aviation, up to that time primarily known for its airmail deliveries and stunt pilots.

In 1935, the airlines knew that someone had to control the growing number of airplanes plying the skies, particularly where they converged for landing, like in Newark, New Jersey. Newark Airport had a departure or an arrival every ten minutes. It was American Airlines that took the lead in designating a "boss," someone in charge who could direct planes to maintain separation by altitude, primarily, as they approached for landing. An agreement between six airlines created a company known as Air Traffic Control, Inc., and it was manned by employees of those companies.

The first facility was set up at Newark (see Figure 4-18), followed by others in Chicago and Cleveland. At first, the controllers sought to track flights within fifty miles of the controlled airport, using blackboards, a large table map, a telephone, and a Teletype. Flight plans were filed by departing pilots who would then keep in contact with their company's radio operator, relaying their position at designated waypoints with their speed and altitude. This information would then be telephoned to the center guarding the destination airport, and the airplane's position would be marked using brass weights that were moved along the table map to represent the airplane's progress. When an aircraft approached one of the staffed centers, directions to the incoming airplane would be issued by the controller to the airline's radio operator by telephone, who would then radio the pilot of the incoming airplane to descend to a certain altitude, to hold at the beacon, or that he was cleared for the approach.

The authority of the controllers was debatable, particularly among the more independent pilots who were used to doing things their own way and in their own time. It was at first considered by the pilots that the controllers' directions to them were advisory only, not mandatory, so that a direction to hold while another aircraft landed might or might not be honored. With air traffic control being taken over by the Commerce Department's Bureau of Lighthouses in 1936, procedures at last began to change. Discipline and self-control were becoming as much a requirement of good piloting technique as airspeed and altitude.

The regulations governing aircraft control adopted by the Commerce Department were actually not known to all airlines and pilots, since they were not required to be published in the Federal Register, a publication whose purpose it was to advise of the adoption of agency regulations. The Supreme Court case of ***Panama Refining Co. v. Ryan***[13] established the proposition that, in order to be binding, a regulation must be published in the Federal Register as notice to all concerned. The next year, 1937, saw the first codification of regulations promulgated by the federal government. They came to be known as the Civil Air Regulations. Not only did these first regulations establish rules governing the movement of airplanes within the designated airspace, they required, for the first time, that the airlines themselves draw up a detailed operations manual, approved by the government, containing procedures for that airline regarding weather, minimum altitudes, approach, departure, and enroute procedures. These Department of Commerce regulations did not apply to airport control towers, however, which remained under local city control until just before the United States entered World War II.

Source: FAA.

FIGURE 4-18 The beginnings of air traffic control—Earl Ward (left) organized the Newark, New Jersey air traffic facility in the mid-1930's. Here he tracks a flight with the aid of a calipar as R. A. Eccles watches. The pointed markers representing aircraft were moved across the map as flights progressed.

Standardization of aircraft procedures was only one aspect of the emerging airline industry that the airlines felt needed the steady hand of

“Before take-off, a professional pilot is keen, anxious, but lest someone read his true feelings he is elaborately casual. The reason for this is that he is about to enter a new though familiar world. The process of entrance begins a short time before he leaves the ground and is completed the instant he is in the air. From that moment on, not only his body but his spirit and personality exist in a separate world known only to himself and his comrades.”

Ernest K. Gann, foreword to *Island in the Sky*

government control. Passenger traffic and airmail carriage had tripled since Black-McKellar, but the airlines were still suffering financially and had, in fact, all lost money each and every year since 1934. The airlines formed their own group, the Air Transport Association, and one of its first acts was to drum up support for and draft a bill creating federal rate and route regulation designed to stabilize the airline industry. The airline industry was demanding to be regulated.

The statute passed by Congress on June 23, 1938, provided a uniform basis of regulation for aviation in the United States and created three independent agencies to discharge the statute's mandate: the Civil Aeronautics Authority, the Administrator of Aviation, and the Air Safety Board. Although control of aviation matters had been removed from the Department of Commerce, with direct reporting to the President, there was a significant degree of overlapping authority among the three agencies. Almost immediately friction developed among the agencies, causing the President to order an investigation into the problems created under the new setup. Within a period of two years from its passage, the Civil Aeronautics Act was amended to resolve the situation inadvertently created by the legislation. Known as the 1940 Amendment to the Civil Aeronautics Act, it dissolved the three agencies originally created by the Act and redistributed their functions between two new agencies that would administer the Act for the next twenty years:

- the Civil Aeronautics Board (CAB) and
- the Civil Aeronautics Administration (CAA).

The CAB was established as an independent board of five individuals, who reported directly to the President, and whose function was primarily to exercise control over air carrier economic regulation, such as rates, routes, and mergers. The CAB was also given the responsibility to investigate aircraft accidents and for safety rulemaking.

The CAA was created as an agency, headed by an administrator, which was placed back within the Department of Commerce. Responsibility for all non-military aviation safety programs, air traffic control, and airway development now was assumed by the CAA. Enforcement became much more stringent, the rules now requiring pilots to obey the controller's instructions. Training centers were established to educate would-be controllers and to standardize their training. Coordination of all controllers followed, with towers and enroute centers falling under the CAA umbrella. The diminishing number of aircraft accidents reflected the wisdom of the new policy, and the public took note.

The Act also provided that anyone wishing to operate an airline business in interstate commerce would be required to apply for and receive a Certificate of Convenience and Necessity, the issuance of which would depend on the results of an appropriate review of the applicant's financial and experience qualifications and hearings conducted by the CAB to determine whether the granting of such an entry into the brotherhood of air carriers was in the public interest. The Act provided a Grandfather clause for all airlines that had provided adequate and continuous airmail service from May 14 to August 22, 1938. In 1938, there were sixteen such carriers in the United States and all of them were granted permanent Certificates of Public Convenience and Necessity to continue operations. But in all of the next forty years, the CAB would not grant entry to another major airline.

Endnotes

1. See Appendix 3 for Max Miller's narration of his survey flight to Chicago across the Allegheny Mountains.
2. Or "dead reckoning." There does not appear to be agreement concerning the use of the term "ded reckoning" or "dead reckoning." "Ded reckoning" is based on contraction of "deductive reckoning."
3. For a more detailed discussion of NACA, see page 29.
4. See Appendix 4 for details of Lindbergh's flight, including hourly log entries.
5. Lindbergh served as technical advisor to Pan American for 45 years.
6. See Appendix 5.
7. Moolman, Valerie, *Women Aloft,* Time-Life Books, 1981.
8. By contrast, the around the world flight in 1938 by Howard Hughes was 14,456 miles in length, incorporating the itinerary New York-Paris-Moscow-Omsk-Yakutsk-Fairbanks-Minneapolis-New York. Except for New York, Paris, and Minneapolis, all stops were above 55 degrees north latitude. Hughes' flight set a new around the world speed record of 3 days, 19 hours, and 8 minutes, beating both of Wiley Post's world records of 8 days and 16 hours in 1931 and 7 days 19 hours in 1933 along a route similar to that flown by Howard Hughes in 1938.
9. See discussion of Pan American on page 36.
10. 98 Ct. Cl. 649 (1942).
11. Pacific Air Transport v. U.S.; Boeing Air Transport v. U.S.; United Airline Transport Corporation v. U.S., 98 Ct. Cl. 649 (1942).
12. The Monroe Doctrine was first expressed by President James Monroe in the State of the Union address to Congress on December 2, 1823. According to this policy, the American continents (North and South America, and including Central America) were to be henceforth free of any further colonization attempts by any European power. This statement of American national interest implied the use of American military and economic power in its enforcement. International adventures by Spain and Portugal triggered this policy.
13. 298 U.S. 388, 55 S.Ct. 241, 79 L.Ed 446 (1935).

CHAPTER 5

World War II

A year after the Civil Aeronautics Act was passed, Hitler invaded Poland to start World War II. Germany, Italy, and Japan comprised the Axis powers arrayed against the Allied forces which were to consist mainly of Great Britain, the United States, the U.S.S.R., Canada, Australia, and the various governments in exile of the conquered countries, notably France. Commercial aviation and aircraft manufacturing would be almost completely devoted to the war effort.

The United States did not immediately enter the war for reasons more fully discussed below. The inertia of pre-war planning for commercial aircraft carried on into 1939 and beyond. Howard Hughes by 1939 was well ensconced as the moving force behind TWA. He had countermanded the stop order on the pressurized Boeing 307 and five of the Stratoliners had been delivered to the airline. But he had bigger plans—plans that would create one of the most impressive airline shapes in the history of commercial aviation. They called it the Constellation.

Hughes knew that Douglas was working on the airplane that was to become the DC-4, and he knew that it was to be unpressurized and subject to the limitations of low altitude flight, particularly weather. On his first transcontinental flight in the H-1 racer in 1935, he had used oxygen to allow flight at higher altitudes, in more favorable winds, and out of most of the weather. Boeing had already produced its first version of a pressurized plane and was committed to the basic design reflected in the B-17 and the commercial, pressurized version in the B-307. Hughes approached Lockheed with the proposal that they design and build the world's fastest, high performance airplane, and that TWA would buy it. Hughes insisted on secrecy, believing that his specifications to Lockheed would produce an airplane that would give TWA a great advantage over the competition.

A young designer-engineer by the name of Kelly Johnson, who was to become a legend as the brains behind such super secret military projects as the U-2 and SR-71 spy planes developed in the famed Lockheed "Skunk Works" in later years, took on the assignment of the design, development and production of the Connie. He and the other designers at Lockheed agreed to Hughes's terms, including the requirement that TWA would get the first forty planes off the line, and that there be absolute secrecy. The deal was made with Hughes Tool Company, not TWA, both to ensure secrecy and because of the fact that Hughes Tool had the money. The project was begun, drawings were prepared, reviewed, revised, and by 1941 about one-half of the original prototype was done.

The war in Europe and Japan's military occupation of Korea, Manchuria, and parts of China caused the War Department to survey U.S. airplane manufacturers' plants with a view to ascertaining production levels in the event of the United States being brought into the hostilities. The Constellation design was thus disclosed. The war atmosphere had also caused the United States to create the War Production Board, whose job was to allocate the industrial and manufacturing resources of the country in a way to best ensure its defense and guarantee the production of essential goods. This higher cause was understood by all concerned, including the secretive Hughes, and by agreement it was determined that Pan American would participate in the Constellation project. Pan Am was the only international air carrier for the United States, and it was not a competitor of TWA. Further, Pan American had the international experience, the routes, the landing rights, and the foreign contacts that could make the best and highest use of the Connie's range and speed. Thus amended, the project went ahead under the auspices of the War Production Board.

The complexity and relatively difficult design shape of the Connie caused General Hap Arnold to stop production of the airplane numerous times during the next two years, in favor of the simpler and relatively inexpensive DC-4. The Connie would not actually fly until December 1943. After the first Connie was rolled out of the Burbank, California plant, as a part of its test flight regimen Howard Hughes and Jack Frye would fly it to Washington, D.C. in a new record time of under seven hours, nonstop.

The Connie would not contribute in any significant way to the war effort. The other four-engine aircraft, the DC-4, would. The DC-4 was a Douglas creation begun in the 1930s on orders from American Airlines and United for domestic service. It was the first production airliner with tricycle landing gear and, with the exception of the five Boeing 307 Stratoliners in service, was the first serious transoceanic aircraft to come available. It came available just as the United States had to have that very type of airplane, and it would fly in 1942. The airplanes came off the production line as the C-54 and some 1,162 of them were built for the military. The first ones did not go to the airlines that ordered them, but to their competitor, TWA, which was flying the southern transatlantic route to Africa for the military. During the war, these planes would log over a million miles a month over the Atlantic, some 20 sorties every day.

The Lend Lease Act

The United States did not immediately enter the war in Europe, partly because of the experience of World War I, and partly because of the geographical fact that two great oceans separated America from the hostilities being conducted in Europe and Asia. The country was politically and emotionally divided between those who favored involvement and those called "isolationists," one of whom was the most prominent man in the world of aviation, Charles A. Lindbergh. Sentiment ran high against becoming once again embroiled in "Europe's wars," which seemed to recur on a more or less regular basis and had done so since time immemorial. On the other hand, England, as the last hope of Western Civilization, stood alone against the totalitarian, fascist Nazi (NSDP) party of Germany. Germany, by 1940, had conquered virtually the entire European continent. A non-aggression pact between Russia and Germany had been concluded in 1939, protecting Germany's eastern flank, and the "Battle of England" raged in the skies over Great Britain between the Luftwaffe and the badly outnumbered Royal Air Force. German submarines prowled the Atlantic unchallenged, conducting unrestricted warfare, sinking prodigious amounts of tonnage of shipping and menacing the ships of the Royal Navy. Supplies, planes, and ships were running low for the forces of freedom, but the United States was diplomatically neutral.

England and the United States were historically bound together—by language, by culture, by common law—and England was calling on the United States, on President Roosevelt, for help. The Roosevelt administration was spiritually and emotionally with the British, and many believed that the defeat of Britain would be tantamount to a defeat of American interests worldwide. Slowly at first, but with increasing resolve, the United States began giving aid, comfort, and supplies to England. None of this was authorized by law. As debate swirled within the country, the country moved ahead with its aid plans without publicity.

The first planes supplied to England, seven Lockheed Hudson twin-engine bombers, were ferried via Gander, Newfoundland, to Aldergrove, Ireland on November 10, 1940, followed by some additional 30 aircraft that winter. On March 11, 1941, the supplying of material became legal with the passage of the Lend-Lease Act, which empowered the President "on behalf of any country whose defense the President deems vital to the defense of the United States, to sell, transfer title to, exchange, lease, lend, or otherwise dispose of, to any such government any defense article … not expressly prohibited." The United States, now legally empowered to do so, began wartime production of defense material and provided it to China, Russia, and to 35 other nations. TWA, the only airline with land-based four-engine aircraft, at government direction set up a training center in New Mexico for instructing American and British pilots how to fly the four engine bombers, the B-24 and the B-17. (TWA had purchased the new Boeing Stratoliner, the 307, in 1940. The 307 was the first pressurized airplane flown by any air carrier in domestic service.) Pan American also contributed to four engine training.

The Army Air Corps formed a ferrying service to deliver aircraft to England and to shuttle ferry pilots back to North America. The industrial capacity of the United States was about to be tapped, and a mighty force it would prove to be. And then on December 7, 1941, all of the fence sitters on the issue of involvement in the European and Asian war were unseated as the Empire of Japan bombed the Naval and Army Air Force facilities at Pearl Harbor. War was declared on Japan by the United States the next day, then Germany and the other Axis Powers declared war on the United States. Global war was on.

The Chosen Instrument

As the sole American airline with operations overseas, Pan American became an important asset of the United States during World War II. Pan American operated flying boats, in part, because of the lack of airfields. Nevertheless, Pan Am had experience in building airfields in remote areas. Roosevelt had secured rights to bases on many of the islands of the Caribbean from the British. He now called upon Pan American to build airfields on these islands as a part of a larger plan to supply the war effort against Germany. Airports would be built down through the Caribbean to South America, along its east coast, for ferrying equipment and supplies across the Atlantic narrows to Africa. The British were engaging the Germans in North Africa, and North Africa would be the location of America's first military engagements in World War II. Although the United States paid for all of the airport construction (some $90 million), Pan American held title to these facilities initially for appearance purposes since the United States was diplomatically neutral prior to its entry into the war. After the war, negotiations caused these improvements to revert to the United States, but with limitations on their use by airlines other than Pan Am.

The Airlines at War

At the beginning of the war, there were only some 365 transport aircraft in the United States. The airplane manufacturing community would shortly begin to produce 50,000 aircraft a year, the largest manufacturing activity in the United States for the duration of the war, and at war's end over 300,000 airplanes would have been produced. America's main contribution was in production, not development, for the existing weapons of war were considered sufficient, at least in the short run, to win the war if only there were enough of them. The P-51 Mustang was the only new development in airplane technology supplied by the United States after the onset of war. All the others produced were pre-war designs, and even the P-51 had a British engine.

The main production effort was, of course, directed toward fighter and bomber aircraft, although over 10,000 DC-3s, designated for the military as C-47s, were built, along with over 1,000 DC-4s (as C-54s). America needed every bit of transport potential it could muster during the years 1941–1945, including railroads and as well as air carriers, and while the railroads enjoyed a resurgence of their former glory during these years, the air carriers came into their own for the first time. Aircrews flew everywhere, either as military or civilian to military or civilian airports, on domestic and overseas routes. Flying northern transatlantic routes became routine, as did South American routes to Africa.

Domestically, the airlines discovered after the onset of hostilities that they had only 165 airplanes to service their routes. The armed forces had commandeered the rest for military purposes. Travel space on the relatively few air carrier aircraft was allocated according to a government imposed "priority system":

- Priority One was for persons traveling under the authority of the President
- Priority Two got military pilots a seat
- Priority Three was other military personnel or civilians on essential wartime business
- Priority Four was military cargo.

The remaining seats, of which there were precious few, went to everyone else. The lexicon of future airline travel was being established too. "Standbys" were those who hoped a priority above them would become a "no-show" so that a seat would become available. To be "bumped" was to have a higher priority passenger show up to take your seat.

American air carriers began to make money for the first time since 1934, and although the high load factor of domestic commercial operations contributed to profitability, the main effort of the airlines during the war was as contract carriers for the military.

The government allocated the airlines' responsibility during the war in logical fashion. Northeast Airlines was given the North Atlantic route as far as Greenland and then Reykjavik, Iceland. Northwest was assigned to the Alaska route, Eastern to the Caribbean and Brazil. American flew to South America and, in the process, caused a radio range to be built along its route from the United States. TWA had its five Boeing 307s, the only four-engine land-based transoceanic aircraft available at the time, commandeered by the military and was given the transatlantic route to Egypt, the most significant long distance route of any airline except Pan American. TWA set up its transcontinental division immediately at the beginning of the war, no doubt with an eye on the postwar period. At first, TWA flew to Africa via the South American route, and later, after Portugal granted landing rights, via the much shorter North Atlantic route by way of Prestwick, Scotland. TWA flew military supplies and equipment, like the other airlines, but it was the preferred carrier for VIPs and, in fact, TWA carried President Roosevelt to the three wartime conferences with Churchill in Casablanca, Tehran, and Yalta.

It is not surprising perhaps, that Pan American, as the only overseas carrier in existence before the war, was counted on as the major civilian arm of the military during the war. Yet, it is noteworthy that Pan American's five divisions, the Alaskan, Pacific, North Atlantic, Caribbean and Africa-Orient, flew half of all contract miles flown by all airlines for the U.S. military. In the process, Pan American began flying landplanes instead of the flying boats that had been its trademark during its early years, thus marking the end of the romantic and adventurous era of the Pan American Clipper. The range and speed of the DC-4 and the airplanes to follow, the availability of the airports that Pan American and others had built around the world, and the relative maintenance costs and requirements of amphibious planes over landplanes sounded the death knell of the flying boat airliner, and Pan American never ordered another one.

Most of the flying done during World War II was not by the commercial airlines but by the military forces created and trained by the government. The exigencies of war, shown once again to be a mighty motivating force, had caused a great technological leap forward in aircraft, engines, and systems. The feats of the non-combat pilots of the military lift branches, some 25,000 of them, during the four-year duration of the war testify to the great advance in air transportation over that short span of time. Feats only imagined a mere four years before were now commonplace. Distances had been covered and heights had been overcome for the first time in the airborne delivery of personnel and goods that would henceforth be considered routine. There was a confidence born not only of victory, but also of accomplishment.

Contrasted to the unspeakable devastation visited on the landscapes and structures of Europe that had been created by the most advanced civilization for the better part of two millennia, the homeland of United States emerged from the war unscathed, and with the mighty industrial plants that had supplied the weapons and material of war intact. America had:

- The pilots
- The planes
- The know-how
- The international presence on the ground
- The financial structure and stability to lead the world into the postwar realms of commercial aviation.

CHAPTER 6

Commercial Aviation in the U.S. post-WW II

Before the war, air travel had begun to catch on, and in 1941 domestic airlines carried four million passengers. With the war over in 1945, air travel quickly picked up again, and by the end of 1945 the airlines had enplaned some 7.5 million passengers. In 1946, the number almost doubled to 12.5 million passengers. The commercial airline fleet before the war provided about 6,200 seats. By 1946, the airlines had tripled capacity to 19,000 available seats. The cost of airline travel had fallen enough to be competitive with first class railroad fares, and the four-hour plane ride between Chicago and New York offered a real choice for any time-sensitive traveler over the sixteen-hour railroad Pullman. It was a new day in commercial aviation.

The development of transport aircraft had progressed rapidly during the war. The expectation was that the pre-war traffic would be promptly reclaimed and then exponentially developed using the new era of airliners. Some of the aircraft were suited to expanding the first class travel begun in the 1930s, particularly on the transcontinental and transoceanic runs. Modern airlines of the postwar era were on the verge of entering the first class travel market long held by the steamship lines and the transcontinental Pullman trains. While it was true that some airlines using DC-3s had offered berths for sleeping on overnight flights, now airliners could fly much higher and faster, and in pressurized and air-conditioned comfort.

TWA launched transatlantic service on December 5, 1945, with a VIP flight to Paris that was completed in the record time of twelve hours and fifty-seven minutes. Pressurized, and with a cruise speed of 280 miles per hour, the Constellation was ready to contribute to the anticipated revolution in transatlantic and transcontinental air travel.

The number of airports used by the airlines more than doubled between 1941 and 1947, from 2,484 to 5,343. Outside the terminals, new sleek aircraft for the first time took on the look of flying in place with their new tricycle landing gear. Interspersed among the ubiquitous DC-3s that appeared to be sitting back on their haunches, these new planes stood tall over them and gave an impression of progress, comfort, and safety. The airplanes were getting larger than the terminals in some places. Mass transit, facilitated by the big airliners, was about to begin.

The DC-4, which had been commandeered by the military upon its appearance in 1942, became available to the domestic fleet in 1946. (See Figure 6-1.) Unpressurized, seating 44 passengers and barely able to muster 200 miles per hour at cruise, the DC-4 was outclassed by the Constellation, yet it became in the late 1940s the four-engine airplane of choice. Its service during the war had proved it to be safe and reliable, something yet to be proved in the Constellation and other advanced aircraft emerging from the war. The tapered lines of the Constellation, its more complex systems, its

Source: Florida State Archives.

FIGURE 6-1 The DC-4 became available to the domestic fleet in 1946.

three vertical stabilizers, and the number of parts required to be stored and available also caused it to be significantly more expensive to maintain than the DC-4. (See Figure 6-2.) The straight lines of the DC-4 proved to be much cheaper to repair, maintain, and to fly.

The availability of these new aircraft increased the airlines' capacity and brought with them options for airline management that had never before been possible. This may be the point in history that the concept of the passenger seat as a "grapefruit," as in a perishable commodity, was articulated as a marketing truism. Every unfilled seat at takeoff was like spoiled grapefruit for that flight; it was forever lost to use. Competitive management thinking recognized that high density seating brought with it pure profit after boarding enough passengers to cover costs.

The scheduled airlines were, and had always been, of one class, and that was first class. When comparisons were made between the airlines and the railroads as to cost, an airline seat was compared to a Pullman berth (these accommodations were seats during day travel and were converted to beds for night travel). The high cost of air travel could be favorably compared to first class rail fares because of the time-distance advantage between comparable points enjoyed by the airlines.

After the war, the CAB began loosening the regulations that bound the air-traveling public to the scheduled airlines (first class service). This allowed aircraft charter, or as some called it, the nonscheduled lines, or "nonskeds." Using DC-3s and then DC-4s, these charter operators flew at off hours, at night, and most importantly, with full airplanes. Not bound to a schedule, these operators were not required to leave the terminal at any particular time. Their schedule was simply dictated by the passenger count. And passengers flocked to them. Soon the nonskeds were going coast to coast and at prices that were thirty percent less than the scheduled airlines. They did the same thing on some international routes, notably to Puerto Rico.

Pan Am's official name had been changed in 1945 from Pan American Airways to Pan American World Airways. Juan Trippe was again ahead of the game and ready for the postwar contest. He saw that by seating five abreast in the DC-4, the passenger count could increase from 44 to 63. But the question remained in what market such a configuration could be put to use. And fares would certainly have to be reduced in order to induce anyone to put up with such crowding. Here, Juan Trippe was about thirty years ahead of his time, ahead of the day of deregulation that would come in 1978.

In 1948, Pan American began flying DC-4s from New York to San Juan, Puerto Rico. Puerto Rico was a relatively impoverished island, and most of its inhabitants could not afford expensive travel of either kind, ship or plane. The low standard of living in Puerto Rico and its mortality rate combined to provide motivation to some people to leave the country. Pan Am tapped this large market for mass air transit in the newly configured DC-4, which had no galley and only one flight attendant. Any Puerto Rican with $75 was given the opportunity to begin a new life in New York, which had a Puerto Rican population of seventy thousand at that time. By 1950, there would be two hundred and fifty thousand Puerto Rican residents in New York City. By 1975, five million of the island's former residents would have migrated to the United States.

Source: Library of Congress.

FIGURE 6-2 The Constellation was much more expensive to maintain than other comparable aircraft.

Domestically, the airlines were losing money to the nonskeds, so they petitioned the CAB for authority to operate a second class of service. "Air coach," as it was called, was introduced by the scheduled airlines in 1948. Some of the crews began to refer to the new passengers as "cattle class." The airlines saw that they could compete with the railroads, not just for first class passengers, but also for coach passengers. Capital Airlines became the first established carrier to offer "coach-class" service, inaugurated on the New York–Chicago route. At a fare of two-thirds the standard, there were few complaints of overcrowding, late night departures, or the lack of a meal service. TWA and American followed suit with their transcontinental service. By the end of 1951, nine domestic carriers offered coach or tourist class service to 34 cities. In 1952, fares were $99 coast-to-coast; $32 between Chicago and New York. Airline passenger traffic doubled in the five years between 1948 and 1952. By 1955, the airlines had passed the railroads for the first time in the number of passengers carried.

For a while the "coach" or "tourist" class flights operated as separate airplanes both domestically and internationally. The smaller international carriers complained to the International Air Transport Association (IATA) that they could not compete with larger airlines since they did not possess the necessary number of aircraft to operate both first class and tourist class airplanes. When IATA authorized them to operate their equipment carrying both first class and tourist in the same airplane, the modern form of aircraft configuration was born. TWA was the first to begin domestic operations with both fare classes on the same airplane after CAB approval.

A new group of air carrier, known collectively as local service lines or feeder lines, completed their first full year of service in 1946. While the CAB would not expand the total number of trunk airlines beyond the sixteen that were grandfathered under the Civil Aeronautics Act of 1938, these smaller carriers received authority to operate on short routes to some 350 small cities. The average distance flown between stops was about 60 miles, and some of the communities served had populations of as few as 3000 people. By the end of 1951, there were 18 local service airlines operating 130 airplanes.

On the Way to the Jet Age

"To put your life in danger from time to time ... breeds a saneness in dealing with day-to-day trivialities."

Nevil Shute, *Slide Rule: The Autobiography of an Engineer*

The last of the big airliners mounting reciprocating engines on their wings were stretched versions of the airliners that had gone before. The DC-6 and the DC-7 were from the DC-4 model with various refinements to go along with the increase in length, breadth, and power. The Super Constellation was 19 feet longer than the original. With increased length came additional seating and with more seating came more revenue. Range was extended so that non-stop service was possible—not only coast-to-coast but transatlantic.

The DC-6 was launched in coast-to-coast service on April 27, 1947, with one stop enroute for fuel. United advertised its service as ten hours total. (See Figure 6-3.) TWA's Constellations could do about the same, advertised as ten hours ten minutes.

Boeing, a late entry to the new postwar aircraft building party, in 1948 introduced the double-decked B-377 Stratocruiser, a four-engine landplane larger than either the DC-6 or Constellation and designed with an emphasis on luxury reminiscent of the Pan Am Clippers. The airplane featured two decks with a cocktail lounge with leather seating located below, accessible by a curved stairway, and with a honeymoon suite in the aft section. Take off performance in the Stratocruiser was marginal, with the DC-6 routinely outperforming it, but it was bigger and faster at 340 miles per hour than any other airliner. It was

Source: Florida State Archives.

FIGURE 6-3 The DC-6 was launched in coast-to-coast service on April 27, 1947 with one stop enroute for fuel.

also expensive, costing over $1.5 million, and its high operation costs did not help matters. It had engine problems (the P & W Wasp Major had 112 spark plugs in 28 cylinders and delivered 3,500 horsepower) and the propellers had a tendency to go flying off on their own. Still, these airplanes were the ultimate in passenger comfort. New York to London was a pleasant affair of 12 hours duration, including cocktails, a five-course dinner, a good night's sleep, and plenty of attention. But the Stratocruiser had the worst safety record of the postwar big planes; six were involved in fatal crashes with the loss of 108 passengers and 28 crew. United unloaded their Stratocruisers early; Northwest kept theirs for years. In the end, the airlines seemed glad to see them go. Figure 6-4 pictures the Boeing 377, the Constellation 049, and the DC-4.

The DC-7 proved to be the first true transatlantic airplane, flying either west or east with a full load. With it, Pan American regained its leadership position over TWA, which was flying the Super Constellation. The DC-7 had engines that were reaching the limits of reciprocating engine

Source: Florida State Archives.

FIGURE 6-4 Boeing 377, Constellation 049, and DC-4.

power possibilities. With four Wright turbo-compound engines providing 3,250 horsepower, each weighing over 3,500 pounds, engine maintenance was a problem; American Airlines reported 10 engine failures a day on average. Westbound DC-7 service to the Pacific coast was advertised as non-stop, but with headwinds the advertised flying time of seven and one-half hours was often missed. Eastbound, American was able to adhere to its scheduled arrivals. The DC-7 made the first non-stop transatlantic crossing in 1957.

But the strain was showing; the limits of the reciprocating engine had been reached. It was time for the jet age.

Jets

It has been said that World War II advanced the airplane by fifty years. Yet it can also be accurately stated that there were really only two revolutionary technological advances to come out of the war—radar and the jet engine—and America had nothing to do with either. Not only were the Americans a bit slow at the beginning, they were even slower at appreciating how great a leap the jet engine represented in the potential for air commerce. This was demonstrated by the reluctance of American aircraft manufacturers and air carriers to pursue jet engine potential after the war.

The idea of turbine engines first manifested itself in England before the turn of the 20th century, when Charles Parsons developed a steam turbine that he applied to the generation of electric energy. He thereafter built a 100-foot vessel, the Turbinia, which he powered by means of a steam turbine to very impressive speeds. The British Navy subsequently built many of its warships utilizing steam turbines for propulsion.

In the early 20th century, attempts to apply the turbine technology to internal combustion engines, or gas turbines, met with disappointment. The design of the gas turbine called for the induction of air (as in a reciprocating engine) that would be compressed prior to ignition (as in a reciprocating engine). In the reciprocating engine, the compression is accomplished by means of the piston, but in the turbine engine, the compression is delivered by means of a series of rotating vanes (the compressor) located in front of the combustion chamber. Although the idea of the gas turbine proved workable, its fuel consumption was four times that of the internal combustion engine, and the economics of the invention simply prevented its adoption into commercial use.

The use of compressors in aircraft engines, however, did find continued use as the combined technology associated with aircraft engines and aircraft designs matured during the early years of aviation. As early as World War I, a turbocharger was fitted to a French aircraft to enable increased engine performance at higher altitudes. Turbochargers use the engine's exhaust gases to propel the turbocharger's compressor, thus making use of the free fuel source of the exhaust to compress air for induction into the engine's cylinders as the aircraft climbs into the thinner air at altitude. This technology was used to good effect in the design of the American fighters and bombers used in World War II. The B-17, for instance, could carry its bomb load to 34,000 feet.

The idea of utilizing the smooth, vibration-free rotation of a turbine, instead of the oscillating pistons of the much more cumbersome internal combustion engine, continued to occupy the minds of engineers and inventors. The first serious work on developing the jet engine, or turbojet, was commenced almost simultaneously in the 1930s in both England and Germany.

In England, Frank Whittle, an officer in the Royal Air Force (RAF), conceived that the application of the principles of the gas turbine might be applied, not to drive a shaft or propeller, but to produce a source of thrust for propulsion. In 1930, he was awarded a patent for his design, which was replete with compressor, combustion chamber, and turbine. Although his ideas and designs appeared to be workable, no assistance was forthcoming directly from the British government. The RAF did allow him to continue his work while on duty, and even assigned him the duty of securing engineering and advanced degrees at government expense, culminating in a master's degree from Cambridge. By 1935, Whittle had almost given up on his dream of producing a prototype jet engine when he was approached by two RAF officers willing to capitalize a company for this work. A company was formed, Power Jets, Ltd., and a workable prototype was achieved. At last, government interest was peaked, and direct funding of refinements of his engine was provided. Shortly, in 1939 and in conjunction with the Gloster Aircraft Company, Whittle was given a contract to produce England's first fighter jet airplane, the Meteor.

In Germany, Hans von Ohain, a physics student at the University of Göttingen, was similarly motivated to develop an efficient compressor for the gas turbine. His efforts led to early disappointments, but tests revealed enough success that his university mentors used their influence to introduce Ohain to Ernst Heinkel, the noted aircraft builder. Ohain found kindred spirits at Heinkel's company, and with that support a workable jet engine, of much simpler design than Whittle's, was incorporated into a newly designed airplane, the He 178, for its test flight in August 1939. The test was completely successful and ultimately produced the Jumo-004, (see Figure 6-5) later incorporated into the Messerschmidt 262 and

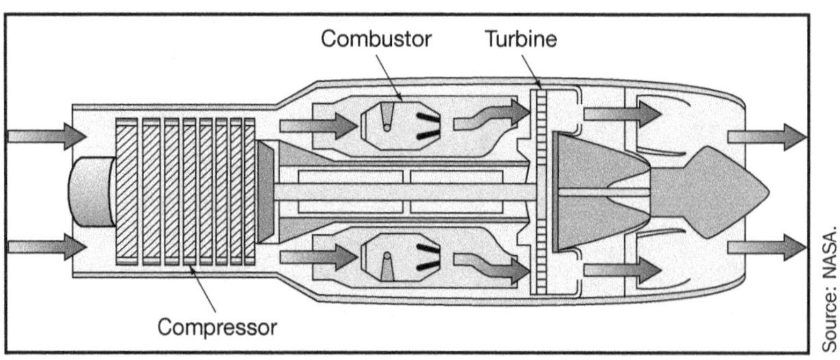

FIGURE 6-5 Germany's jumo-004 engine.

Heinkel 280. These airplanes made limited but impressive appearances over the skies of Europe during World War II.

In October 1942, Hap Arnold, aviation pioneer and general officer in the Army Air Corps in World War II, on learning of Whittle's work, had one of his engines flown to Washington. General Electric Company, the most experienced turbine producer in the country, was given the assignment of producing an American model along the lines of Whittle's design. Bell Aircraft was appointed to build the airplane, and in 1943 the XP-59, known as the Airacomet, flew as the first U.S. jet aircraft. Arnold also knew of the German jet program and had kept his superiors fully informed. The result was a government decision in 1943 to begin initial studies by the four largest aircraft manufacturers in the United States to test the feasibility of jet bombers. From this initiative came the North American B-45 Tornado and the vastly superior Boeing B-47 Stratojet.

Boeing had engineered and put in service in 1944 a very large wind tunnel capable of design testing shapes at speeds close to the speed of sound. Engineers had been dealing with drag at lower speeds for decades, but compressibility was a phenomenon that had only manifested itself as aircraft gained speeds approaching the speed of sound. Drag also appeared to increase exponentially as the speed of sound was approached. The Boeing tests concentrated on these phenomena and, unlike the B-45, were to result in a sweepback wing design never before incorporated in production aircraft. This design had the effect of delaying the onset of compressibility and of raising the speed at which the exponential increase in drag occurred. In short, it facilitated faster flight. Interest in production jet aircraft in the United States was in military aircraft, as evidenced by the B-47 medium bomber program, followed by the B-52 heavy bomber program. The first fully operational jet fighter, the Lockheed P-80, appeared in production in 1945 with its straight wings. The swept wing North American P-86 (designation changed to F-86) appeared in production in 1948 and set the world speed record that year at 671 miles per hour. These and the other jet aircraft of the time incorporated the conventional turbojet engine with all of its drawbacks and shortcomings. Chief among these drawbacks was fuel consumption, followed by high initial cost and frequent maintenance requirements. Airline chieftains were wary of the new, noisy technology. While there was little long-term operational data on jet engines available at the time, what was known was daunting. The consensus among American airline executives was that the jet was too risky, too unproven, and too expensive to be seriously considered as an addition to the fleet.

In the late 40s, a secret design concept for an improved jet engine, known as a "twin-spool turbojet" was in the works in military circles. The prototype of this engine would be known as the "J-57," and it was first tested aloft in 1951. After successful tests, it would be incorporated in a new generation of military airplanes, and it ultimately would make the difference the airline industry needed to consider turbojet propulsion in American commercial aviation. But in the early 1950s, the J-57 was still a military secret.

The British Comet

After World War II, commercial aviation interests in England conceded that the state of British aircraft technology and production was woefully behind that of the United States. The British, by necessity, had concentrated their efforts on fighter aircraft during the war, while the United States had been able to also pursue transport development. The British concluded that they could never catch up with the Americans in existing technology, but they saw a chance at leveling the competitive playing field with the United States by using the conventional turbojet in a new series of passenger transports.

In 1949, the de Havilland Comet began flight-testing with a design expectation of speeds of 480 miles per hour at flight levels of 35,000 feet. (See Figure 6-6.) While pressurized aircraft had been flying since the late 1930s, no airliner had faced the stresses that would be imposed at this projected altitude. The de Havilland Comet completed testing and entered into service on

Source: © Hulton-Deutsch Collection/CORBIS.

FIGURE 6-6 The Comet entered into service on British Overseas Airways Corporation in May 1952.

British Overseas Airways Corporation (BOAC) in May 1952, to the thrill and applause of the world. The first turbojet airliner halved flight times over BOAC's world routes. To the surprise of many, the Comet made money even though its operating costs were three times that of the DC-6, even charging regular fares. The difference was that the Comet flew virtually full on all flights, proving that high density seating was commercially feasible, at least on the vibration-free Comet. Next, Air France" inaugurated jet service on some of its routes with the Comet. In the United States, it was still DC-7 and Superconstellation piston engine service.

The Comet had three serious accidents in 1953. The third one involved the airplane coming apart in the air, possibly due to a design flaw, but it had occurred in connection with suspected thunderstorm penetration and was written off to the expected result of thunderstorm force. A fourth accident on January 10, 1954, grounded all seven of the Comets. This fourth Comet was lost over the Mediterranean Sea as it climbed above 26,000 feet. Its wreckage fell into the sea and was not immediately available for study. A commission formed in England to study the accidents came up with some fifty fixes to be incorporated into the Comet fleet. These adjustments were made and the Comets resumed service.

The wreckage of the Mediterranean crash was recovered and taken to the Civil Aviation Investigation Branch in England for analysis. As the investigation was proceeding, another Comet disappeared on April 8, 1954 on a flight from Rome to Cairo as it climbed to 35,000 feet. The fleet was again grounded and an all-out investigation was ordered to resolve the cause. It was fully appreciated that the future and reputation of the English aircraft production industry was now at risk, as was the entire future of commercial jet transportation.

The Royal Aircraft Establishment at Farnborough, headquarters for British aeronautical research, was given the task of solving the mystery of the Comets' crashes. A test procedure was contrived to expose the fuselage to a lifetime of pressurization and depressurization cycles, but at a rate 40 times faster than would normally occur. On June 24, 1954, the Comet's fuselage failed, developing a structural crack at the corner of one of the square windows, and expanding away down the fuselage. This indicated that the Comets likely had exploded, not unlike a bomb, due to the interior pressure of the aircraft. In August 1954, the last section of the doomed Rome to Cairo Comet was recovered. The investigators' conclusion of the cause of the crash was confirmed as the Comet's fuselage disclosed an almost exact duplication of the test results. The Comet 1 never flew again. Two later iterations of the Comet never flew commercially. The official findings of the British government's inquiry included that "more study both in design and by experiment" was needed to secure an economically safe life of the pressure cabin. These requirements were not met until 1958, at which time the Comet 4 made the first transatlantic jet commercial flight, on October 4, 1958. By then, the British advantage had been lost, and the United States aircraft production community was just getting started.

America Catches Up

The U.S. aircraft industry took its cue from its potential customers. The industry was obviously not interested in designing and building airplanes unless and until a market existed for them. The concept of the turboprop (the turbine jet engine used to drive a propeller) was considered the likely next commercially successful form of propulsion, and research and development efforts were stepped up both in military and commercial circles, particularly at Lockheed.

There was one other possibility. Boeing's reputation as a builder of military aircraft, mainly bombers and tankers, was unequalled. But Douglas and Lockheed were far ahead of Boeing in the commercial transport field. Boeing had taken a back seat to Douglas and Lockheed in every commercial airliner contest thus far—the B-247 ran second to the DC-3, the B-307 lost out to the Constellation and to WWII, and the B-377 could not compete with the DC-7 and the Super Constellation. Boeing had built the C-97 piston engine tanker for the Air Force, but it was not adequate for fueling the new jet bombers, the B-47 and the B-52. The Air Force, Boeing reckoned, would be in the market for a new jet tanker.

Boeing decided to take the gamble. On April 22, 1952, Boeing's board of directors authorized the expenditure of one-fourth of the company's total net worth, $15 million, to develop a prototype. Neither the airlines nor the military had actually expressed an interest in purchasing such an airplane, nor had any appropriation been secured in Congress for replacing the C-97. Boeing officially designated the project the model 367-80, and it was known internally at Boeing as the "dash 80" thereafter. But the designation that the world would come to know was the "707."

Boeing had the largest and the only state-of-the-art wind tunnel for testing aircraft shapes. The development of this wind tunnel, in fact, had been responsible for the adoption of the swept wing design first incorporated in the B-47. (See Figure 6-7.) The 707 design was adapted through wind tunnel testing for over 4,000 hours. Redundant systems, overlapping structural components, multiple strength round windows, plug type doors for better pressurization seals, and spot welds set the standard for the jet aircraft production industry to come. The exterior shell of the 707 was engineered before the Comet disasters. Boeing decided that the skin of the aircraft would be aluminum, of a thickness that turned out to be 4 1/2 times as thick as the Comet's (the Comet's exterior shell was so designed in order to save weight). Boeing also incorporated a new alloy, known as titanium, that was as light as aluminum but stronger than steel to bolster the strength and fatigue resistance of the 707's skin. Then the engineers put the design through 50,000 pressurization cycles with no evidence of metal fatigue.

Source: Florida State Archives.

FIGURE 6-7 A KC-97 refueling a B-47.

Tests revealed that the positioning of the aircraft's engines on pylons slung underneath the wings provided the best lift efficiency and had the added benefit of allowing easy access to the engines for maintenance. The wings were also designed to carry 17,000 gallons of fuel, thus allowing for non-stop transcontinental range.

In 1952, Douglas was investigating the feasibility of jet powered airliners but only went so far as to construct a full scale wooden mockup of what was to become the DC-8. The problem in the industry, with both manufacturers and carriers, was one of confidence. None of the industry leaders could seem to project a solution to the financial impact of the cost of construction of the high-flying jet airliner and its cost of operation, particularly from the fuel standpoint. The J-57 was still not available. Projections of cost per aircraft approximated $4 million, contrasted to the $1.5 million price tag of the DC-7. Aviation industry leaders were not convinced that jets were commercially viable. C.R. Smith of American Airlines was of the opinion that in order to justify going to jetliners, the cost of operating them should be no higher than the cost of operating the DC-6. Of course, no one knew what the costs of operation of a jet fleet would be. Fuel consumption could be projected, but some costs, such as parts, maintenance, and engine life (TBO) would have to await experience. Among the Americans, only Juan Trippe dissented.

Trippe had been interested in the de Havilland Comet when it first came out, and had placed orders for three of the airplanes subject to their specifications meeting the United States Civil Aeronautics Administration requirements. There was, in fact, some concern that the CAA would not grant the Comet an airworthiness type certificate based on CAA reservations (prophetic, as it turned out) about the square corners of the windows in the aircraft. The CAA had recommended oval windows but de Havilland appeared to be satisfied with its design, citing design safety tolerances much in excess of expected stresses. Subsequent events would tragically vindicate the CAA's position, but without U.S. approval, Trippe was left in the age of piston aircraft. By the time that the problems with the Comet had been rectified in the redesign of the Comet 4, in 1958, the 707 was light years ahead of the old Comet design. The Comet, for instance, had seating for 67 while the 707's capacity was 130.

Boeing's gamble paid off when, in March 1955, it received its first order for the 707, not as the anticipated passenger airliner, but as the first jet tanker ordered by the Air Force. The first 707 was rolled out of its hangar at Renton, Washington, in May 1955, and completed its maiden flight on July 15, 1955. The 707 prototype would undergo flight testing for the next three years before being placed in commercial airline service. The largest aircraft then in commercial service was the Boeing Stratocruiser and the 707 was 15 times more powerful, twice as fast and almost twice as big. Douglas, now convinced of the feasibility of building civilian jet aircraft, announced that it would complete its design and begin production of the DC-8.

Pan American, alone among the American carriers, seemed interested in jets, despite their projected economic indicators. Trippe had seen how the public had abandoned piston-powered airplanes in droves for the Comet, and it was his purpose to be the first to supply the high-flying, vibration-free, 500-mile-per-hour airplane of the future to America. Just as he had been among the first to abandon the wood and wire airplane and put the Fokker trimotor all metal cantilevered monoplane in service in 1928 (Key West to Havana), the first to inaugurate extended over-water service in the great Clipper amphibians, the first to offer his airline passengers hotel accommodations in his own hotels at their destinations, and among the first to switch to pressurized aircraft, he was now the first to order the first U.S. produced commercial jet airliner. He did so against the prevailing industry tide in October 1955 with the announcement that he had ordered twenty Boeing 707s and, to the great delight of Douglas who did not actually have a real airplane in existence, twenty-five Douglas DC-8s. At a total capital outlay of $269 million, Pan American had committed to the largest airplane acquisition in the history of the industry.

The 707 and the DC-8 were so similar in appearance that it was hard to distinguish between them. But there were real differences to the potential customers, the airlines. First was cabin width, then length, then seating capacity, then the engines. The airlines seemed to prefer the DC-8 design. Before very long, Douglas had twice the orders for DC-8s than Boeing did for 707s. Boeing began making modifications, first to widen the fuselage to a dimension one inch wider than the DC-8, then to increase its length, wingspan, and range. Soon, it had another version of the original 707, and this time the airlines liked it. In 1955, Douglas outsold Boeing, only to be put in second place at the end of 1956. The airlines were now getting caught up in the idea of the jet age, and orders began to pour in. United States airlines bought, but so did foreign airlines. Eastern, Delta, KLM, SAS, Japan Air Lines, and Swissair all bought Douglas. American,

Continental, Western, TWA, Air France, Sabena, and Lufthansa went with Boeing. Boeing had finally broken the old jinx of second best. All told, Boeing would win the numbers competition against Douglas by almost 2 to 1. The most satisfying event, though, might have been the selection by the President of the United States of the Boeing 707 as the first jet Air Force One, in 1959.

Lockheed declined to enter the competition, concentrating on the turboprop as its best guess of where the future of commercial aviation lay. Lockheed's contribution was to be the Electra, which in 1957 became the first propjet put in service by U.S. airlines. Only 169 planes were produced, some for the Navy, designated as the P-3. Convair submitted its CV-880 in 1959 but was unable to compete with the Boeing and Douglas jetliners, losing some $270 million for its efforts.

On October 26, 1958, Pan American became the first American air carrier to inaugurate scheduled jet service with the 707 on its New York to Paris flight. National Airlines was next on December 10, 1958, with a 707 leased from Pan Am that was put on the New York-Miami run to mark the first domestic jet service. Eastern was flying the same route with Lockheed Electras, and immediately began losing out to National. American followed domestically by putting the 707 to work on the transcontinental route, then TWA. United was out of action awaiting the delivery of the first DC-8s, still in the production phase. Eastern could not seem to accept that the jet age had really arrived, and was woefully late in acquiring its first jets, much to its economic disadvantage against its competitors.

On any competitive route in the late 1950s, jets trounced the piston airplanes. The flying public loved jets, and this translated into filled passenger seats. The load factor went up dramatically on jet routes, and their capacity was almost twice that of the DC-7. The airlines were surprised to find that the reliability of the new jet engines greatly reduced failure concerns which had become commonplace with the great turbo-compound piston engines, and that replacement parts and maintenance costs were much lower than expected. Time Between Overhaul (TBO) was a federally-mandated life expectancy of the piston engine used in commercial service in the late 1950s, and it was about 800 hours. The FAA found that jet engines could greatly exceed this limit, and gradually raised the TBO for jets to 4,000 hours. This artificial limit was ultimately discarded entirely in favor of a progressive maintenance schedule designed around the few critical components of the jet engine.

Passenger-mile costs proved to be about the same as on the DC-7. Although fuel consumption in the 707 was much higher than the DC-7, the actual passenger miles per gallon for the 707 was 42 compared to the 59 passenger miles per gallon for the DC-7. The lower cost for jet fuel (kerosene) compared to high-octane gasoline offset this slight difference. The economics of commercial jet travel were working out after all, and the flying public embraced the jet age.

The Next Jets

Boeing had upgraded the 707 in 1959 with the new J-75 engine. The DC-8 was flying. Big jets were flying long distances and setting records, and the public was enthralled. Governments the world over were buying these jets and setting up their own airlines. Flying in jets was a prestigious activity.

The government of France had been eclipsed in the jet design and production market. Its aviation representatives took note of something that was not in production and not even on the drawing boards—a medium-range jet that could carry 60 passengers up to 1,200 miles. This was the airplane for the European market, and presciently, was to become the airplane for the deregulated market of the future. This was the Regional Jet.

In response to a government-sponsored competition, Sud Aviation in Toulouse, France, came up with a novel idea in aircraft construction. They placed the aircraft's engines on the side of the fuselage near the tail of the aircraft instead of under the wings. They called this prototype the Caravelle

Source: National Air and Space Museum, Smithsonian Institution (SI 82-14081).

FIGURE 6-8 The Caravelle was a prototype with the aircraft's engines on the side of the fuselage near the tail of the aircraft.

Image Courtesy of Corel.

FIGURE 6-9 Boeing 727.

(see Figure 6-8), a name given small sailing ships during the age of exploration. Production began, and in 1956 Air France contracted for the first twelve airliners to come off the line.

British European Airways, the government-owned airline, flew many of the same routes on the Continent using turboprops. Given the proven popularity of jets, already evident in the 1950s, Britain realized that it must build its own short-haul aircraft in order to compete. Its entry was the Hawker-Siddeley Trident, which incorporated the Caravelle aft-engine innovation but added a third engine housed within the vertical stabilizer and aft fuselage. The aircraft designers placed the horizontal stabilizer at the top of the vertical stabilizer, out of the way of the jet exhaust, an arrangement that provided more stability at low airspeeds.

Meanwhile, Boeing was testing the aft-engine concept with an aft-mounted engine attached to its 707 prototype, and it was pondering the viability of such an aircraft in the domestic market. Douglas had designed the DC-9, with its aft-mounted engines, in response to a request by United Airlines, but no other carrier expressed interest, and the design was put on hold. The airlines specified an aircraft with two or three engines, for cost effectiveness, that could operate from shorter runways like LaGuardia. Boeing's engineers were first to conclude that a three-engine airplane with a T-tail was the most likely airplane to succeed, borrowing from the Trident design, which had proven out in Boeing's tests. They designated the new airplane the Boeing 727 and incorporated the new Pratt & Whitney JT8D turbofan, with up to 17,500 pounds of thrust, as the power plant. Turbofans evolved from turbojets as early as 1960, mainly in response to complaints about the noise produced by straight jets, both while in taxi and airborne. The JT8D was not only quieter, it was more economical to operate than any other engine at the time.

The 727 was an aesthetically pleasing airplane. (See Figure 6-9.) It was said that building the 727 would have been warranted even if it couldn't fly. It utilized the same basic fuselage as the 707 and incorporated a new flap design that, at slow airspeeds, increased the wing area by twenty-five percent; thereby greatly reducing the aircraft stall speed. This reduction in speed enabled the 727 to operate from shorter runways, just as specified by the airlines.

The first production model of the 727 flew late in 1962 and immediately began to surpass its design criteria. It was faster, its fuel consumption was less, and its payload was greater. Short landing and takeoff was proven in operation, and its superb handling made it one of the most trusted and respected aircraft flying. Concerns arising from a series of four crashes occurring in 1965 were alleviated when it was determined that they were all caused by pilot error in allowing the airplane to descend at a rate from which recovery was difficult. These accidents established that the 727, in spite of its easy handling characteristics, had to be flown by the numbers, like most jets. The performance of the 727 would go on to earn it

Image Courtesy of Corel.

FIGURE 6-10 Douglas DC-9.

Image Courtesy of Corel.

FIGURE 6-11 Boeing 737.

a reputation as the most successful commercial transport aircraft in the history of aviation. By the early 1980s, Boeing had delivered or contracted to deliver almost 2,000 of the very unique airplanes.

After Douglas had placed its DC-9 plans on hold in the late 1950s, the emergence of the short-to-medium range aircraft market caused Douglas to dust off its DC-9 blueprints. In April 1961, Douglas announced that it would begin production of the DC-9. Although Douglas had no orders placed at the time of its announcement, within a month Delta disclosed its contract to purchase 15 of the new jets. Boeing did not respond to the DC-9 until 1965, the same year the first DC-9 went into service. (See Figure 6-10.) Then Boeing unveiled its plans for the 737. The 737 was not a sleek airplane, having a width equal to the 727 and 707 but not the length—it was shorter even than the DC-9. Lufthansa Airlines was instrumental in the design of the 737 because they were first to order the airplane, insisting that it carry 100 passengers, ten more than the DC-9. The 737 entered service in 1968. (See Figure 6-11.)

Sales of the 737 were initially depressed primarily because the Air Line Pilots Association (ALPA) took the position that ALPA crews would not fly the 737 with only two flight crew members; demanding that a flight engineer be included in the cockpit. ALPA was playing catch-up from its earlier failure to require 3-man crews in the DC-9. This requirement made the 737's operating costs too high to be competitive, so the airlines largely rejected the airplane. ALPA abandoned its 3-crew position in 1974, partly because of worldwide recession based on the fuel crisis that year, and partly because of the untenable and obvious featherbedding aspects of its 3-crew position. Airlines then started buying the 737.

For the first time, feeder airlines began to buy the short-to-medium range jets and to bring jet service to the hinterlands of America. Piedmont, North Central Airlines, and Allegheny Airlines were able to expand their service, and in the later years of regulation, beginning in the early 1970s, these airlines were able to secure routes to destinations previously unavailable to them. These jets made routes between small airports—like Tri-Cities, Tennessee to Chicago, or to Washington, D.C., or to New York—convenient and profitable. The feeder lines preferred one class service and gave the world a glimpse of the age of deregulation to come. But first, the jumbo jets had to fly.

The Really Big Jets

In 1962, Lockheed won an Air Force contract to build the largest cargo plane ever conceived. The aircraft specified by the government included power plants of four 21,000-pound thrust turbofans, a range of 4,000 miles, and a useful load of 71,000 pounds plus fuel. When complete, the aircraft would be known as the C-141 Starlifter, and it would have shortcomings. Chief among these was the fact that the C-141 did not have the design volume required to house the cargo load specified. Already recognized by the Air Force was the need for a larger airplane. The Air Force had put out for competition the design of what was to be known as the C-5A, a truly mammoth creation. Lockheed won this competition too, even though Boeing's entry was a serious contender in the competition and, on reflection, perhaps the best of the three entries.

Second place Boeing decided to convert its design and engineering effort to commercial passenger use. Juan Trippe, ever on the cutting edge, had indicated an interest in such an aircraft. Boeing showed that its cargo plane could be modified to accommodate 450 passengers, at nineteen feet, five inches in width, and two hundred thirty-one feet in length. The JT9D turbofan, a high bypass ratio jet engine with 41,000 pounds of thrust, was chosen to power the aircraft. This airplane would also fly faster than previous models, at 625 miles per hour, and would be known as the 747. (See Figure 6-12.) Juan Trippe had long ago concluded that the key to making money in the airline business was to fill the airplanes with paying customers, like he did with the DC-4 in the late 1940s in the San Juan to New York migration. Now, this was a dream come true. He signed a letter of intent to purchase twenty-five of the "wide bodies," as they were to be known.

Boeing, just as it had during the design phase of the first American jet transport, the 707, took its safety responsibilities seriously. The "carnage factor" of a crash of such a large aircraft was daunting, and only increased Boeing's commitment to safety in the design stage. A "safety committee" was formed to review every aspect of the new aircraft. Concerns of the committee ran the gamut of engineering and construction, from hydraulics, to wing loads, and even to coffee pots. The airplane was so huge that Boeing did not even have a facility large enough to build it, so a new plant had to be constructed at Everett, Washington. It was the largest factory in the world.

The 747 first flew on February 9, 1969. Once again, Pan American was the first to place yet another new prototype airliner in service, this time the 747 Clipper Young America out of JFK for Europe. The 747 had initial problems, mostly because of its size. For instance, baggage facilities were overloaded at destination, causing delays; cabin attendants were overwhelmed by the number of drinks, meals, and related requirements caused by the passenger count; the lavatories seemed inadequate for the needs of passengers; and so on. Each of the concerns was addressed, resolved, and the 747 gradually became a favorite of the flying public. The upper deck of the 747, complete with its cocktail lounge atmosphere for first-class passengers, which was sometimes converted to a restaurant, and its piano manned by a professional pianist, was reminiscent of the lower deck of the Stratocruiser of the 1940s.

In 1967, Lockheed completed its design for its wide-bodied entrant into the field, known as the L-1011 Tristar. (See Figure 6-13.) Lockheed utilized the fuselage tail-mounted engine of the original Trident, together with two wing-mounted

Source: Florida State Archives.

FIGURE 6-12 The Boeing 747 flew faster than previous models.

Image Courtesy of Corel.

FIGURE 6-13 Lockheed L-1011.

engines, for its combined power plant, and it could accommodate 300 passengers.

Circumstances, primarily financial, had required Douglas to merge with McDonnell Aircraft of St. Louis. The new company was known as McDonnell-Douglas. Its submission to the wide-body contest was the DC-10, which bore a marked similarity to the L-1011. Both aircraft had three turbofans, one mounted under each wing and one tail mounted. The L-1011's rear engine's intake was built into the vertical stabilizer above the top of the fuselage, with the engine mounted at the rear of the cabin. In the DC-10, the third engine was mounted through the vertical stabilizer, with the intake and exhaust in a direct line fore to aft. (See Figure 6-14.)

In 1970, European aircraft builders, funded by their national governments and comprising a loose consortium of French and British interests that were later joined by the Germans, established their own aircraft production company, Airbus Industrie. The purpose, as they said, was "to reduce dependence on foreign equipment, facilitate survival of a struggling European aircraft industry and address a market opportunity not being met by the Americans." This consortium designed the Airbus 300, a 300-passenger entry actually built by Sud Aviation in Toulouse, France. The A300 had only two turbofan engines, either Rolls-Royce or General Electric, but for various reasons the A300 was slow to materialize. The A300 did not fly until 1972, over three years after the 747, and over one year after the DC-10.

Image Courtesy of Corel.

FIGURE 6-14 DC-10.

Problems related to structural integrity were encountered by the DC-10 shortly after its inauguration. First, in June 1972, an American Airlines DC-10 out of Detroit suffered a decompression incident when a baggage door, located on the lower deck, blew off and collapsed the supporting deck of the passenger section above. Hydraulic lines had been designed and built to run the length of the aircraft through the floor or deck between the upper and lower compartments, and when the floor collapsed, some of these lines were severed, causing serious control problems for the flight crew. Only the ingenuity and skill of the crew allowed the stricken craft to be brought in for a safe landing.

The baggage doors were not the plug-type doors designed into many jet aircraft, but were dependent on latch mechanisms that, upon investigation, were found to be defective. An aircraft directive mandating corrective action was issued and the modifications were performed with the exception of two airplanes.

One of these was found and fixed; the other was not. On March 3, 1974, the airliner that had been overlooked, a Turkish Airline DC-10 from Paris to London, suffered a similar baggage door failure with a similar floor collapse. This time the crew was unable to fly the aircraft, which crashed, taking all 346 lives aboard.

In the DC-10, no further baggage door incidents occurred, nor were any other serious failures experienced for six years. Then on May 25, 1979, as an American Airlines DC-10 climbed out from Chicago O'Hare after takeoff, the left engine separated from the wing pylon, causing the aircraft to roll inverted and nose down, a condition that the crew was unable to correct at such a low altitude.

The wide-bodied experience of the American producers could be said to have been only marginally successful. Ultimately, McDonnell-Douglas sold 300 DC-10s while Lockheed sold only 244 L-1011s. By 1982 when production of the L-1011 was halted, Lockheed is said to have lost some $2.5 billion on the project.

Airbus, on the other hand, had managed to design a product that would crack the American airline market from Europe for the first time. The A300 had only two engines. This was of some concern initially for transoceanic flight, but it translated directly into reduced operating costs. Secondly, the A300 had incorporated composite, lightweight materials in its structure, adding to its cost effectiveness. As Air France began in 1974 to operate the A300 around the world, the airplane soon began to sell in the European and Asian airline market. Korean Air Lines, Lufthansa, Indian Air Lines and South African Airways bought the A300. Frank Borman of Eastern arranged a six-month trial of the A300 for its New York to Miami route, without any commitment to buy the airplane. This was an unprecedented deal, amounting to a manufacturer loss-leader arrangement whereby Airbus, in effect, loaned its airplane to Eastern on a trial basis. It turned out to be a brilliant stroke by Airbus that resulted in Eastern placing an order for 23 of the aircraft at the price of $25 million a copy, in April 1978.

With the Airbus 300, a trend began in airliner construction of wide-bodied, twin-engined, and lighter weight airplanes that still endures. Boeing contributed the 767 in 1983, using weight-saving composite materials and an advanced wing structure. Since the Boeing and Douglas face-off in the 1930s, beginning with the introduction of the 247 and the DC-1, the history of commercial airliner production competition had been an altogether American affair. Now, with the emergence of Airbus Industrie, combined with the shrinking number of American aircraft manufacturers, the contest is becoming not only international, but also specifically European versus American.

CHAPTER 7

Aviation and Federal Legislation

The Post Office Department, through the development of airmail, was directly responsible for the beginning of commercial air transportation and of the airline industry. Therefore, the Post Office can be credited with being the father of United States commercial air transportation.

Benjamin Franklin, who established the Postal Service at the birth of the United States, recognized the importance of such a service. He said that a progressive nation requires a good communication system. His policy for the Post Office was that it should assist development of all new forms of transportation, which would, in turn, provide better mail delivery. Examples of this were the mail subsidies paid to the early stagecoach lines, which depended upon mail pay to stay in operation; only then were they able to provide passenger service. Their real income was from the transportation of mail. The famous Pony Express was established solely because of the lucrative mail payment offered to the contractors.

Government assistance in the early railroads is further evidence of the federal policy to encourage new forms of transportation. In this case, the government granted land to the railroads to lay their tracks, and then loaned or gave them money for equipment.

By 1925 the Post Office had conducted airmail service long enough to prove the practicability of such a commercial, nonmilitary use of the airplane. It was generally understood that the Post Office operation would only be temporary, and that as soon as it was feasible, carrying mail by air would be turned over to private industry.

In 1918, it was estimated that it would be at least five years before this transfer could take place. By the mid-1920s the Post Office was ready to ask private industry to take over the airmail operation.

The story of private companies beginning to carry airmail is also the story of the inception and growth of the commercial airline industry.

The last flight of the Post Office operation took place on September 9, 1927. Beginning in June 1927, the Post Office pilots were released as the newly formed airlines took over the airmail routes. There were 43 pilots at that time in the Postal Service, and an additional 600 employees in ground jobs. All of these had contributed to more than 12 million airmail miles having been flown.

The entire cost to the United States government for the airmail service, from its beginning in 1918 until the last flight in 1927 was $17.5 million. During these nine years about $5 million in airmail postage was sold to the public. The difference of $12.5 million cannot be counted as a loss, since it was a very small price to pay for the establishment of commercial air transportation and the United States airline industry. This industry has repaid the debt many times over, not only in taxes, but in valuable service to both the government and to the public.

Airmail Act of 1925 (The Kelly Act)

Civil air transportation got its start in the United States when on February 2, Congress passed the Airmail Act of 1925. It is also known as the Kelly Act, because it was Representative Clyde Kelly of Pennsylvania who sponsored and led the congressional fight for its passage.

This Act made possible the awarding of contracts to private contractors for the transportation of airmail. Therefore, the purpose of the law was to provide for private airmail contractors, as clearly indicated in the title of the Kelly Act: "An Act to Encourage Commercial Aviation and to Authorize the Postmaster General to Contract for the Mail Service."

One important feature of the Kelly Act was that the amount of compensation paid to an air carrier could not be more than four-fifths, or 80%, of the revenue derived from the sale of airmail postage for such mail. This feature was a compromise with those in Congress who were opposed to the federal government paying for the development of the airline industry. The remaining 20% was for the ground handling expense of the airmail. In essence, the operation of the airmail under the Kelly Act would not cost the government—this feature proved to be the main weakness in the Kelly Act.

By the middle of 1925 Postmaster General Harry S. New advertised for bids on eight airmail routes. These routes were to be short feeder routes in order to allow the airlines time to gain experience; after the airlines had proved themselves, the transcontinental routes would be awarded. Only those companies with proper experience and equipment were eligible to bid.

On September 15, 1925, bids for the eight routes were filed with the Postmaster General, and they were opened on October 7. There were 10 companies which submitted 17 bids. Three were dispensed with due to the lack of proof of financial responsibility. As a result, only five routes were immediately awarded. By the beginning of 1926, 12 contract airmail (CAM) routes had been awarded. (Table 7-1.)

The first of these awards went to companies supported by men of various backgrounds and personalities. Winners of the initial five contracts were National Air Transport (owned by the Curtiss Aeroplane Co.), Varney Air Lines, Western Air Express, Colonial Air Transport and Robertson Aircraft Corporation. National and Varney would later become important parts of United Air Lines (originally a joint venture of the Boeing Airplane Company and Pratt & Whitney). Western would merge with Transcontinental Air Transport (TAT), another Curtiss subsidiary, to form Transcontinental and Western Air (TWA). Another company, later to join American, was Robertson Aircraft Corporation, which was founded by two brothers, one of whom gained experience during World War I. One of their three employees was Charles A. Lindbergh. Robertson would become part of the Universal Aviation Corporation, which in turn would merge with Colonial, Southern Air Transport and others, to form American Airways, predecessor of American Airlines. Juan Trippe, one of the original partners in Colonial, later pioneered international air travel with Pan Am—a carrier he founded in 1927 to transport mail between Key West, Florida, and Havana, Cuba. Pitcairn Aviation, yet another Curtiss subsidiary that got its start transporting mail, would become Eastern Air Transport, predecessor of Eastern Air Lines. Wall Street promoter Clement M. Keys was Chairman of the Board and President of Curtiss Aeroplane and Motor Company. Keys succeeded in getting CAM 3 for National Air Transport, later to join United Airlines. The Los Angeles-Salt Lake City award went to Western Air Express. Western was formed by journalist Harry Chandler, publisher of the *Los Angeles Times.* Awards to Henry Ford of routes 6 and 7 attracted the most publicity because of his influence on the business community.

Florida Airways Corporation received CAM 10 between Atlanta and Jacksonville, Florida. Its president was Reed Chambers who sought his former flying companion from the 94th Aero Squadron, Eddie Rickenbacker, to assist Florida Airways' growth.

Route #	Company	Route	Began Operating
CAM 1	Colonial Air Lines	New York to Boston	June 18, 1926
CAM 2	Robertson Aircraft Corp.	Chicago to St. Louis	April 15, 1926
CAM 3	National Air Transport	Chicago to Dallas	May 12, 1926
CAM 4	Western Air Express	Los Angeles to Salt Lake City	April 17, 1926
CAM 5	Varney Speed Lines	Elko, Nevada to Pasco, Washington	April 6, 1926
CAM 6	Ford Air Transport	Detroit to Cleveland	February 16, 1926
CAM 7	Ford Air Transport	Detroit to Chicago	February 15, 1926
CAM 8	Pacific Air Transport	Los Angeles to Seattle	September 15, 1926
CAM 9	Charles Dickenson	Chicago to Minneapolis	June 7, 1926
CAM 10	Florida Airways Corp.	Atlanta to Jacksonville	September, 1926
CAM 11	Clifford Ball	Cleveland to Pittsburgh	April 21, 1927
CAM 12	Western Air Express	Pueblo, Colorado to Cheyenne, Wyoming	December, 1926

TABLE 7-1 The First Contract Airmail Routes

CAM 11 had the most interesting award. It went to Clifford Ball who was the Hudson-Essex automobile dealer in McKeesport, Pennsylvania. As a result of a real estate transaction, Ball acquired land near Pittsburgh that pilots had been using as an airport. When they could not pay the rent, he seized their aircraft as payment. Since he had an airport and airplanes, Clifford Ball bid and won the Pittsburgh-Cleveland route.

These first contract airmail routes awarded under the Kelly Act were feeder lines branching off into the main transcontinental route. Because of economic conditions and a desire to reduce costs in the Department, it was then decided that bids would be given on main routes.

Although Colonial Air Lines received the first airmail contract, the honor for being the first in operation goes to Henry Ford, who owned Ford Air Transport. In the early days of commercial aviation the name Ford was known in aviation equally as well as in the automobile industry. As early as 1923 Ford became interested in an all metal, internally braced wing, monoplane designed by William B. Stout. Ford bought Stout's factory and put three engines on the aircraft instead of one. The result became known as the Ford TriMotor, or more affectionately, the *Tin Goose*. It was such a strong and dependable aircraft that it became the leading passenger aircraft for many years. The aircraft designated 4-AT was unique in that it was covered in corrugated metal; considered state of the art when it was produced. Ford built 199 of them over seven years beginning in June 1926 at a price of $50,000 each. This type of aircraft was later used to found air transportation in South America. In fact, some of them are still in operation today, a record unequaled by any other type of airplane.

In April 1925, Ford, using his single-engine airplanes, began flying a scheduled route from Detroit to Chicago to Cleveland for the exclusive use of the Ford Company. Thus Ford not only operated the first airmail route by a private contractor, but also is credited with having one of, if not the first, air freight operations as well as with having one of the first airlines to carry passengers.

Under the Kelly Act a contract airmail carrier was paid up to 80% of the postal revenue of the airmail he carried. In trying to avoid subsidy Congress created a problem in computing payment to the carrier. Because the airmail was handled both by the Post Office and by the airline, each letter had to be counted before it was loaded aboard the airplane so that the postage could be apportioned between the Post Office and the air carrier. Under this system, an irrational distribution of airmail payments occurred.

The Postmaster solved these inequalities by the first amendment to the Kelly Act, approved June 3, 1926. New contracts were provided for at rates not to exceed $3.00 per pound for the first 1,000 miles and thirty cents a pound for each additional 100 miles. Again compensation was based upon nonsubsidy; this was still a weakness.

On January 25, 1927, the bid for the San Francisco to Chicago portion of the route was awarded to the Boeing Airplane Company and Edward Hubbard. Several months later the joint contract was turned over to Boeing Air Transport. Boeing Air Transport bid $2.89 per pound on the Chicago-San Francisco route; their nearest competitor bid $5.09 per pound. At that rate and with the aircraft available, Boeing critics predicted substantial losses. When the contract was awarded, William Boeing went to Pratt & Whitney and requested that they build an air-cooled Wasp radial engine for his Boeing 40-A. By June 30, 1927, 25 Boeing 40-A's were ready. The airplane carried two passengers in addition to its airmail load. During its first two years Boeing Air Transport carried over 1,200 tons of airmail and approximately 6,000 passengers. The Post Office and Boeing urged lower rates to stimulate the use of airmail. At the time the western sector was awarded, no acceptable bid was received for the route east of Chicago. This was determined later by awarding a contract to National Air Transport on March 8, 1927.

In February 1927, an airmail postage rate of ten cents per half ounce between any two points in the United States became effective and provided a great impetus to the public in the use of airmail. In August 1928, rates were reduced to five cents for the first one-half ounce and ten cents for each additional ounce.

The first amendment to the Kelly Act did away with the difficult method of counting each letter carried, and the second amendment, passed in May 1928, did away with assuring the government against losses. Thus subsidy quietly came into the airline industry. The second amendment also reduced the airmail postage to five cents an ounce, and a 95% increase in airmail traffic resulted. The airlines could actually receive more money for carrying the airmail than the cost of the postage on the mail carried. To the young profit-starved airlines, this presented a golden opportunity. Now that they were being paid by the pound to carry the airmail, several of the airline companies began sending large quantities of airmail, as private citizens, destined to go over their own routes. They sent thousands of letters stuffed with paper, even telephone directories and spare parts. One airline contractor mailed two tons of lithograph material from New York to Los Angeles. It cost him over $6,000 in postage; but he was paid $25,000 by the Post Office to carry it, a neat profit in anyone's estimation.

Obviously, the opportunity for these inequalities had to be corrected; and so the third amendment to the Kelly Act so completely changed it that it was, in reality, the second major law passed by Congress having to do with airmail carried by the private contractors. This piece of legislation was called the Airmail Act of 1930.

Air Commerce Act of 1926

When the Kelly Act provided that the airlines would carry the airmail, it was apparent that they were neither financially able, nor large enough, to provide for the maintenance and operation of the airway system organized for the Post Office operation.

Therefore, on the heels of the Kelly Act, Congress passed another major piece of legislation known as the Air Commerce Act of 1926, the purpose of which was to promote air commerce. This law charged the Federal Government with the operation and maintenance of the airway system as well as all aids to air navigation, and to provide safety in air commerce generally through a system of regulation.

It was unlike the other mail legislation acts. It left no doubt that the Federal Government was strongly in the picture of growth and development of aviation by way of aiding and encouraging air transportation.

The function of safety regulation was to be carried out by the Department of Commerce; therefore, the Bureau of Air Commerce was established.

Among the safety regulations provided were the requirements of registration and licensing of aircraft and the certification and medical examination of pilots. Civil penalties were allowed to enforce these requirements. This was the start of what later became the Civil Aeronautics Administration, and today the Federal Aviation Administration.

The Kelly Act and the Air Commerce Act provided the foundation of civil air transportation in the United States.

The late 1920s brought many new ventures in aviation. Four major groups developed:

- United Aircraft and Transport Corporation
- North American Aviation
- Aviation Corporation of Delaware
- Transcontinental Air Transport

Airmail Act of 1930 (McNary-Watres Act)

The Airmail Act of 1930, also known as the McNary-Watres Act, was passed April 29, 1930, in an attempt to unify the industry. Up until this time, the newly organized airlines were not interested in carrying passengers. During the 1930s the airline airplanes were referred to as mail planes. Today, when an airliner passes overhead, it is thought of as a passenger airplane, even though airmail is aboard.

An important event which vastly effected development of the airline industry, took place in November 1928, when Herbert Hoover was elected President and shortly thereafter appointed Walter Brown as the new Postmaster General. Brown initiated passage of the McNary-Watres Airmail Act of 1930. (See Figure 7-1.)

Brown served under Hoover as Assistant Secretary of Commerce in the Coolidge Administration and had served as Hoover's campaign manager. He was previously chairman of the Republican National Committee.

When Brown assumed the office of Postmaster General there existed 44 small airline companies lacking capital and dependent upon government airmail contracts.

FIGURE 7-1 Three-engined airplane of the 1930s meets the three-engined airplane of the 1960s.

Brown felt that the responsibility to encourage commercial aviation was not being carried out by competitive bidding although airmail was being carried at the lowest cost. Brown defined the airline problems as:

1. Being unwilling to invest in new equipment,
2. Operating obsolete aircraft,
3. Demonstrating questionable safety performance resulting from cost cutting, and
4. Maintaining marginal operations with no growth.

To Brown the solution was to eliminate competitive bidding and use airmail pay to support those corporations strong enough to encourage commercial aviation. To accomplish his solution Brown needed legislation. This legislation was the Airmail Act of 1930. The law received difficult passage because Brown failed to consult Congressman Clyde Kelly. Kelly had gone on record supporting new legislation. Brown's oversight resulted in Kelly's vigorously opposing the law and succeeded in making critical amendments to it.

The new legislation authored by Brown himself contained three important provisions—

1. Operators would be paid for flying mail according to the space available for mail in their airplanes. As much as $1.25 per mile was paid for large aircraft like the Ford TriMotor.
2. The Postmaster could extend or consolidate routes at his discretion.
3. Routes would be awarded to the lowest responsible bidder who had owned an airline operating on a daily schedule for at least 250 miles over a six-month period.

Brown's goal was to build a viable and safe commercial airline system. Once put into place, the airmail contract system subsidized large operations at the expense of smaller air carriers. This gave substantial incentives to airlines to acquire large multiengine airplanes and to develop passenger routes. The new system actually did result in a strong commercial aviation system early in the 1930s. No other Postmaster, before or since, has held such power over the United States air carrier system as did Walter Brown. Brown alienated many airlines by his insistence on forced mergers.

It was under Brown's administration that the second phase of the airline industry's development took place. Brown became a most controversial figure and even today is both violently criticized and praised for the part he played in the development of civil air transportation and airline history.

Upon taking the office of Postmaster General, given to him for being partly responsible for Hoover's nomination, he began a thorough study of the problems of aviation. He felt the airlines should be able to develop profitably without the help of airmail payments. He also believed that competition was necessary in the airline industry to stimulate growth.

Brown was almost 50 years ahead of time in his thinking—The Airline Deregulation Act of 1978 was passed for the purpose of encouraging an air transportation system which relies on competitive market forces to determine quality and price. In a meeting with airline leaders, Brown encouraged them to develop a profitable passenger business to become independent of the subsidy of airmail contracts. Postmaster Brown's goal was a system of three self-supporting transcontinental airline systems, and so he urged the airlines to form three large transcontinental companies.

The Airmail Act of 1930 gave Brown dictatorial power over the airline industry; therefore, commercial air transportation took on the characteristics of a federally regulated industry. The law gave Brown extensive power—he could grant an airmail contract without competitive bidding. He stated at the time that if the airlines were ever to become self-supporting they had to carry passengers and compete with each other. He felt that the airline industry needed leadership and that he was the one to give it. Postmaster Brown arranged that his office extend or consolidate routes in the public interest. This resulted in several smaller companies joining together to qualify for an airmail contract.

The new law brought about a change in the method of computing mail pay rates, paying the airlines for space instead of weight. This encouraged the airlines to purchase larger aircraft, which in turn gave rise to passenger traffic. In addition to computing the new airmail rates on space, the McNary-Watres Act also provided for variables such as bad terrain, bad weather, night flying, radio equipment, and multiengine aircraft.

Brown met with the contracting airlines to discuss the new airmail contracts and plan for passenger service. It became clear that the airlines were not willing, as yet, to move in this direction. This further forced Brown into assuming the responsibility of leadership of the airline industry and using the far reaching powers granted to him under the McNary-Watres Airmail Act of 1930.

Meeting with the airlines, Brown outlined his philosophy and his plan for development of a national air transportation system. Brown disliked the reckless competition that was going on, but even more so he disliked a monopoly. Brown's solution was regulated competition. He made it apparent that he believed an airline should be a large, sufficiently financed company, and that the transcontinental airmail routes must be conducted by one company. Since there was no large single airline, mergers of several smaller companies resulted. This action contributed to the beginning of the corporate entities of airlines that we know today, such as United Air Lines, American Airlines, and Trans World Airlines. Many smaller companies were left out. To them it meant no mail contract, and no contract meant going out of business. Brown got the large transcontinental airline companies he wanted. The four years Brown was Postmaster General resulted in the emergence of a consolidated airline route pattern and the formation of the great air carrier companies of today.

Brown has been called both good and evil because he acted in this direct fashion. In retrospect, however, it becomes obvious that the airline industry development would have been much slower were it not for his accomplishments.

The complaints of the small airlines were heard in Washington. With a presidential election coming up, Brown's actions made a good basis for political discussion. Hoover lost to the Democratic candidate, Franklin D. Roosevelt, who upon taking office, named James A. Farley of New York Postmaster.

When Brown left office there were 34 established airmail routes. The per mile cost of airmail was 54 cents, down from $1.10 when he entered office. The result of his reign was good if no attention were paid to those who had been trampled upon.

Roosevelt's New Deal program created many investigations, one of which was to examine mail contracts subsidizing the Merchant Marine. This congressional investigation committee was chaired by a second-term Senator from Alabama, Hugo Black, who later became a Supreme Court Justice.

One small airline that lost its bid on an airmail route was the Ludington Line. Although Ludington was making a profit flying passengers between New York, Philadelphia, and Washington, the bid was won by Eastern Air Transport. But Ludington Line was innovative. In 1930 they introduced a concept not duplicated again until 1960; the operation consisted of flights leaving New York and Washington every hour on the hour.

In 1932, word was circulating through Congress for the cancellation of the airmail contracts awarded in 1930. In early 1933 a House committee expressed its feelings that certain interlocking relationships had prevented the development of aviation and resulted in wasteful expenditures of public funds. It was the committee's recommendation that the powers of the Postmaster be restrained.

On February 9, 1934, as a result of the Black Committee investigation, James Farley issued an

The Case of Walter Brown

The investigating committee, with the cooperation of small carriers who lost contracts, found that apparent favoritism had been used by Walter Brown.

Senator Black found little interest in investigating the Merchant Marine, but airmail contracts added fuel to a growing political career. The committee hearings took on the form of a prosecution of Walter Brown. Cross-examination and witnesses developing full statements were not acceptable to Chairman Black. Black's line of attack focused on two points. First, small operations were forced out of business by Wall Street seeking a monopoly. Second, enormous profits were made through stock manipulations.

The main outcome was to crucify Walter Brown by saying he had favorites among the airlines and by charging him with collusion with the larger airlines. His methods drew the attention of the Black Committee. Brown's defense was that he acted in a manner according to his beliefs for the public interest.

The new Postmaster General, James Farley, maintained that the routes had been awarded at a series of secret meetings held in May and June of 1930, called the Spoils Conferences.

The main point was that all but two of 20 contracts went to United Aircraft and Transport Corporation, Aviation Corporation of Delaware, and the North American Aviation Group of General Motors.

Senator Black's committee overlooked the issue of how much growth and progress took place during Walter Brown's handling of airmail contracts. Little time was spent on how United Aircraft and Transport Corporation, Eastern Air Transport, and American Airways had improved passenger service, safety, and aircraft development.

The airlines were not entirely without fault. Expense account favors were utilized in entertaining government officials. Jobs and loans to postal officials were offered. Relatives of senators and congressmen were hired as consultants. However, as far as Walter Brown was concerned, no testimony or evidence was ever recorded that showed that he benefited financially or otherwise. Walter Brown acted as he felt was best for the United States. If a small carrier's contract was awarded to a large carrier, Brown made it a condition that the large carrier would buy the small carrier at a fair price.

Black went to the White House in January 1934 to present to the President the findings of his committee. Roosevelt said he would study the matter. Black then met with the United States Solicitor General, and both men went to Postmaster General Farley. Farley felt many of Black's statements were for publicity. He asked the Solicitor General to verbally sum up the volume of evidence, and he was eventually convinced.

The Postmaster General and Solicitor General met with the President and again discussed the findings. Farley wanted to advertise for new bids and then cancel the existing contracts. Roosevelt accepted Black's view for immediate cancellation. The Democrats accused Brown with nine charges of favoritism, collusion, and misconduct of office.

Brown subsequently retired to his home in Ohio and became the dean of Toledo politics and the head of the Republican Party there. He was vindicated in 1941. He died at the age of 91 in 1961.

order which canceled all domestic airmail contracts on charges of collusion. The Army, he said, carried the airmail originally in 1918 and they could do it again. General Benjamin Foulois, an aviation pioneer and Chief of the Air Corps, was called upon to do the job.

The airline pilots at the time of the cancellation were experienced even for a fledgling industry. Most of the pilots had over 4,000 hours of flying time and some had as much as 9,000 hours. They had day and night flying time and knew their routes well. They were used to the particular weather associated with the route and their aircraft were equipped with radio for air-to-ground communication.

In contrast, the Army assigned open-cockpit airplanes to carry the airmail and their pilots had been restricted to flying 170 hours a year. The Army aircraft were not equipped for weather or night flying, and only 500 men were assigned to the entire operation, which had been handled by several thousand airline employees.

In taking over the airmail on February 19, 1934, the route mileage was greatly reduced. The Army was given ten days to get the schedules in operation. The Air Corps itself was struggling for recognition and had not been given adequate funds to accomplish their growth.

The ill-prepared and poorly equipped Air Corps had tragic results in its effort to carry out Farley's order. Three days before they were to take over the airmail routes three pilots were killed on their way to begin their assignment, and another crashed but escaped serious injury.

The record for the first week's operations was equally tragic. Five Army pilots were killed and six more were seriously injured. Eight airplanes were ruined, and there was $300,000 worth of property damage. A few days later two more pilots were killed and one was critically injured.

The needless slaughter went on into March of 1934. Finally, after 66 crashes, 12 deaths, and governmental cost of nearly $4 million, President Roosevelt ordered the Army to cease flying the airmail on June 1, 1934, after nearly six months' operation.

At the time of the cancellation of airmail contracts, the airlines were receiving from 42.5 cents to 54 cents per mile. It cost the Army $2.21 during its operation. It was obvious that for economy, service, and safety, the airlines could do a much better job.

As a result, all the former airmail contractors were given three-month contracts to transport airmail. Before the time expired it was apparent that airmail legislation had to be completely revamped.

It was neither the President nor Black who received the criticism for the failure of the Army in carrying the airmail; it was James A. Farley who signed the cancellation order.

In March 1934 Roosevelt asked Congress to provide adequate legislation for returning airmail to private companies.

The Last Industry Airmail Flight—1934

The day before the Army took over airmail, the airlines demonstrated their progress and ability to carry the mail, as well as to show the ridiculousness of the postmaster's order. A new Douglas airliner, the DC-2, with two famous pilots at the controls, Jack Frye of Trans World Airlines and Eddie Rickenbacker of Eastern, flew the last load of mail from Los Angeles, California to Newark, New Jersey and set a new transcontinental flying record of 13 hours.

Frye and Rickenbacker wanted everything perfect for their transcontinental flight. It was their desire to make the last trip carrying airmail in less than 15 hours. Scheduled to fly on board were several journalists to record the event. The flight was to land in Newark before midnight to prove that air transportation had progressed to a point of flying an accurate schedule.

The new DC-2 took off on schedule at 9:00 p.m. from Los Angeles; en route to Albuquerque the flight averaged 230 miles per hour. Donald Douglas had guaranteed 180 miles per hour for the new aircraft. The first stop was Kansas City for ten minutes to refuel, then on to Pittsburgh. Weather reports received en route indicated a storm over Pittsburgh; therefore, Columbus was chosen as the second refueling point. After take-off, they passed the storm and landed at Newark 13 hours and two minutes after leaving Los Angeles.

It was not until 1941 that Brown was vindicated when the United States Court said that the charges of collusion against him were untrue. Therefore, the cancellation of the airmail contracts, the deaths of the pilots, the property damage, and the high expense were all unnecessary—except as a period in the historical development of United States air transportation. This action of the court made it possible for the airlines, whose airmail contracts had been cancelled by Farley, to claim and receive about $2.5 million dollars in damages.

Airmail Act of 1934 (Black-McKellar Act)

The remedial act meant to prevent the happenings of Brown's administration again was passed in June 1934 and is known as the Airmail Act of 1934, or the Black-McKellar Act.

Under the provisions of this act airmail contracts were to be granted upon competitive bidding, as was the case before 1930. However, no contract would be awarded to any company who had been involved in the supposed collusion. The three air carriers involved in the alleged collusion did show up at a meeting called by Farley on April 20, 1934, to issue new contracts. The three airlines had disguised themselves by changing their names slightly; American Airways became American Airlines, Eastern Air Transport became Eastern Airlines, and Transcontinental & Western Air just added the suffix "Inc." Furthermore, the control of the industry was placed in the hands of three agencies of the Federal government. This joint administration was divided between the Post Office, the Interstate Commerce Commission, and the Department of Commerce.

The Post Office awarded all airmail contracts and enforced the regulations governing airmail. The Interstate Commerce Commission set the rates of mail pay to the contracting airlines as well as reviewing all rates periodically. The Department of Commerce had established the Bureau of Air Commerce (the ancestor of the present Federal Aviation Administration) whose function was to provide safety in air transportation and to maintain, operate, and develop the airway system.

The old airmail contracting airlines were awarded most of their previous airmail routes. To ensure regaining these routes, some airlines submitted extremely low bids; several were only a fraction of a cent per mile.

By midsummer the airlines were financially weakened because their airmail income was less than one-half the former amounts. The depression and the publicity of the Black hearings also added to the difficulty. The situation reached a point where the airlines almost suspended operations.

Many of the smaller airlines were successful in obtaining routes; some lasted only a year or two. By World War II almost half of these companies had merged or suffered financial collapse and no longer existed. This seemed to confirm Walter Brown's philosophy.

The Act of 1934 also created a Federal Aviation Commission to study aviation policy headed by Clark Howell of Atlanta. The commission submitted its report on February 1, 1935, and recommended that a separate organization be created to control the airline industry. Senator Pat McCarran, during debate on the Black-McKellar Act, suggested this same course of action.

The Commission, in discussing the reorganization of regulatory agencies, came to two possible choices: control by the Interstate Commerce Commission, or control by a new independent agency. The White House was in favor of an independent agency.

At about this time the airlines were forming a trade association, known as the Air Transport Association, whose function it would be to speak in unity for the airline industry.

A minor provision of the Act of 1934, seemingly unimportant at the time, required the separation of airline companies and aircraft manufacturers. It was considered a matter of safety that airplane users not be controlled by those who controlled the airplane builders. Had this been allowed, an airline might be required to buy an airplane from a manufacturer owned by the same holding company. This action resulted in the final organization of the present day airline corporate names. This provision of the Airmail Act of 1934 required North American Aviation to divest itself of its airline companies. Through a loophole in the law, General Motors was allowed to keep Eastern. However, Eastern was in poor financial condition.

The responsibility for building Eastern was given to Eddie Rickenbacker. As a result, Eastern's condition substantially improved so that John Hertz, of Hertz Rent-A-Car, offered General Motors $3 million for Eastern Airlines. Rickenbacker went to the chairman of General Motors, Alford Sloan, and asked for an opportunity to match the Hertz offer. After receiving that option he set about securing financing for the deal. Rickenbacker called upon friends on Wall Street, and financing was acquired. The completion of the sale took place in Eastern's hangar at Newark airport. Rickenbacker also had obtained a half million dollars for operating funds.

Both the air transport industry and the Federal government learned many valuable lessons during the early months of 1934. Among these was the conviction that a healthy air transport service was essential to the nation's social, economic, and political strength, and that the time was at hand to revise the existing air legislation. It took four years of discussion before satisfactory legislation was finally enacted.

To do a good job for both the airlines and the public, the Federal government decided it must have regulatory power over the airline industry and that it must regulate air commerce in the same way it regulated interstate and foreign commerce. The authority by which it conducted these functions stemmed from the United States Constitution.

By 1938, the airline industry was an established and recognized segment of the American way of life. The airlines had airmail contracts, and they were developing passenger service. The economic and business condition of the entire United States was improving after the disastrous years of the depression. Aircraft manufacturers had begun providing aircraft to the airlines for passenger service.

All seemed well with the industry—in fact, too well. Uneconomic passenger route duplication and rate cutting began to surface.

Civil Aeronautics Act of 1938 (McCarran-Lea Act)

The airlines approached the Federal government and suggested working out a basic set of guiding principles. The three agencies under the Airmail Act of 1934 were overlapping in authority, causing further hardship on the airlines.

After several months of study by the industry leaders and government officials, a completely new all-encompassing law was passed affecting civil aviation in the United States. The new law found its basis in Article I, Section 8, of the United States Constitution known as the Commerce Clause. The progression of authority was the Interstate Commerce Commission Act of 1887, and the essentials of the new law came from the then recently passed Motor Carrier Act of 1935.

The Civil Aeronautics Act, introduced by Senator Pat McCarran and Representative Clarence Lea, passed the Congress on June 23, 1938. The McCarran-Lea Act amended or repealed all major existing and previous legislation having to do with aviation.

The mechanics of the Civil Aeronautics Act were simple. All air transportation regulation, both economic and safety, was administered by three separate agencies.

These original three agencies were:

1. The Civil Aeronautics Authority, composed of five members who were to establish aviation policies by legislation of safety and economics of air transportation.
2. The Administrator of Aviation, appointed to carry out the safety policies of the Authority, while the Authority provided for the execution of the economic regulations concerning the air transport industry.
3. The Air Safety Board, formed as an independent group of three people responsible for the investigation of aircraft accidents.

For the first time the airline industry could look upon a firm regulatory system, making it possible to plan for future development.

After the functioning of the Authority, the Administrator, and the Air Safety Board began, it became obvious that some duplication of authority existed and that an adjustment was necessary. Therefore, President Roosevelt in April and June of 1940 proposed changes that became effective on June 30, 1940, and became known as the 1940 Amendment of the Civil Aeronautics Act.

The result of the organizational changes was the reshuffling of functions into two agencies instead of three. The first as the Civil Aeronautics Board, which was to be an independent group of five people reporting to the President. Its function was to exercise legislative and judicial authority over civil aviation, as well as executive control in the area of air carrier economic regulation. The investigation of aircraft accidents also became a responsibility of the Civil Aeronautics Board.

The second agency created by the 1940 Amendment was the Civil Aeronautics Administration, headed by an Administrator, whose function was solely to be the execution of safety regulations. This was to include enforcement and promotion of aviation safety as well as safety in the entire operation of the airway system. The Civil Aeronautics Administration was placed under the Department of Commerce.

The Grandfather Clause

Included in Title IV of the economic regulation of the Civil Aeronautics Act of 1938 was section 401e, which stated that the carriers who had provided adequate and continuous airmail service from May 14 to August 22, 1938, the effective date of the Act, would receive a permanent Certificate of Convenience and Necessity for those routes. This clause was a grandfather clause, and the first airline to receive this award was Delta Air Corporation. (Table 7-2.) Only a violation of the Act can cause an airline to lose its grandfather rights.

FIGURE 7-2 A DC-3 of the 1930s superimposed on the outline of a Boeing 747 of today.

American Airlines	Eastern Airlines
Braniff Airways	Inland Airlines
Chicago and Southern Airlines	Mid-Continent Airlines
Continental Airlines	National Airlines
Delta Air Corporation	Northeast Airlines
Northwest Airlines	United Airlines
Pennsylvania-Central Airlines	Western Air Express
Transcontinental & Western Air	Wilmington-Catalina Airlines

TABLE 7-2 Airlines that Received Permanent Certification by the Grandfather Clause

Of these 16 carriers, six have faded away through mergers or reorganization. Chicago and Southern, and Northeast merged with Delta; Inland merged with Western; Mid-Continent merged with Braniff; Pennsylvania-Central became Capital, which later merged with United; Transcontinental and Western changed their corporate name to Trans World Airlines. Additional carriers were merged and others are no longer operating because of bankruptcies.

Civil air transportation continued under the Civil Aeronautics Act for a period of 20 years, and the Act provided the practical machinery for the development of civil air transportation. It formulated continuing public service objectives and provided the means whereby these objectives could be established. The purpose of the Act was not only to regulate civil air transportation, but also to develop and encourage its growth.

Because of the growing importance of the air transport industry to the economy and the national defense of the United States in the later 1930s, the writers of the Civil Aeronautics Act agreed that air transportation must be kept a strong and dynamic force.

Time alone proved the truth of their wisdom.

The Federal Aviation Act—1958[1]

The Civil Aeronautics Act of 1938 was basically a constitution of aviation upon which the aviation industry could expand and develop. It was general in character and left the interpretation of the specific nature to the Civil Aeronautics Board in the matter of economic regulation, and to the Civil Aeronautics Administration in the area of adherence to air safety.

Since 1938 and development of the air industry, particularly after World War II, it became apparent that changes had to be made. Amendments and revisions were passed to take care of minor current adjustments; however, the main points of contention became more severe. Many plans were forthcoming, ranging from overhauling and modernizing the Civil Aeronautics Act, to a completely new method of aviation regulation. Some people proposing changes had been responsible for the writing and enactment of the Civil Aeronautics Act in 1938; a few felt that aviation regulation had not been carried out according to their intended meaning. Some of their arguments, no doubt, were justified. It is sufficient to state that no major change came about and the Civil Aeronautics Act remained in effect until the passage of the Federal Aviation Act in 1958.

The Federal Aviation Act of 1958, however, is not entirely new; it is largely a reenactment, in amendment form, of the prior Civil Aeronautics Act. The economic regulatory provisions set forth therein are unchanged. Therefore, pre-1959 judicial and administrative decisions interpreting and applying predecessor provisions of the Civil Aeronautics Act are of precedential value with respect to the Federal Aviation Act. The Federal Aviation Act is the old Civil Aeronautics Act in practically every respect except one—the expanding of power in the area of air safety. The Civil Aeronautics Administration became the Federal Aviation Agency. Whereas, the Civil Aeronautics Administration was under the Department of Commerce, the Federal Aviation Agency was organized at an independent level answerable only to Congress and the President. With the enactment of the Department of Transportation in 1966, the Federal Aviation Agency became the Federal Aviation Administration and was placed under the Secretary of Transportation.

The Federal Aviation Agency was granted expanded authority in air safety. It would not be hampered organizationally or budgetwise as had been the case of the Civil Aeronautics Administration. The Federal Aviation Agency was able to make long-range plans and to implement these plans without interference.

The Federal Aviation Act of 1958 worked virtually no changes of substance into the economic regulatory provisions, but made several revisions in the safety program. While the Board retained its duties in the fields of air carrier economic regulation and aircraft accident investigation, the Board's safety rule-making powers were transferred to the administrator of the Federal Aviation Agency, with the result that the latter official then promulgated the very regulations and standards which he administers and enforces. The Board's role in safety rule-making is presently limited to participation as an interested party in FAA proceedings.

A second important revision of prior law concerns procedure in certificate suspension and revocation cases (Section 609). Whereas under the former law, only the Board could suspend or revoke

Mid-Air Collisions

The circumstances which led to the Federal Aviation Act of 1958 were a series of mid-air collisions, some of which involved military aircraft. These highlighted the need for improved air traffic control.

- The first of these accidents occurred on June 30, 1956, over the Grand Canyon. The aircraft involved were a Trans World Airlines Constellation and a United Airlines DC-7. Both aircraft fell into the Canyon.

TWA Flight 2 left Los Angeles International Airport at 9:00 A.M. bound for Kansas City; and three minutes later, bound for Chicago, UAL Flight 718 took off from the same runway. At 10:31 A.M. Flight 2 and Flight 718 collided over the Grand Canyon.

The probable cause, according to the official Civil Aeronautics Board Accident Report, was that the pilots did not see each other in time to avoid the collision. It is not possible to determine why the pilots did not see each other, but the evidence suggests any one, or a combination of the following factors: intervening clouds, visual limitations, preoccupation with normal cockpit duties, attempting to provide passengers with a more scenic view, physiological limits to human vision, or insufficiency of en route air traffic advisory information due to inadequacy of facilities and lack of personnel in air traffic control.

- The second accident involved a DC-7B owned by Douglas Aircraft Company and a USAF F-89J near Sunland, California on January 31, 1957. The DC-7B crashed into a junior high school playground killing three students and injuring 70 others. The F-89J crashed in the Verdugo Mountains.

The probable cause was the high rate of near head-on closure at high altitude, which, together with physiological limitations, resulted in a minimum avoidance opportunity during which the pilots did not see the other's aircraft.

- The third accident involved a United Airlines DC-7 and a USAF F-100F near Las Vegas, Nevada, on April 21, 1958.

The probable cause of this mishap, according to Civil Aeronautics Board investigation, was a high rate of near head-on closure at high altitude; human and cockpit limitations; and the failure of the Nellis Air Force Base and the Civil Aeronautics Administration to take every measure to reduce a known collision exposure.

These accidents and other occurrences caused Congressional action.

in the first instance; the new act contemplated initial action by the Administrator, subject to the certificate holder's privilege of appeal to the Board. Apart from these matters, the Federal Aviation Administrator has substantially all the powers and duties of his predecessor under the 1938 Act, plus a clearer authority to allocate the navigable airspace between military and civilian users.

Whereas the Civil Aeronautics Board had been responsible for the enactment of air safety legislation, and such regulations carried out by the Civil Aeronautics Administration in their executive capacity, the power of air safety legislation was transferred to the Federal Aviation Agency. The Civil Aeronautics Board retained the responsibility of aircraft accident investigation and economic regulation of the air carriers.

All ongoing air safety research and development was consolidated and placed in the care of the Federal Aviation Agency. Therefore, the work of such agencies as the Airways Modernization Board, the Air Coordinating Committee, and the National Advisory Committee for Aeronautics could be coordinated at one location and under one agency, the Federal Aviation Agency.

The area of economic regulation of the air transportation industry was not altered in any way. The Federal Aviation Act was passed hurriedly in the closing days of Congress in 1958. No attempt was made to change or modify any economic regulation of the air transport industry.

In the study of present Federal regulation of air transportation, it must be remembered that such regulation is now much as it was in 1938 except in the area of air safety including air traffic control.

Although the Federal Aviation Act is the law of the land in aviation matters, it is very broad and in many cases loose in definition. This fact is not a weakness, for the Civil Aeronautics Board and the Federal Aviation Agency were to interpret and apply the law within the framework of the Act.

The Federal Aviation Act was passed by Congress on August 23, 1958. That part of the law dealing with the creation and organizational structure of the Federal Aviation Agency became effective on that date. The main body of the law was placed in effect on December 31, 1958.

The Airline Deregulation Act of 1978 substantially changed the Federal Aviation Act of 1958.

Officially the Federal Aviation Act is known as Public Law 85-726, 85th Congress, S.3880. The introductory paragraph very clearly states the purposes of the new law as:

1. Continuing the Civil Aeronautics Board (disbanded 1/1/85).
2. Creating a Federal Aviation Agency (Administration).
3. Providing for the regulation and promotion of Civil Aviation.
4. Providing for the safe and efficient use of the airspace.

Structurally, the Act is divided into 15 main divisions called "Titles." Each title is divided into various sections which have several subsections. Reference is made to any part of the Act by stating the title number and section combined; thus 411 means the eleventh part of the fourth title.

Title II deals with the organization, functions, and general powers of the Civil Aeronautics Board which ceased to exist on January 1, 1985. Title IV, Air Carrier Economic Regulation, was the direct responsibility of the Civil Aeronautics Board and indicated the majority of the work which they performed. Title VII, Aircraft Accident Investigation, was a function of the Civil Aeronautics Board, but it was transferred to the National Transportation Safety Board under the Department of Transportation.

The organization, powers, and duties of the Federal Aviation Administration, headed by Administrator, are outlined in Title III. The main responsibilities of the Administrator, and therefore the Federal Aviation Administration, are stated in Title V, Nationality and Ownership of Aircraft, as well as Title VI, Safety Regulations of Civil Aeronautics.

The remaining titles of the Act, although important, deal with miscellaneous aviation matters.

Section 101—Definitions

Except for the minor changing of words to a more modern connotation, the definitions had not been altered until the passage of the Airline Deregulation Act on October 24, 1978. In 1938 the writers of the Civil Aeronautics Acts wanted to be sure the terms and words were generally understood by all; therefore, the definitions represent an effort at standardization. There are now 41 definitions.

The statutory definitions are of great importance since application of the substantive regulatory provisions of the Act depend upon them to a large extent, and since the definitions also include certain substantive grants of power. Thus, the definition of "air commerce" largely fixes the scope of the Federal Aviation Administrator's jurisdiction in the safety field because the substantive provisions of Title VI, Safety Regulations of Civil Aeronautics, are couched in terms of "air commerce." It should be observed that the definition of "air commerce" is exceedingly broad. It includes not only what is traditionally regarded as interstate and foreign commerce (physical movement of aircraft across state and international boundaries and the transportation within a single state of persons or property moving to or from another state), but also includes all transportation of mail, operations, or navigation within the limits of any federal airway, and any operation or navigation, which may endanger safety in air commerce. This definition is generally regarded as conferring federal jurisdiction in safety matters with respect to all flights of aircraft, since all flights, whether commercial or private, and irrespective of whether they cross state lines or operate within a federal airway, represent a potential danger to interstate commerce. Moreover, counsel for the Board and the Administrator, and thus far some courts, have considered the definition to be so broad as to vest in the Federal government exclusive jurisdiction in safety matters, thus preempting the states from enacting air safety regulations under their general police powers.

On the other hand, the definition of "air transportation," which fixes the scope of the Board's regulation under Title IV, Air Carrier Economic Regulation, is considerably narrower than the definition of "air commerce." "Air transportation" includes only common carriage by aircraft "in commerce between" the states, etc., and the carriage of mail by aircraft. This definition is believed to confer jurisdiction over the common carriage of persons and property only in those areas traditionally regarded as constituting interstate and foreign commerce. Unlike in the safety field, the definition does not preempt the entire economic field to the exclusion of the states, and the courts have in certain context so held. Thus, while the Board's economic jurisdiction was believed to extend to all common carrier movements of aircraft across state lines and within the territories, and to movements of aircraft within a state but engaged in interstate transportation of more than a minimum amount, there was no direct economic jurisdiction over purely intrastate transportation. Air transportation of mail, whether interstate or purely intrastate, falls within the definition of air transportation, except for certain special transportation defined elsewhere.

Another important term is "air carrier" defined as any U.S. citizen "who undertakes, whether directly or indirectly or by a lease or any other arrangement, to engage in air transportation." This definition serves to fix the Board's jurisdiction over "indirect air carriers." The definition provides further that the Board "may by order relieve air carriers who are not directly engaged in the operation of aircraft" from the various regulatory provisions of the statute. This provision has been used to establish the classification of Air Freight Forwarder and to exempt both them and other indirect air carriers such as Railway Express (an "air carrier" as to air express) from the Certificate of Convenience and Necessity and other provisions of the Act.

Mention should also be made of the term "aeronautics," which is defined as "the science and art of flight." Since the control provisions of the statute (Sections 408 and 409) confer jurisdiction over relationships between air carriers and persons engaged in a "phase of aeronautics," this definition permits considerable breadth in the application of the statute. It has been held to include manufacturing companies, leasing companies, and a host of other persons engaging in activities closely related to air carrier operations.

The examples listed are for the purpose of illustration, and only those which bear special attention will be discussed.

Administrator—the Administrator of the Federal Aviation Administration.

Board—the Civil Aeronautics Board.

Aircraft—any contrivance now known or hereafter invented, used, or designed for navigation of or flight in the air.

Civil aircraft—any aircraft other than public aircraft.

Civil aircraft of the United States—any aircraft registered as provided in this Act.

Public aircraft—an aircraft used exclusively in the service of any government or of any political subdivision, but not including any government-owned aircraft engaged in carrying persons or property for commercial purposes.

Citizen—anyone, a partnership, or corporation, who is a citizen of the United States or its possessions.

Persons—any individual, firm, co-partnership, corporation, company, association, joint-stock association, or body politic; and includes any trustee, receiver, assignee, or other similar representative thereof.

Foreign air carrier—any person, not a citizen of the United States, who undertakes, whether directly or indirectly or by lease or any other arrangement, to engage in foreign air transportation.

Charter Air Carrier—an air carrier holding a certificate of public convenience and necessity authorizing it to engage in charter air transportation.

Charter Air Transportation—charter trips, including inclusive tour charter trips, in air transportation rendered pursuant to authority conferred under this Act under regulations prescribed by the Board.

Interstate air commerce, overseas air commerce, and *foreign air commerce,* respectively—the carriage by aircraft of persons or property for compensation or hire, or the carriage of mail by aircraft, or the operation of aircraft in the conduct of a business or vocation.

Interstate air transportation, overseas air transportation, and *foreign air transportation,* respectively—the carriage by aircraft of persons or property as a common carrier for compensation or hire or the carriage of mail by aircraft.

Airman—any individual who engages, as the person in command or as pilot, mechanic, or member of the crew in the navigation of aircraft while under way; and any individual who is directly in charge of the inspection, maintenance, overhauling, or repair of aircraft, engines, or propellers; and any individual who serves in the capacity of aircraft dispatcher or air traffic control tower operator.

Ticket agent—any person, not an air carrier and not a bona fide employee of an air carrier who, as principal or agent, sells or offers for sale any air transportation, or negotiates for, or holds him or herself out by solicitation, advertisement, or otherwise as one who sells such transportation.

Airport—a landing area used regularly by aircraft for receiving or discharging passengers or cargo.

Landing area—any locality, either land or water, including airports, which is used or intended to be used for the landing and take-off of aircraft, whether or not facilities are provided for the shelter, servicing, or repair of aircraft, or for receiving or discharging passengers and cargo.

Federal airway—a portion of the navigable airspace of the United States designated by the Administrator as a federal airway.

Appliances—instruments, equipment, parts, and accessories, which are used in the navigation, operation, or control of aircraft in flight, and which are not parts of the aircraft, aircraft engines, or propellers.

Spare parts—pans, appurtenances, and accessories of aircraft (other than aircraft engines and propellers), of aircraft engines (other than propellers), of propellers, and of appliances, maintained for installation or use in an aircraft, aircraft engine, propeller, or appliance, but which at the time are not installed or attached.

Propeller—all parts, appurtenances, and accessories thereof.

Mail—United States mail and foreign-transit mail.

Navigable airspace—airspace above the minimum altitudes of flight prescribed by regulations issued under this Act, and shall include airspace needed to insure safety in takeoff and landing of aircraft.

Navigation of aircraft or *navigate aircraft*—the piloting of aircraft.

Operation of Aircraft or *operate aircraft*—the use of aircraft, for the purpose of air navigation. Any person who authorizes the operation of aircraft, whether with or without the right of legal control (in the capacity of owners, lessee, or otherwise) of the aircraft, shall be deemed to be engaged in the operation of aircraft within the meaning of this Act.

Possessions of the United States—(a) the Canal Zone, but nothing herein shall impair or affect the jurisdiction which has heretofore been, or may hereafter be, granted to the President in respect of air navigation in the Canal Zone; and (b) all other possessions of the United States. Where not otherwise distinctly expressed or manifestly incompatible with intent thereof, references in this Act to possessions of the United States shall be treated as also referring to the Commonwealth of Puerto Rico.

United States—the several States, the District of Columbia, and the several Territories and possessions of the United States, including the territorial waters and the overlying airspace thereof.

Predatory—any practice which would constitute a violation of the antitrust laws as set forth in the first section of the Clayton Act.

The Declaration of Policy, Sections 102 and 103 state the guiding principles which the Civil Aeronautics Board and the Federal Aviation Administration follow to accomplish their three basic objectives of:

1. Economic regulatory control of air carrier activity.
2. Promotion of maximum progress in aeronautics.
3. Promotion of air safety.

Since the Deregulation Act the Department of Transportation has assumed the responsibilities of the Civil Aeronautics Board, therefore, any reference to the Board in the following Sections of the Federal Aviation Act of 1958 are now meant for the Department of Transportation.

Sections 102 and 103 are a preface to every section of the Federal Aviation Act pertaining to either the Civil Aeronautics Board or the Administrator. Therefore, consideration must be given to the effects of these two sections for complete understanding.

Section 102—Declaration of Policy: The Board Factors for Interstate, Overseas and Foreign Air Transportation

This section, originally a part of the Civil Aeronautics Act of 1938, is analogous to the National Transportation Policy contained in the Interstate Commerce Act and provides that the Board shall consider, among other things, the specific factors there set forth "as being in the public interest, and in accordance with the public convenience and necessity." Since, throughout the substantive provisions of the Act, the Board is required as a prerequisite to various actions to find either "public interest" or "public convenience and necessity," Section 102 serves in part as a statutory definition of these terms. It speaks, however, only in terms of air carriers and air transportation. Counsel for the Board have contended that this section, among others, permits the Board to award operating authority to air carriers and to take other regulatory action with respect to air transportation without regard to the possible adverse competitive impact that such action might have upon surface carriers.

In performing its responsibilities under the Act with respect to interstate and overseas air transportation, the Board shall consider the following, among other things, as being in the public interest and in accordance with the public convenience and necessity:

1. Maintaining safety as the highest priority in air commerce.
2. Preventing any deterioration in established safety procedures.
3. Availability of economic, efficient, and low-price air carrier service without discrimination or deceptive practices.
4. Rely upon competition to provide needed air transportation and to encourage well-managed air carriers to earn adequate profits.

5. To develop and maintain regulation responsive to the public need of air transportation for the commerce, Postal Service, and national defense of the United States.
6. Encourage air service at major areas through secondary airports consistent with local authorities when endorsed by state entities encouraging air carriers whose responsibility is to provide service at secondary airports.
7. Prevent unfair, deceptive, predatory, or anticompetitive practices in air transportation and avoid:
    a. Unreasonable industry concentration, excessive market domination, and monopoly.
    b. Any other conditions that would allow an air carrier(s) to unreasonably increase prices, reduce services, or exclude competition.
8. Maintain a system of scheduled air carrier service for small communities with direct federal assistance where appropriate.
9. Encourage, develop, and maintain an air transport system relying on competition to provide efficiency and low prices.
10. Encourage entry into markets by new and existing air carriers, and the strengthening of small air carriers to assure a competitive air carrier industry.
11. Promote, encourage and develop civil aeronautics and a viable privately owned U.S. air transport industry.
12. Strengthen the competitive position of U.S. air carriers to at least assure equality with foreign air carriers.

Factors for All-Cargo Air Services. In addition to the declaration of policy above, the Board in performing its duties with respect to All-Cargo Air Service shall consider the following, among other things, as being in the public interest:

1. Encourage and develop All-Cargo Air Service by private enterprise responsive to the needs of shippers, the commerce of the United States, and national defense.
2. Encourage and develop an integrated transportation system relying upon competition to determine quality and price.
3. Provide service without discrimination, deceptive practices, or predatory prices.

Section 103—Declaration of Policy: The Secretary of Transportation

In this exercise and performance of his powers and duties under this Act the Secretary of Transportation shall consider the following among other things, as being in the public interest:

a. To regulate air commerce to promote and develop safety.
b. To promote, encourage, and develop civil aeronautics.
c. To control use of the navigable airspace of the United States and to regulate both civil and military operations therein.
d. To consolidate research and development of air navigation facilities.
e. To develop and operate an air traffic control system for both civil and military aircraft.

This section was originally written in 1958 at the time the Federal Aviation Agency was created by the legislation of the Federal Aviation Act. It parallels the previous section concerning the Civil Aeronautics Board in that it specifies several factors which the Administrator shall consider as being in the public interest.

As previously stated, the Federal Aviation Act created the Federal Aviation Agency to replace the Civil Aeronautics Administration. The name of Agency was changed to Administration in 1966 when the Department of Transportation Act was passed. The Administrator is charged with encouraging civil aviation and air commerce in the United States. In exercising his authority, he must consider the needs of the national defense and of commercial and general aviation, as well as the public right to use the navigable airspace of the United States.

Specifically, the Administrator is authorized to prescribe rules governing the safe and efficient use of the navigable airspace; to provide for necessary air navigation facilities; to promulgate air traffic rules; to grant exemptions from his own regulations when the public interest requires it; to veto federal nonmilitary expenditures for airports and air navigation facilities if such are deemed by him as being not reasonably necessary for use in air commerce, or not in the interest of national defense; to make recommendations to the Secretary of Transportation concerning weather services for air commerce; to disseminate and exchange information relative to civil aviation; to engage in longrange planning, research, and development with respect to the utilization of the navigable airspace; and to provide for improved control of air traffic.

Section 104—Public Right of Transit

This section of the Act recognizes the right enjoyed by free people; it is a precious right that is not abundant in many countries on earth. To a person born in the United States, Section 104 is taken for granted, because it recognizes the public right of freedom of the United States citizen to use the navigable air space of the United States.

The declaration of a public right of freedom of transit through the navigable airspace of the United States has been the subject of considerable controversy in lawsuits involving low-flying aircraft. "Navigable airspace" is defined as airspace above the minimum altitudes of flight prescribed by regulations issued under this Act, and it includes airspace needed to insure safety in take-off and landing of aircraft. The Administrator prescribes such altitudes in the exercise of his safety rule-making powers. Prior to the passage of the Federal Aviation Act, when the Board was the body which promulgated the minimum altitude regulation, the Board's position in the courts was that the freedom of transit provision relieved aircraft operators from technical claims of trespass, but that it did not serve to extinguish a landowner's rights in the case of an actual "taking" of property occasioned by low-flying aircraft. In such latter circumstances, contended the Board, the landowner would have his state law remedies against the person causing the damage.

The Secretary shall consult with the Architectural and Transportation Barriers Compliance Board established under Section 502 of the Rehabilitation Act of 1973, prior to issuing or amending any order, rule, regulation, or procedure that will have a significant impact on the accessibility of commercial airports or commercial air transportation for handicapped persons.

Section 105—Federal Preemption

a. No state or interstate agency can enact or enforce any law or rule relating to rates, routes, or services of any air carrier having authority to provide interstate air transportation except for air transportation conducted within the State of Alaska.
b. Nothing here shall be construed to limit the authority of any State or interstate agency as the owner or operator of an airport served by certificated air carriers.

Section 106—Report on Subsidy Cost-Sharing

Not later than January 1, 1980, the Secretary of Transportation, shall submit a comprehensive report to the Congress on the feasibility and appropriateness of devising formulas by which States and their political subdivisions could share part of the costs being incurred by the United States under sections 406 and 419 of this Act. Such report shall include any recommendations of the Secretary for the implementation of such cost sharing formulas.

Section 107—Safety Study

Congress intends that the Airline Deregulation Act of 1978 will not result in any diminishing of safety attained.

Not later than January 31, 1980, and each April 1 thereafter, the Secretary of Transportation shall prepare and submit to the Congress and the Board a comprehensive annual report on the extent to which the implementation of the Airline Deregulation Act of 1978 has affected, during the preceding calendar year, or will affect, in the succeeding calendar year, the level of air safety. Each such report shall, at a minimum, contain an analysis of each of the following:

1. All relevant data on accidents and incidents occurring during the calendar year covered by such report in air transportation and on violations of safety regulations issued by the Secretary of Transportation occurring during such calendar year.
2. Current and anticipated personnel requirements of the Administrator with respect to enforcement of air safety regulations.
3. Effects on current levels of air safety of changes or proposals for changes in air carrier operating practices and procedures which occurred during the calendar year covered by such report.
4. The adequacy of air safety regulations taking into consideration changes in air carrier operating practices and procedures which occurred during the calendar year covered by such report.

Based on such report, the Secretary shall take steps necessary to ensure that the high standard of safety in air transportation is maintained in all aspects of air transportation in the United States.

Section 108—Report on Air Carrier Marketing of Tours

The Board shall prepare and submit a report to the Congress which sets forth the recommendations of the Board on whether this Act and regulations of the Board should be amended to permit air carriers to sell tours directly to the public and to acquire control of persons authorized to sell tours to the public. The report shall evaluate the effects on the following groups of allowing air carriers to sell tours:

1. The traveling public.
2. The independent tour operator industry.
3. The travel agent industry.
4. The different classes of air carriers.

Civil Aeronautics Board

A rare instance of a government agency being legislated out of existence took place on January 1, 1985, when the Civil Aeronautics Board ceased to exist.

The Airline Deregulation Act of 1978 provided for a slow phasing out of its authority. "Sunset Provisions" (Title XVI of the Act) provided for termination of the Civil Aeronautics Board and its legislative foundation provided by Title II of the Federal Aviation Act of 1958. This Act is a revision of the Civil Aeronautics Act of 1938 which established the powers of the Civil Aeronautics Board. These powers were economic in nature and dealt with the rates that the air carriers charged for their services and the routes over which these services were offered.

When the Civil Aeronautics Board was created, the United States air carrier system was a struggling, subsidized collection of private companies earning little or no profit. When the Deregulation Act was passed the air carrier industry had become one of the world's most efficient and carefully controlled competitive route networks that served high density markets as well as small cities.

The legislative demise of the Civil Aeronautics Board left a competitive free-for-all air carrier industry marked by disruptions in air service (especially to small cities), lower fares in high density markets and an industry of leaner but more efficient air carriers. Occasional bankruptcies also occurred.

Because the efforts of the Civil Aeronautics Board will long remain and be remembered, a brief presentation of its organization and functions is presented below.

The Civil Aeronautics Board derived its authority from, and was organized in accordance with, the Federal Aviation Act of 1958 as amended by the Department of Transportation Act of 1966 and the Airline Deregulation Act of 1978. It was an independent federal agency comprised of five members appointed for six-year terms by the President with the advice and consent of the Senate, with no more than three members appointed from the same political party. Each year the President designated one member as Chairman and another as Vice Chairman.

Each member of the Board had to be a citizen of the United States, and no member could have any pecuniary interest or own any stock or bonds in any civil aeronautics enterprise. Also, no member of the Board could engage in any other business, vocation, or employment. Three members of the Board constituted a quorum.

The Board exercised its power independently. Its decisions were not subject to review by any executive department or agency, except for the approval of the President required in Board decisions granting or affecting certificates for overseas and foreign air transportation, and foreign air carrier permits.

In general, the Board performed two major functions:

1. Regulation of the economic aspects of domestic and international United States air carrier operations and of the operations of foreign air carriers to and from the United States.
2. Participation in the establishment and development of international air transportation.

Economic Regulations. The Board's responsibilities changed dramatically under the Airline Deregulation Act of 1978. Routes, rates, mergers, foreign air carrier permits, and all other economic matters were amended.

International Civil Aviation. The Board consulted with and assigned the Department of State in the negotiation of agreements with foreign governments for the establishment or development of air routes and services between the United States and foreign countries.

The Board's function dealing with aviation safety was transferred to the Department of Transportation several years before the Board went out of existence.

The Board's authority over Certificates of Convenience and Necessity (CCN), rates, and the transportation of mail ceased to be in effect on December 31, 1981. The Act of 1978 provided that the authority of the Board relating to the air carriers filing tariffs; filing of accounts, records, and reports; and complaints and investigations ceased on January 1, 1983. The finale occurred on January 1, 1985, when Title II ceased.

Functions of the Board Members

The Board members were charged with carrying out the duties and responsibilities of the Board under the Act and the statutes. Action initiated pursuant to the Board's own initiative or by any document authorized or required to be filed with the Board originated in or was referred to the appropriate organizational unit for study and recommendation to the Board. The Board had the authority to delegate any of its functions to a division of the Board, an individual Board member, an Administrative Law Judge, or any employee or employees of the Board.

In addition to duties as a member of the Board, the Chairman served as presiding officer at meetings of the Board, determined the order in which day-to-day matters received attention of the Board, and acted as Spokesman for the Board before committees of Congress. The Chairman was responsible for the executive and administrative functions of the Board.

The Board's staff was composed of the following Offices and Bureaus:

Managing Director
Public Affairs
Management Systems
Community and Congressional Relations
Secretary

Civil Rights
Human Resources
Administrative Support Operations
General Counsel
Comptroller
Administrative Law Judges
Carrier Accounts and Statistics
International Aviation
Domestic Aviation
Economic Analysis
Compliance and Consumer Protection

Federal Airport Act of 1946

Orville Wright, in a statement before the House of Representatives, Committee on Interstate and Foreign Commerce, October 12, 1925, said:

> The greatest present drawback to the use of aircraft for civil purposes, such as commerce, mail travel, and sport, is the lack of suitable airports.

Prior to 1946, the airports of this country were financed and operated by state, county, or municipal governments. The cost of building and maintaining these airports was very high, and because of this the development of airports was slow. It was the purpose of the Federal Airport Act to give the United States a comprehensive system of airports, administered by the Civil Aeronautics Administration. Small communities which needed airports to help develop their social and economic structure, theoretically were supposed to benefit from this program.

Congress appropriated $520 million over a seven-year period to aid in the development of this airport system; however, the money had to be spent on operational facilities (runways, etc.). This federal aid program was organized so that the Federal government would pay as much as 50% of the cost and that the airport (sponsor) would pay the balance, or at least 50% of the cost. As far as large cities were concerned, this was excellent, because they could issue and sell bonds to pay their share of the cost. However, for small cities even 50% of the cost of development was too much of a burden. The purpose of the legislation was failing, since small cities did not benefit.

In 1953, President Dwight D. Eisenhower suggested that the Act be amended to include a provision that the money could be spent on public buildings (terminals) as well as operational facilities.

In 1958, Congress passed a bill which would continue the Act until 1963 with an increase in appropriations. However, the Eisenhower administration disagreed, and the Undersecretary of Commerce for transportation said that the construction, maintenance, and operation of civil airports was primarily a matter of local responsibility. The President vetoed the bill by saying, "I am convinced that the time has come for the federal government to begin an orderly withdrawal from the airport grant program."

The Senate Committee on Interstate and Foreign Commerce then stated that, "The committee is convinced that the capital investment required to bring airport facilities up to the present and future needs of this Nation is far beyond the financial capacity and capabilities of local communities with continued effective assistance and encouragement from the federal government."

Congress later passed, and the President approved, a bill which extended the Act through 1961.

For an airport or governmental unit to be available for such money, it is necessary that the airport be in the National Airport Plan. Under the Federal Airport Act, the Administrator of the Federal Aviation Administration is directed to prepare such a plan. In formulating the plan the Administrator shall take into account the needs of both air commerce and private flying, technological developments, probable growth, and any other considerations he may find appropriate.

The money which is appropriated for use is divided into two areas: 75% is distributed for projects among the several states and 25% is placed in a discretionary fund for the Administrator to use as he sees fit for airport construction. One-half of the 75% is apportioned to the States, based on population, and the other half is based on land areas. The discretionary fund allows the Administrator to choose the projects regardless of their location.

As a condition to his approval of a project, the Administrator shall receive in writing a guarantee that the following provisions will be adhered to:

1. The airport will be available for public use without unjust discrimination.
2. The airport will be suitably operated and maintained.
3. The aerial approaches will be cleared and protected, and future hazards will be prevented.
4. Proper zoning will be provided to restrict the use of land adjacent to the airport.
5. All facilities developed from federal aid shall be made available to the military.
6. All project accounts will be kept in accordance with a standard system.
7. All airport records will be available for inspection by an agent of the Administrator upon reasonable request.

On May 17, 1966, the Senate Commerce Committee gave approval to a bill to continue for an additional three years a program authorizing $75 million annually in federal matching funds for airport construction. This extended the Federal Airport Act to June 30, 1970. This program was expanded in 1970 to include a separate Airport and Airway Trust Fund.

The airlines and the air industry can develop only as rapidly as the facilities they use; the Federal Airport Act has aided this development.

Airport and Airway Development Act of 1970 (Airport Development Aid Program)

The Second Session of the 91st Congress passed an act to provide for the expansion and improvement of the airports and airway systems of the United States. In order to do this Congress also provided for the imposition of an airport and airway user charge. In declaring the policy of this law they stated that the nation's airport and airway system was inadequate to meet the current and projected growth in aviation; and in order to meet the demands of interstate commerce, the postal service, and the national defense, substantial expansion and improvement were necessary to meet the proper requirements of the future.

The Secretary of Transportation is directed by this Act to prepare and publish a national airport system plan for the development of public airports. The plan shall show the type and estimated costs of airport development necessary to provide a system of public airports adequate for the needs of civil aeronautics. In developing this plan the Secretary is directed to consider the relationship of each airport to the rest of the transportation system, including all modes of intercity transportation.

One section of this Act provides for the Secretary to make available grants of funds to planning agencies to develop airport master plans.

In order to develop a proper airway system and adequate airports, the Secretary may authorize grants for the purpose of acquiring, establishing, and improving air navigation facilities; and for existing airports served by air carriers certificated by the Civil Aeronautics Board. Included in this group of airports are those general aviation airports, which serve to relieve congestion at airports having high density traffic serving other segments of aviation. Thirty million dollars was made available for the fiscal years 1971 through 1975 for airports serving segments of aviation other than air carriers.

The Secretary is authorized to apportion funds in the following manner for air carrier airports:

a. One-third of the funds will be distributed to the various states, one-half of this amount distributed according to the geographic size of the state and the other half according to the size of the population.
b. One-third of the funds will be distributed to the sponsors of airports served by certificated air carriers in the same ratio as the number of passengers enplaned at each airport of the

sponsor bears to the total number of passengers enplaned at all such airports.

c. One-third will be distributed at the discretion of the Secretary.

The Secretary is authorized to apportion the funds for general aviation airports in the following manner:

a. 73.5% to the states based on the distribution related for air carrier airports.

b. 1.5% for Hawaii, Puerto Rico, Guam, and the Virgin Islands.

c. 25% to be used at the discretion of the Secretary.

The most controversial section of this Act appears to be a section which amends the Federal Aviation Act of 1958 by adding Section 612. This section is titled "Airport Operating Certificates." The Administrator is empowered to issue airport operating certificates to airports serving air carriers certificated by the Civil Aeronautics Board and to establish minimum safety standards for the operation of these airports.

Any person desiring to operate an airport serving certificated air carriers must file with the Administrator an application for an airport operating certificate. If, after an investigation, the Administrator finds the applicant properly equipped and able to conduct a safe operation, he shall issue an airport operating certificate. The details for this procedure of airport certification may be found in the Federal Aviation Regulation Part 139.

It shall be unlawful for any person to operate an airport serving certificated air carriers without an airport operating certificate, or in violation of the terms of that certificate.

Title II of the Airport and Airway Revenue Act of 1970 established new and increased taxes which are imposed on the users of the airport and airway system. These taxes include the following:

1. A 10% tax on airline tickets for domestic flights.
2. A $6.00 charge on airline tickets for most international flights beginning in the United States.
3. A 6.25% tax on air freight waybills.
4. A 7¢ tax on fuels used in noncommercial aviation.

Criticism of the Act of 1970

The operation of the Act of 1970 has been severely criticized because the money collected for the aviation trust fund is placed in the general budget of the nation. Any expenditures are counted against the budget total which usually has a ceiling set by the President.

Critics state that about half of the money collected has not been used for aviation safety projects but rather to make the general budget deficit appear to be smaller than it is. At the end of fiscal year 1991 the surplus in the trust fund was over $15 billion.

Legislation has been proposed to remove the trust fund from the general budget, thus allowing expenditures from the aviation fund to be freed from the restrictions of the general budget.

The Government has proposed, and will probably approve, the elimination of several airports from the group of those that receive grants under the Airport Development Aid Program (ADAP). This action has been called "airport defederalization."

Airport interests are outspoken on the fact that surplus money is being used to off-set the federal deficit rather than to aid airports as the Act intended. At the same time, airports are to be removed from the list of recipients of the trust fund money.

5. A basic annual registration tax of $25.00, plus an added charge (2¢ per pound for piston aircraft, and 3–4¢ per pound for jets) applicable to aircraft over 2,500 pounds.

As a result of the Airport and Airway Development Act of 1970, the Federal Aviation Administration adopted a Federal Aviation Regulation, titled Airport Aid Program. The new regulation prescribes the policies and procedures for administering the Airport Development Aid Program and the Planning Grant Program. Both of these programs were established by the Airport and Airway Development Act of 1970. The new regulation, Part 152, incorporates the general requirements of Part 151 which governed the old Federal Aid Airport Program and remained in effect for that program until all of its projects were completed.

Airport Improvement Program

Currently, federal assistance for airports is provided by the Airport Improvement Program, which awards grants to airports proportionally based on formulas considering use (enplaned passengers or pounds of cargo) and location (land area and population of the State). A separate discretionary fund also exists, but disbursements are largely driven by congressional mandate.

The adequacy of the Nation's airports and airways continues to be a priority issue. Among the most prominent problems is the sufficiency of airport capacity, or the ability of airports to accommodate demand for takeoffs and landings at certain times. The Federal Government takes an interest in airport capacity in order to meet military, commercial, and safety needs. It accomplishes this by its management of air travel through the Air Traffic Control System and by meeting airport capital assistance needs through the Airport Improvement Program.

The Airport Improvement Program is the Federal Government's capital grant funding program that helps airports sustain or increase their capacity through facility expansion and improvement. Grant allocation is based on the type of project and airport involved and other funding goals outlined in authorizing legislation. Capital grants address, at least partially, the problem of airport capacity; other difficulties, however, arise from implementation of the Airport Improvement Program.

Congress has continued to support the Airport Improvement Program largely to meet national needs; numerous issues, however, remain unresolved. Pressure to minimize the federal deficit, as well as questions about the effectiveness of capital spending in addressing airport capacity problems suggest that increased funding for the Airport Improvement Program is unwise. Other viewpoints, however, indicate that the need for a safe and effective aviation system, as well as the need to confront noise and other ongoing problems may be sufficient incentive to repel attacks on this program's budget. Legislation by Congress indicates that it is satisfied with the Airport Improvement Program and that curtailing its growth is not a likely outcome in the near future.

The Airport Improvement Program (AIP) (49 U.S.C. Chapter 471) has provided federal grants for airport development and planning since the passage of the Airport and Airway Improvement Act of 1982 (P.L. 97-248). Airport Improvement Program funding is usually spent on projects that support aircraft operations including runways, taxiways, aprons, noise abatement, land purchase, and safety, emergency or snow removal equipment. Funds obligated for the Airport Improvement Program are drawn from the Airport and Airway Trust Fund, which is supported by user fees and fuel taxes.

The September 11, 2001 terrorist attacks on New York and Washington led to passage of the Aviation and Transportation Security Act (ATSA) (P.L. 107-71). ATSA broadened the range of security activities and projects that are eligible for AIP grants.

On April 5, 2000 President Clinton signed the Wendell H. Ford Aviation Investment and Reform Act for the 21st Century (FAIR21; P.L. 106-181). Two years in the making, this $40 billion multiyear Federal Aviation Administration (FAA) reauthorization bill included AIP authorizations of $3.2 billion for FY2001, $3.3 billion for FY2002, and $3.4 billion for FY2003. The Act also increased the Passenger Facility Charge (PFC) ceiling to $4.50 per boarding passenger.

Raising the ceiling on the Passenger Facility Charge had been one of the most contentious policy issues related to Airport Improvement Program. The PFC is essentially a local tax on each boarding passenger that is levied by an airport with federal approval. The ceiling had been set at $3 since 1990. During the reauthorization airports had supported eliminating or raising the ceiling while airlines had argued for no change.

FAIR21 includes "point of order" provisions that were designed to assure that all trust fund receipts and interest are spent annually and increased the likelihood that Airport Improvement Program will be fully funded at the authorized level. To date, excepting across-the-board recissions, this has been the case: Airport Improvement Program has received the fully authorized amounts of $3.2 billion, $3.3 billion, and $3.4 billion for FY2001, FY2002, and FY2003 respectively.

Noise mitigation spending is closely linked to airport capacity policy because airport noise levels are a major factor in local resistance to airport improvement projects. FAIR21 increased the set aside for noise mitigation from 31% to 34% of Airport Improvement Program discretionary funds.

As reauthorization of Airport Improvement Program approaches some issues including, noise mitigation, the appropriate range of Airport Improvement Program and Passenger Facility Charge projects, and the scope of airport development needs, are reemerging. Along with these issues are newer issues such as the impact of the use of Airport Improvement Programs funds for security purposes on the funding for Airport Improvement Program's more traditional uses.

Most Recent Developments

The September 11, 2001 hijacking of four airliners from three airports and the enormous loss of life from their use as weapons has had an impact on the Airport Improvement Program (AIP). In the aftermath of the attack, many airports are seeking to use Airport Improvement Program funds for airport security improvements. Congress responded to the attacks by passing the Aviation and Transportation Security Act of 2001 (ATSA) (P.L. 107-38). ATSA includes provisions that broaden Airport Improvement Program eligibility to cover airport security costs incurred because of Federal Aviation Administration mandated security changes imposed in response to the September 11 attacks.

On February 20, 2003, President Bush signed H.J.Res. 2, the Consolidated Appropriations Resolution (P.L. 108-7). The Act included $3.4 billion less and across-the-board rescission of 0.65% (roughly $20 million). President Bush's FY2004 budget proposal would hold the AIP budget steady at $3.4 billion.

Background and Analysis

The Airport Improvement Program (AIP) provides federal grants to airports for airport development and planning. Airport Improvement Program funding is usually limited to improvements related to aircraft operations, typically for planning and construction of projects such as: runways, taxiways, aprons, noise abatement, land purchase, as well as security, safety, or emergency equipment. Commercial revenue producing portions of terminals (such as shop concessions or commercial maintenance hangars), automobile parking garages, and off-airport road construction are examples of improvements that generally are not eligible for Airport Improvement Program funding. Airport Improvement Program money cannot be used for airport operational expenses or bond repayments.

The Airport Improvement Program is one of five major sources of airport capital development funding. The other sources are tax-exempt bonds, passenger facility charges (PFCs), state and local grants, and airport operating revenue. Different airports use different combinations of these sources depending on the individual airport's financial situation and the type of project being considered. Small airports are more likely to be dependent on Airport Improvement Program grants than large- or medium-sized airports. The larger airports are also much more likely to participate in the tax-exempt bond market or finance capital development projects with the proceeds generated from Passenger Facility Charges.

The Passenger Facility Charge is a local tax imposed, with federal approval, by an airport on each boarding passenger. Passenger Facility Charge funds can be used for a somewhat broader range of projects than Airport Improvement Program grants and are more likely to be used for "ground side" projects such as passenger terminal and ground access improvements. Passenger Facility Charges can also be used for bond repayments.

This issue brief discusses the Airport Improvement Program and its complement, the Passenger Facility Charge (PFC). After a brief history of federal support for airport construction and improvement, the report describes Airport Improvement Program funding, its source of revenues, funding distribution, the types of projects the program funds, Airport Improvement Program and Passenger Facility Charge policy issues, and the allowable use of Airport Improvement Program funds for airport security purposes.

Founding Legislation

Prior to World War II the federal government limited its role in aviation to maintaining the airway system, viewing airports as a local responsibility. Some federal monies were spent on airports during the 1930s (about $150 million) but only as part of federal work relief activities. The national defense need for a strong system of airports during World War II led to the first major federal support for airport construction. After the war, the Federal Airport Act of 1946 (P.L. 79-377) continued federal aid under the Federal Aid to Airports Program, although at lower levels than during the war years. In the 1960s substantial funding also went to upgrade and extend runways for use by commercial jets. Congestion, both in the air and on the ground at U.S. airports, was seen as evidence by some that past federal support for airports had not been sufficient to maintain adequate airport capacity.

Airport and Airway Development and Revenue Acts of 1970 (P.L. 91-258)

Congress responded to the congestion problems and capacity concerns at airports by passing two Acts. The first, the Airport and Airway Development Act, dealt with the spending side of federal aid to airports. It established the forerunner program of the Airport Improvement Program, the Airport Development Aid Program (ADAP), and set forth the program's grant criteria, distribution guidelines, and first five years' authorization. The second Act, the Airport and Airway Revenue Act of 1970, dealt with the revenue side of airport development. This Act established the Airport and Airway Trust Fund (also known as the Aviation Trust Fund). Revenues from levies on aviation users and fuel were dedicated to the fund.

Airport and Airway Improvement Act of 1982 (P.L. 97-248)

This Act created the current Airport Improvement Program. Although the Airport Improvement Program maintained the ADAP's approach of using grants-in-aid to support an integrated national system of airports, it did make some significant changes in the operation of the program. The program differences included altering the funding distribution among the different categories of airports, extending aid eligibility to privately owned general aviation airports, increasing the federal share of eligible project costs, and earmarking a portion of total funding for noise abatement and compatibility planning.

The Airport and Airway Trust Fund

The money that goes into the Aviation Trust Fund comes from a variety of aviation user fees and fuel taxes. These tax revenues are authorized through

September 30, 2007, by the Taxpayer Relief Act of 1997 (P.L. 105-34). Revenue sources include:

- 7.5% ticket tax;
- $3.00 flight segment tax;
- 6.25% tax on cargo waybills;
- 4.3 cents per gallon on commercial aviation fuel;
- 19.3 cents per gallon on general aviation gasoline;
- 21.8 cents per gallon on general aviation jet fuel;
- $13.40 international arrival tax;
- $13.40 international departure tax;
- 7.5% tax on second party sales of airline award miles (normally "frequent flyer" awards); and
- 7.5% ticket tax at rural airports.

Over much of the life of the trust fund, these revenues plus interest on the trust fund's unexpended balances brought more revenue into the fund than was being paid out. This has led to the growth in the end-of-year unexpended balance in the trust fund. There are outstanding commitments against these unexpended balances, so not all of the unexpended balance would actually be available in any given year. Nonetheless, these unexpended balances (somewhat inaccurately referred to by some as a surplus) have been large enough relative to the Federal Aviation Administration budget to make their existence controversial.

The scenario of an unexpended trust fund balance, that grows substantially larger each year, ended in FY2001. Most observers believe the drop in demand for air travel that began during 2001, due to the recessionary economy and potential passengers' fear of flying following the September 11 attacks, significantly constrained the revenues available from the trust fund. The end of year balance for the trust fund dropped from a high of $14.5 billion for FY2001 to $12.6 billion for FY2002. Estimated balances for FY2003 and FY2004 are $12.3 billion and $10.6 billion, respectively.

Airport Improvement Program Funding

Airport Improvement Program spending since FY1982 is illustrated. From FY1982 to FY1992 annual spending (obligations) increased from $412.5 million to $1,954.5 million. From FY1982 to FY1992 the obligation limits increased every year except for FY1986, when it dipped by $28.6 million below the FY1985 level. For FY1993–FY1997 spending was reduced as part of overall deficit reduction. Airport Improvement Program spending declined in FY1993 and FY1994 before leveling off at about the $1.5 billion level during FY1995–FY1997.

Obligations for FY1998 rose to $1.7 billion. The FY1999 omnibus appropriations act (P.L. 105-277) provided obligational authority for $1.95 billion. However, the money was released intermittently as a series of partial year authorizations were passed. On October 1, 1999, with the beginning of the new fiscal year 2000, the Airport Improvement Program went into abeyance, causing the suspension of Airport Improvement Program grant distribution. The enactment of FAIR21 allowed the FY2000 Airport Improvement Program funds to be distributed.

For FY2000 appropriations, the enacted appropriations legislation (P.L. 106-69) again provided for $1.95 billion. However, the Consolidated Appropriations Act for FY2000 (P.L. 106-113), called for an across-the-board cut of 0.38% from all discretionary budget authority and obligation limitations. Another $45 million is to be obligated to pay for the administration of the Airport Improvement Program. This allowed FAA to obligate just over $1.85 billion for airport grants in FY2000.

For FY2001, the Department of Transportation Appropriations Act funded Airport Improvement Program at the authorized level of $3.2 billion. This was an increase of nearly 70% over the FY2000 enacted funding. The Administration had proposed $1.95 billion for Airport Improvement Program. Following passage of the FY2001 Department of Transportation appropriations bill, the FY2001 Consolidated Appropriations Act (P.L. 106-554) provided for a government-wide rescission that reduced the amount available for Airport Improvement Program by roughly $7 million.

In the aftermath of the September 11 terrorist attacks on New York and Washington, $175 million in FY2001 supplemental appropriations (available until expended), included in the Department of Defense Appropriations Act (H.R. 3338; H.Rept. 107-350), were made available for Airport Improvement Program to help reimburse airports for the costs of post-September 11 security mandates imposed by law or Department of Transportation. The FY2002 DOT Appropriations Act (P.L. 107-87) provided for the fully authorized funding of $3.3 billion for Airport Improvement Program. In addition, the FY2002 Emergency Supplemental Appropriations Act (P.L. 107-206) includes a provision for $150 million for emergency grants, contingent on a request by the Administration for their use, to airports to offset the costs of Federal Aviation Administration security mandates. The Bush Administration, however, has announced that it will not utilize any of the contingent emergency funds provided for in P.L. 107-206.

On February 4, 2002, President Bush submitted his FY2003 budget to Congress. The budget requested the fully authorized $3.4 billion for Airport Improvement Program. Of this amount, the President requests that $83 million be provided for the Essential Air Service Program (EAS). On July 26, 2002, the Senate Committee on Appropriations recommended $3.4 billion (S. 2808; S.Rept. 107-224), as did the House Committee on Appropriations on October 7, 2002 (H.R. 5559; H.Rept. 107-722). Both of these bills rejected the use of Airport Improvement Program funds for EAS. The FY2003 Consolidated Appropriations Resolution (P.L. 108-7) provided for the fully authorized $3.4 billion less the 0.65% across-the-board rescission (roughly $20 million). (For more see CRS Report RS31308, *Appropriations for FY2003: Department of Transportation and Related Agencies*, coordinated by David Randall Peterman and John Frittelli).

Airline Deregulation Act of 1978

The Airmail Act of 1925 provided for the encouragement of the air carrier industry. The Civil Aeronautics Act in 1938 established economic and other regulations upon which the industry matured and developed.

Human nature is such that not everyone is satisfied when legislation is enacted. Laws passed by Congress for the benefits of the public are compromises at best. Many factions and individuals representing the aviation industry, government, and the general public continued to express dissatisfaction after the Civil Aeronautics Act was passed in 1938 and again after the Federal Aviation Act became law in 1958. Dissent and criticism with federal aviation regulation continued with increasing force until the 1970s. As early as 1975 a law was proposed known as the Federal Aviation Act of that year. It was not passed, but opposition grew regarding the economic regulation of the aviation industry. Almost every opinion was heard—from complete abolishment of economic regulation to total control. Some people were in favor of operation of the industry by the government. Somewhere in between these diverse ideas lies a compromise.

The idea of retaining economic regulation on the one hand, and allowing the air transportation industry greater freedom on the other persisted and merited further consideration.

In the early 1970s, many academic economists questioned the need for economic regulation of air carriers. As a result, President Ford began to press for deregulation. Then President Carter appointed Alfred Kahn as Chairman of the Civil Aeronautics Board. He moved the Civil Aeronautics Board quickly toward deregulation in areas of pricing, entry, and exit.

In 1975, Senator Edward Kennedy began an investigation of the regulatory practices of the Civil Aeronautics Board and the effects of these practices upon the air carrier industry. He was seeking answers to such fundamental questions as, why were rates so high and why was there so little route competition.

Kennedy was concerned that between November 1973 and January 1975 the standard coach fare rose 16% and the average yield rose 20%. The solution to these problems was rather simple, he said. By allowing new entry and pricing freedom, air carriers would be forced to operate more efficiently, and consumers would reap the benefit of competition in the form of lower fares.[2]

As early as 1975, the Civil Aeronautics Board began adjusting for deregulation. For the following four years until 1979 the Board engaged in five major domestic regulatory actions. These were:

1. Scheduled operating rights and new entry
2. Charter rules and entry
3. Pricing
4. Agreements, antitrust and control relationships
5. Air taxis, commuters, and air freight forwarders

The result of industry comment, congressional study, and rhetoric by almost every side involved during the years would require a volume many times larger than this text. The final result was the Act by Congress was signed into law by President Carter on October 24, 1978, known as the Airline Deregulation Act of 1978.

The new law consists of amendments to the Federal Aviation Act of 1958. Those Titles substantially changed were:

I General Provisions
IV Air Carrier Economic Regulation
VIII Other Administrative Agencies
X Procedures
XI Miscellaneous
XVI Sunset Provisions (This was a new part added to the Act.)

In addition to these changes the Airline Deregulation Act amended certain sections of the Airport and Airway Development Act of 1970.

The purpose of the Airline Deregulation Act of 1978 is given in its opening paragraph and states as its purpose to amend the Federal Aviation Act of 1958, "to encourage, develop, and attain an air transportation system which relies on competitive market forces to determine the quality, variety, and price of air services, and for other purposes."

The Deregulation Act of 1978 caused many changes, the most important of which are briefly summarized below.

1. Elimination of most of the economic regulations which were a barrier to competition.
2. Provided for the guarantee of future air service to small communities.
3. The ease of air carriers, both new and established, to enter new market areas.
4. The protection of air carrier employees if bankruptcy occurs within ten years or a 75% reduction in employment in one year due to the effects of deregulation.
5. The elimination of the Civil Aeronautics Board and its authority over domestic routes and fares which then became subject to U.S. anti-trust laws.

Much has been said about deregulation over the past several years and much more will be said about it in the future. Many people say that it is the best thing to ever happen to the United States air transportation industry, while others take the view that it is the most disastrous.

The Airline Deregulation Act of 1978 was a milestone occurrence in the deletion of a governmental controlling agency. It was part of the movement that changed the trucking, brokerage, oil, gas, telecommunications, financial, electric and gas utility industries.

Many observers agree that airline deregulation has been a success. The majority of passengers have enjoyed the benefits that its proponents expected. However, it has also caused many problems including airport congestion and a limited reemergence of monopoly power. With it came the exploitation of a minority of customers. It is a mistake to regard these developments as failures of deregulation when, in fact, they are manifestations of its success.

Deregulation—What Is It?

What does it mean? What brought it about? How does it affect the air carrier industry? What does it mean to the flying public? The dictionary defines "regulation" as the act of governing or directing according to rule, bringing under control and fixing the amount or rate by adjusting. This definition describes the United States air carrier industry from 1938 to 1978. The prefix "de" means separation, reversing or undoing an action or freeing from a rule. Placing the "de" in front of "regulation" reverses its meaning. Therefore, "deregulation" denotes the freeing from governing rules which control the amount or rate. Specifically, deregulation of the air carrier industry provides for the removal of rules which control the routes the air carriers fly and the rates which they charge the public. Directly, deregulation allowed for the removal of Civil Aeronautics Board control thus permitting open competition, primarily in routes and rates. Deregulation in the United States is also prevalent in such industries as trucking, railroads, communications, and banking.

Since 1938 the Civil Aeronautics Board had held tight economic control over the air carriers. It was the Board that decided which cities would get service and when they would get it. The Board decided what air carriers would serve these cities and what prices they would charge. Also, it was the Board that decided which air carriers could merge and when and under what conditions they could do so. Some mergers may even have been forced by the Board. The Civil Aeronautics Board made so many decisions for the air carriers that about the only decisions they did not make were what types of aircraft were used, and how frequently service was provided.

Many reasons were given to support the Civil Aeronautics Board's tight control. The main reason, it was believed, that without regulatory control, the air carrier industry would become embroiled in cut-throat competition which would force some companies out of business and result in an industry structure of a few large companies with a considerable amount of power. A few small weak air carriers might remain, fighting for the leftover "crumbs."

These reasons provided the basis for opposition to deregulation. Opponents of deregulation said that it would leave the air carrier industry in ruin and that even safety would be jeopardized. One valid point was that small communities might lose their air carrier service. It was decided that

Economic Regulation Needed?

During the mid-1970s, Kennedy and other Congressmen started to listen to the questions that economists had been raising for many years about the need for close regulation of the air transportation industry. They argued that there was no longer a need for detailed economic regulation, if, in fact, there ever had been a need. They theorized that air transportation was like any other business and that air carrier managements should be allowed to make all, or almost all, of the decisions affecting them.

As examples they pointed to Intrastate Air Carriers, under no federal economic regulations, that were operating successfully in California, Florida, and Texas. These air carriers were making large profits charging low rates, while the regulated air carriers were earning little or no profit while charging higher rates. This seemed to prove the economic theories of competition—supply and demand and pricing.

These questions concerning the need for economic regulation could not bring forth any valid answers.

this point had merit and therefore the Airline Deregulation Act contains protection for small communities.

One reason Congress began listening to arguments for deregulation was the growing public mood that government was getting to be too large and too intrusive.

The decision to deregulate the air carrier industry was premised upon the fundamental economic principle that the air carrier industry is inherently competitive. Congress concluded that the industry sells a basically homogeneous product and that there are no serious barriers to entry at the firm level. This reflected a substantial change from prior government regulatory principles. For 40 years the air carrier industry existed in an environment that was virtually exempted from antitrust laws. Actually, the Civil Aeronautics Board, through its control over air carrier routes and rates, was acting as a cartel agent directed by Congress to promote and to protect the industry.

In a system of limited entry and rigidly controlled prices, no real competition could exist. Over the years the Board had established a policy of parallel growth for the various air carriers, and the aviation industry market shares remained relatively constant for 40 years.

No doubt, the Civil Aeronautics Board believed it was promoting competition in certain city-pair routes. However, the air carriers had no incentive to seriously compete in individual markets or at the national level. The system that had evolved was one of sharing. Each air carrier got its share of an ever increasing population.

During the debate over deregulation, Congress found that this system created few, if any, incentives for operating efficiently and that it led to high costs and higher rates. Congress concluded that deregulation would be in the public interest.

Deregulation has affected more than just the air carrier industry. It has also affected the travel agency industry, the operation of airports, the methods of selling and distributing aviation fuel, and even the development of aircraft technology.

The debate on air carrier industry deregulation which had begun in early 1975 ended with the passage of the Airline Deregulation Act in October 1978. The case for reduced government intervention in air carrier pricing and services decisions had won.

Three major interrelated arguments were advanced by the proponents of deregulation and accepted by Congress. They were:

1. That supply and demand would be in balance; that air carrier service would be provided at the level of demand in all markets that could be profitably served. Transitional problems and economic air service would be provided for by means of explicit subsidy.
2. That improvements, which would be reflected in lower costs and therefore lower prices, would come from improved efficiency.
3. That the air carrier industry would profit sufficiently to provide for the cost of capital replacement and expansion.

The air carrier industry today does not resemble the same industry of the 1970s; many areas have been affected. The ultimate restructuring of airline organizations and capitalization has been widespread through the industry.

Comments

Deregulation has been criticized since its inception. There are many opposed to it as well as those who claim that it was the best thing that happened to the Air Transportation Industry. Many of the problems associated under the days of regulation are claimed under deregulation. Regulation was criticized for obstructing creativity and expansion in the air carrier industry. Now that deregulation has been in effect for more than a couple of decades, its critics claim that the industry is being obstructed and expansion stifled. The opponents to deregulation give three classic examples. The first is the telephone industry since the court ordered the breakup of the Bell Telephone system. Telephone users now must have a “provider” in addition to paying a server.

This means that one company provides the telephone system while another transmits the messages. There is intense competition for your telephone business. Advertisements received by mail and TV commercials encourage the user to switch services. Each company promises better service at cheaper rates, however, upon close examination this is not always the case. Information assistance was once provided free but now information is provided at a nominal cost. When the information service is used the dialer is asked if they want to be connected. If the answer is yes then an additional charge will be levied to dial the number.

A second example of problems resulting from deregulation is to be found in the electrical generating industry. Deregulation is given as the blame for the outages and problems in the state of California; although in fairness this may or not be entirely true. A closer examination of each individual circumstance is necessary.

A third example presented by the opponents to deregulation is the trucking industry. Cutthroat competition brought on by deregulation has turned those knights of the road into road warriors.

Summarily, deregulation in any shape or form, wherever it is found does have its problems but so did regulation. Deregulation has not been found to be the panacea of regulatory problems. Inefficiencies, many of which existed during the regulations years, can be found in one way or another under deregulation. One critic was quoted as saying "somebody is profiting and the rest of us are paying for deregulation."

Thomas Jefferson once said that "the art of life is the art of avoiding pain." This could well be applied to the present problem of deregulation versus regulation.

Labor Relations

The changing pattern of air carrier service brought about by deregulation has made it difficult for labor unions representing air carrier personnel to negotiate contracts that will keep their members within the mainstream of the economic spiral.

Unions and air carrier management officials differ in their opinions about the impact of deregulation. Unions attribute layoffs of their members to deregulation; while management says they are necessitated by economic conditions. Negotiations between unions and airline management have become difficult and many strikes have resulted. Labor demands for higher compensation have not usually been met due to higher airline operation costs and decreased profits; in some cases bankruptcies have occurred.

Essential Air Service

The Deregulation Act contains a provision for air service to small communities. It provided that a minimum level of essential air service to all certificated cities be guaranteed. The Board was directed to solicit applications from air carriers to perform the service and, when necessary, to pay a specific cost based subsidy for the service if it would not be provided without subsidy.

Not every community benefits to the same degree under deregulation. Some have different kinds of air service than they had before deregulation. Different air service does not mean less or bad service.

Several years before deregulation, the Civil Aeronautics Board had exempted commuters from some regulations so that they could develop air services to better meet the needs of small communities.

Under deregulation, an air carrier must give notice before eliminating service in a city. The Department of Transportation must then determine if the elimination will cause service to fall below what is considered essential for the community in order for it to have access to the national air transportation system. If it is determined that the city would be left without essential air service, the Department of Transportation may require the air carrier to continue service to that city until a replacement is found. In nearly every case so far, commuter air carriers have been eager to serve the city and to do so without the benefit of subsidy. In very few cases is subsidy being provided. For example, in Jamestown, North Dakota, Northwest Airlines was providing one flight each day using large jet aircraft. When Northwest gave notice of its intention to vacate Jamestown, the Board determined that essential air service to that town would be two flights a day westward to Bismarck and two flights eastward to Minneapolis. Air Wisconsin replaced Northwest with three daily flights to Minneapolis and two to Bismarck using a 19-seat aircraft on a non-subsidized basis. In the first month of the new service traffic doubled. This illustrates that the flexibility of frequent flights with small aircraft very often serves the need of the small town better than one flight a day using large aircraft.

In Providence, Rhode Island, American Airlines moved out; but TWA, Air New England, and Pilgrim moved in. Providence soon set a record for passenger boardings. This evidence indicates that if a market exists and the community responds, air service will be provided.

Airports

Deregulation has caused problems not only in the air carrier industry; it has also brought about massive difficulties with airport facilities. Due to the route freedom allowed by deregulation, air carriers found it difficult to obtain airport terminal facilities for loading and unloading passengers and cargo. It may be that airport space will come to dictate air carrier operations at all major cities and many other cities as well.

The established air carriers have made large investments to expand their terminal facilities and are now being challenged by new companies whose equal access is provided for by law under deregulation. The problem is complicated by the trend to reduce federal funds for building and expanding airport facilities. Airport authorities will have to devise standards for the utilization of airport terminals.

The Federal Government operates Washington National Airport through the Federal Aviation Administration, and it has taken the lead in devising new procedures for allocating airport "slots" after the air carriers serving the airport and others who wanted to do so were unable to negotiate a satisfactory solution.

Other airport problems brought about by deregulation are: increased noise from aircraft and highway traffic, increased road congestion, and airport parking limitations. Section 105 of the Federal Aviation Act of 1958 was amended by the Airline Deregulation Act of 1978. It prohibits any attempt by a local airport or government to control the markets served by the airport.

Some new airport leases may contain provisions allowing the airports to buy back underutilized areas and reassign them to other air carriers who have a greater demand. This would assure more efficient use of gates and other terminal facilities. New leases at Miami International Airport, for example, allow the air carriers a preferential, but not nonexclusive, right to the use of gate space.

Postal Service

When the Civil Aeronautics Board ceased to exist on January 1, 1985, its authority over the mail rate also ceased. From that time on, it has been the responsibility of the U.S. Postal Service to determine the rates to be paid air carriers for transporting the mail. This has been accomplished by competitive bidding, and the Postal Service contracts for air transportation when and where it is needed.

This change has had the largest effect on the air carriers. Historically, the Civil Aeronautics Board has used airmail rates as a method of subsidizing some air carriers. The Postal Service has no interest in subsidizing air carriers.

International

The United States Government, together with other governments, has begun to question the conventional system of international and foreign air transportation and to suggest that a competitive system would lead to a healthier industry.

Domestically, the competition encouraged by deregulation has led to a realignment of the industry. Internationally, the elimination of restrictions on competition should also lead to greater efficiency and traveler benefits. However, few countries appear willing to change as the U.S. has, opting for a competitive regime including multiple designations, dual disapproval pricing, and third country matching.

Pro-competition policies have been endorsed by the Congress and the President in the International Air Transportation Competition Act passed on February 15, 1980. This Act encourages pro-competitive arrangements. It dictates a resolute yet flexible approach to bilateral negotiation. The Act's Declaration of Policy states that the U.S. shall consider in the public interest the placement of maximum reliance on competitive market forces. The U.S. recognizes that there are substantial differences between countries and markets. For example, the grounds upon which the U.S. is willing to compromise with a small and/or isolated country may differ substantially from what might be considered liberalization by a large country even though the air route from each country to the U.S. is the same or very similar. Domestic flexibility in pricing and freedom of entry and exit have permeated the international scene which now appears to be more open minded toward a competitive system. The U.S. has negotiated more liberal agreements with a number of countries and has achieved liberalization with others. As a result, increased rate competition in international markets has been achieved and the public has benefited. The entire structure of international aviation fares has been made more complicated by the newly negotiated bilateral agreements.

Some important markets are still restricted. The U.S. will have to continue to negotiate for more liberal agreements to bring the benefits of competition to more international travelers. The bilateral agreements that were negotiated before deregulation exchanged route rights for route rights. The new agreements give foreign air carriers route rights in exchange for foreign market access by U.S. carriers. Market access is difficult to enforce because foreign countries may seek to deny the U.S. air carriers' rights indirectly through intimidation—making the other country's breach of the bilateral agreement more difficult to detect. Preserving the United States international aviation system requires that the U.S. take steps which are necessary to assure that its bilateral agreements do not become one way deals.

Summary of Deregulation Changes

1. During 1979 and 1980 new domestic route automatic entry was allowed, multiple permissive awards were made, simplified procedures for obtaining route certificates were provided, and reasonable domestic fares were implemented.
2. As of December 31, 1981, most regulation of domestic route authority expired, including Civil Aeronautics Board authority of air carrier and route selections.
3. On December 31, 1982, authority over domestic passenger fares was terminated. Board authority over domestic mergers and interlocking relationships were transferred to the Department of Justice.
4. On December 31, 1983, a report was sent to Congress on the success or failure of deregulation. This report consisted of before and after comparisons and recommendations for the future of the Civil Aeronautics Board and its remaining programs.
5. On December 31, 1984, authority over foreign air transportation was transferred to the Department of Transportation and authority over domestic mail rate determinations transferred to the Postal Service. Air transportation subsidy programs went to the Department of Transportation.
6. On the same date, authority over agreements and foreign matters transferred to the Department of Transportation. All other programs and the Civil Aeronautics Board ceased to exist.
7. Title II of the Federal Aviation Act of 1958 concerning the Civil Aeronautics Board ceased to be effective on January 1, 1985.

Congress must be sensitive to any losses of public service, the possible financial failure of any large air carrier, continued service suspensions, and any failure of the commuter air carriers to provide necessary service to small communities.

Few can doubt the economic principle embodying the view that the competition process inevitably leads to the most efficient and socially desirable utilization of resources. One fundamental conclusion to be drawn from the benefits of deregulation is a direct link between a competitive environment and the increases in industry productivity.

The impact of deregulation continues to alter the commercial aviation industry. October 1988 marked the tenth anniversary of the Airline Deregulation Act. During the decade a number of significant structural and operational changes in the commercial aviation industry have occurred. The industry went through three distinct phases:

1. Expansion (from 30 large air carriers in 1978 to 105 in 1985)
2. Consolidation (11 mergers and 16 buyouts of smaller regionals/commuters in 1986–1988) to 61 active carriers at the end of 1988
3. Concentration (4 largest carriers accounted for 60.4 percent of traffic in 1988—up from 52.5 percent in 1978)

The regional/commuter airlines experienced similar changes, with the number of carriers increasing from 210 in 1978 to 250 in 1981, then declining to 170 in 1988. The fundamental character of the regional/commuter industry has changed significantly since the mid-1980s. These changes include the relative size and sophistication of airline operations, the carriers involved, the aircraft fleet mix and the industry's relationship with the large commercial air carriers. The overall number of regional commuter airlines declined by two-thirds between 1981 and 2003 from 250 air carriers to 75. The coming years may well see more new air carriers that can fill the void left by the larger regional air carriers as they abandon their short-haul markets in pursuit of longer-haul markets.

One reason for the decline in the number of air carriers over the past two decades has been the dramatic growth in the number of code sharing agreements with the major air carriers thus making it difficult for air carriers without such agreements to effectively compete.

Airlines changed the structure of the routing system from predominantly linear operations to a system of hub and spoke. The development of connecting hub airports led to high frequencies in peak hours at major air carrier airports and has significantly increased the demand for FAA services at these airports. The U.S. deregulation experience has become a model for the rest of the world. As the current movement toward worldwide deregulation progresses, it opens the possibility of the creation of multinational carriers throughout the world. The U.S. experience with code-sharing agreements between large air carriers and regionals/commuters suggests that smaller regionals/commuters benefit from working relationships with a larger airline. In future years, the same could hold true for competition in international markets.

In 1978 when Congress deregulated the commercial aviation industry, it argued that the U.S. commercial airline industry had matured and no longer needed the protection afforded them through economic regulation. With the passage of the Airline Deregulation Act of 1978 several expectations were widely held regarding the benefits of deregulation:

1. There would be improved service to the traveling public;
2. The public would be offered lower fares;
3. U.S. carriers would achieve higher profits; and
4. The resultant commercial airline industry would be more competitive.

The following review by the Federal Aviation Administration shows the results of the past in light of the expectations held earlier. In June 1978, U.S. certificated airlines (the designation prior to deregulation) served a total of 470 airports in 48 states (no service in Delaware and South Dakota) and the District of Columbia. By December 1988, however, the number of airports served by large U.S. air carriers totaled only 232, less than half the number served in 1978. Only two states (Florida and Washington) had service to more airports in 1988 than in 1978. Five states (Indiana, Maryland, New York, Ohio, and Rhode Island) and the District of Columbia received service to the same number of airports in both periods. All of the other states had service to a smaller number of airports in 1988.

Included in the total of the 232 airports served in 1988 are 16 airports which did not receive certificated air carrier service in 1978. This means that over the 10-year period since deregulation, a total of 254 airports have lost certificated air carrier service. At first glance it would appear that deregulation has resulted in some 254 cities receiving less service today than they had in 1978. However, the fact is that few, if any, of these cities actually lost all their service. In fact, many of the cities are receiving air service that is superior, at least from a marketing standpoint, to the service they received in 1978. In 1978, service to many of these cities was subsidized and generally consisted of one or two flights daily. In addition, service generally was not at the most convenient travel times. Most of the service was provided by local service or regional carriers, operating smaller aircraft similar to the aircraft operated by the commuter/regional airlines of today.

Many communities are served by commuters/regional carriers that operate new technology turboprop aircraft with the comfort and amenities of larger air carrier jet aircraft. With the advent of commuter code-sharing agreements and schedule tie-ins with the larger air carriers, most communities receive more frequent service as well as an increase in the number of available destinations. This is because the commuter/regional flights are timed to connect at the larger hub airports with connecting flights to many different destinations. Despite a cutback in the number of cities served, air carrier activity has increased substantially (up 26.2 percent) during the 10-year period since deregulation. This is largely the result of a proliferation of hub airports supporting carriers' hub-and-spoke route systems. Prior to deregulation, some hub-and-spoke networks did exist, although they were not as numerous as the number in existence today. This was due to the difficulty of obtaining new routes. Deregulation changed all this by making it easier for a carrier to enter more markets and to fly any routes it chooses.

Saturation of the larger hub airports, and the attendant congestion and delays, led to the establishment of secondary hub airports. One of the major disadvantages associated with the development and expansion of the hub-and-spoke route system is that it has resulted in a greater number of air carrier operations being scheduled into relatively few airports. The only possible result of such behavior is readily apparent and predictable. That is, increased airport congestion and an increase in the number and length of delays in the National Airspace System.

The proponents of deregulation may view these shifts as a restructuring of pricing policy to better reflect the actual costs of providing service. Critics of deregulation may say that short-haul travel is now subsidizing long-haul service. Whichever view is held, the fact remains that there has been a dramatic shift in pricing policies since the passage of the Act.

On average, "real" fares have fallen since 1978, but this phenomena has not been uniformly distributed among all city-pair markets. Some fares are significantly lower than they were prior to deregulation: e.g., Los Angeles to Phoenix—actual fare down 30%; "real" fare down 58% since 1979. On the other hand, some fares are significantly higher than they were prior to deregulation: e.g., Indianapolis to St. Louis—actual fare up 297% since 1979.

Whether the fare an individual traveler pays is higher or lower than would have been the case without deregulation ultimately depends on: where one lives; which cities one wishes to fly between; the number of competing carriers in those markets; whether one is able to plan his or her travel in advance; whether one is able to travel at non-peak hours or days of the week; and often, whether one is willing to accept substantial cancellation penalties. Discount fares are available in most, if not all, markets. The individual air traveler must seek them out.

Conceivably, the final phase of the deregulation process may have begun in December 1987 when United Airlines announced a marketing merger with British Airways, whereby the two carriers would begin code-sharing operations in the Seattle-Chicago-London and Denver-Chicago-London markets. In addition, the agreement called for the two carriers to share facilities at the Seattle/Tacoma, Chicago O'Hare, and New York Kennedy airports. Since that time, a number of agreements between U.S. and foreign flag carriers have been proposed and/or put into effect. Most notable among them is the agreement between the Texas Air Corporation and Scandinavian Airways.

What makes these agreements particularly important is the movement toward deregulation and merger/consolidation among Europe's 21 national carriers. In December 1988, the European Community's transport ministers agreed to start freeing up European skies, a move which could possibly lead to the creation of transnational "megacarriers" inside Europe. The European Common Market became completely deregulated in December 1992. When these factors are considered along with the current U.S./foreign flag agreements, it opens the possibility of the creation of multinational "megacarriers" throughout the world. The U.S. experience with code-sharing agreements has shown that, while the smaller regionals/commuters have benefited from a working relationship with a larger airline, those without such a relationship find it difficult to compete. In future years, the same could hold true for competition in international markets.

The moves today toward regulatory reform around the world are limited essentially to commercial matters (fares, rates, capacity, market entry and frequency). There is no question that governments might relax their control over things like safety, airworthiness requirements, licensing of flying and maintenance staff, air traffic control, and similar matters.

Some countries view their airline as an arm of the state, designed to serve broad national interests. Generally, airlines seek, with varying degrees of success, both to provide essential public service and to support national goals, while at the same time at least breaking even if not achieving profit.

In a far-reaching domestic experiment, cargo deregulation within the U.S. was introduced in December 1977, followed by passenger deregulation in November 1978. In parallel, the U.S. began to seek liberalized bilateral agreements for international services on various routes. These new agreements place emphasis on more flexible fare structures, services to new destinations, direct or point-to-point service, and increased numbers of competing airlines.

In Britain, there is increasing emphasis on airline competition and in the European Economic Community a growing interest is being shown in how airlines organize their operations and the level of fares and rates. However, there is a recognition in Europe that, while greater competition is desirable, it should be an evolutionary process and should not prejudice the public service character of airline operations, avoiding any disruption which could have harmful effects for the consumer.

There is a considerable difference between applying deregulatory principles domestically within a single country and seeking, as the U.S. has done, to extend them throughout the global system. The latter cannot be done unilaterally, since it involves the susceptibilities of numerous countries in differing stages of development. Ironically, one effect of the U.S. international deregulatory thrust has not been to "get governments off the back of aviation"—for which there is considerable sympathy—but rather to increase government involvement in airline affairs.

This has come about because, where the multilateral airline tariff coordination process becomes hamstrung or inoperable, governments have found themselves getting into the details of fare and rate setting in bilateral negotiations. Thus deregulation can create a situation internationally which is the exact opposite of that of the U.S. domestic scene—more government involvement in airline commercial activities rather than less. Where there are differing regulatory philosophies this could mean clashes between governments over issues capable of solution between airlines multilaterally at the tariff coordination level.

One "spin-off" effect of deregulatory policies and the push for cheaper air travel is an emphasis on point-to-point fares. Internationally, this could lead to fragmentation of the unique interlining system pain-stakingly built up over the years to enable passengers and cargo to move easily and simply from one country to another. The flexibility of the interlining system also enables passengers to change their plans en route—a feature of particular importance to business travelers—and to arrange or rearrange their schedules on short notice.

The successful operation of this system requires standardization of airline paperwork, procedures and practices, and a pattern of agreed tariffs so that each airline involved in an itinerary knows what share to expect of the total fare. Point-to-point fares cannot be accommodated, it is claimed, because they are calculated with margins that prohibit any division of revenue. The challenge for the airline industry—and for governments—is to find the right balance between the demands for cheap point-to-point travel and for fullscale interlining facilities, because there are clearly markets for both.

What Remains Regulated

International: Among the Civil Aeronautics Board functions shifted to other parts of the government was the responsibility for awarding landing rights and other privileges in foreign countries to U.S. carriers. All international air service is governed by bilateral agreements between nations. The agreements specify such things as the cities each nation's airlines may serve, the number of flights they may operate, and the range of fares they may offer. Bilateral negotiations involving the United States are led by the State Department with active Department of Transportation participation. When the United States secures new international rights, it is the task of the Department of Transportation to decide which U.S. airlines will get those rights. When more than one airline is interested in a new route, an administrative law judge will hold public hearings to allow each airline to make a case for why it is best suited to operate the new service. The judge then makes a recommendation to the Secretary of Transportation, who makes the final decision.

Mergers

Another Civil Aeronautics Board function assigned to the Department of Transportation was the review and approval of airline mergers. The Department of Transportation must decide if a merger will result in reduced service and less competition—in short, whether it is in the best interests of the traveling public. In addition, the Department of Transportation must consider the effect of the merger on employees of the two airlines. Department of Transportation has the power to impose "labor protective provisions" (LPPs) as a condition of the merger. LPPs are designed to guarantee adequate severance and relocation pay to employees. They also establish procedures for merging seniority lists, which is particularly important to flight crews, who bid on work assignments based on years of service with the airline. LPPs were last imposed by the Civil Aeronautics Board when it approved the merger of Mohawk Airlines into Allegheny Airlines (later renamed USAir). In all recent mergers, LPP issues have been resolved through collective bargaining. The Department of Transportation has never imposed them.

Essential Air Service

A third Civil Aeronautics Board function the Department of Transportation assumed following deregulation was the responsibility for maintaining air service to small communities. With carriers free to go wherever they want, Congress anticipated that some of the lightly traveled routes would lose service, and it did not want that to happen. So it established the Essential Air Service program, which provided subsidies to carriers willing to provide services that otherwise would be money-losers. The Department of Transportation administers the program determining subsidy levels and soliciting bids from carriers.

Safety

The government continues to regulate the airlines on all matters affecting safety. It has performed this regulatory role since 1926, and continues to do so through the Federal Aviation Administration, a part of the Department of Transportation since the department was founded in 1967. The Airline Deregulation Act ended government regulation of airline routes and rates, but not airline safety.

Effects of Deregulation

Airline deregulation has had a tremendous effect on the airline industry. Almost immediately following passage of the act, carriers began revamping their route structures for greater efficiency and moved into markets they had not served previously.

Hub and Spoke

A major development in the decade following deregulation was the widespread development of hub and spoke networks, which existed on a very limited basis prior to 1978. Hubs are strategically located airports used as transfer points for passengers traveling from one community to another in the region surrounding the hub. They also are collection points for passengers traveling to and from the immediate region and other parts of the country or points overseas. Airlines schedule "banks" of flights into and out of their hubs several times a day. Each bank typically entails dozens of planes arriving within minutes of each other so that, once on the ground, the passengers on those flights can transfer conveniently to the planes that will take them to their final destination. Total ground time for each bank usually is about 45 minutes.

Airlines developed hub and spoke systems because they enable them to serve far more markets than they could with the same size fleet if they offered only direct, point-to-point service. At a hub, travelers can connect to dozens, sometimes hundreds, of flights to different cities, and often can do so several times of the day. An airline with a hub and spoke system thus has a better chance of keeping its passengers all the way to their final destination rather than hand them off to other carriers, and travelers enjoy the convenience of staying with a single airline.

The carriers also found that with hub and spoke systems they could achieve higher load factors (percentage of seats filled) on flights to and from small cities, which in turn lowered unit operating costs and enabled them to offer lower fares. A city of 100,000 residents, for example, is unlikely to generate enough passengers to any single destination to fill more than a handful of seats aboard a commercial jet. However, it many very well generate enough passengers to a dozen different destinations to fill a high percentage of seats. Operating a jet into a hub where passengers can connect to dozens of different cities, therefore, makes economic sense for small-city markets.

Hub and spoke systems also have disadvantages, which by 1993 were leading some large carriers to reduce their hub operations and increase direct service. To work efficiently, hubs require a large number of ground service personnel to transfer baggage, assist passengers with connections, and perform other tasks. The additional ground personnel needed to handle each bank of flights are largely idle between banks—an inefficiency that can increase unit costs. In addition, when bad weather moves in on a hub, a carrier's whole system can be affected as planes are delayed and/or re-routed.

Most of the major airlines, nonetheless, continue to maintain hub and spoke systems, with hubs in several locations across the United States. Geographic location, of course, is a prime consideration in deciding where to put a hub. Another is the size of the local market. Airlines prefer to "hub" at cities where there already is significant "origin and destination" traffic to help support their flights.

New Carriers

Deregulation did more than prompt a major reshuffling of service by existing carriers. It opened the airline business to newcomers, and dozens of them stepped through the door, just as Congress intended. In 1978, there were 36 carriers certified under Section 401 of the Federal Aviation Act for scheduled service with large aircraft

Positive	Negative
1. Airline fares have increased at a rate far lower than that of inflation. Average fares (measured as yield/revenue earned per mile) have risen 59.8% in the period from 1977 through 1990. During the same period the consumer price index has risen an estimated 120.3%.	1. Fares may have gone down initially, but they have risen dramatically as industry concentration has increased. Fare increases were particularly apparent in 1990 according to some estimates. Fare increases have been much greater in certain markets for reasons that are not always apparent or rational to the traveler.
2. More discount fares are available than ever before; 91.4% of all airline traffic (measured on a revenue passenger mile basis) during 1990 was carried at a discount. This compares with a discount level of only 48% in 1978. The average discount in 1990 was 65.4% of the average full fare.	2. Hub-and-spoke systems have resulted in the creation of concentrated hubs where there is little competition between airlines. A recent study by the General Accounting Office indicates that air fares for trips originating at these hubs are higher on average than fares for trips originating at unconcentrated hubs.
3. According to a 1990 survey by the U.S. General Accounting Office, fares to small-city airports, long thought by many to be well above those to larger cities, were only 3% higher than those charged at major airports.	3. Non-stop flights between many destinations have disappeared as a result of the industry transition to a hub-and-spoke route system. The circuitous routes now required for travel between certain destinations are time consuming, inconvenient, and add significant time-related costs to air travel.
4. Airlines and airline passengers have saved large sums as a result of deregulation. One 1989 estimate is that deregulation has provided travelers and airlines with $14.9 billion of annual benefits.	4. Rural areas have been adversely affected by greatly reduced flight schedules and much higher fares. Some rural destinations feel that they have not only not received any benefit from deregulation, but have been effectively disenfranchised from the national air transportation system as well.
5. Changes in industry route structure resulting primarily from the adoption of extensive hub-and-spoke route structures by most carriers have led to a significant increase in flight frequencies between most destinations. As a result most travelers have significantly greater flexibility and opportunity to choose when and where they want to fly.	5. The airline industry has not paid adequate attention to consumer issues. Too many flights are delayed or canceled, flights are too crowded, too much baggage is misplaced and too many flights are overbooked. Inadequate enforcement by the Department of Transportation has allowed airlines to treat their passengers in a manner that would have been unthinkable prior to 1978.
6. Airlines have been very innovative in their transition to a deregulated environment. The airlines have shown an ability to develop new marketing tools, such as frequent flyer programs, that are very popular with travelers. In addition they have developed new and beneficial route structures. All of these activities would likely have been prevented by continued Civil Aeronautics Board regulation.	6. The fare system created over the last 10 years is confusing to most travelers. Only the most knowledgeable can decipher the system and even for these individuals there is no guarantee that they are not paying more for the same transportation as the person who will sit next to them. This situation raises numerous questions about equity as far as the traveler is concerned.
7. Average airline industry earnings, which have always been below those of other industries, showed improvement prior to the onset of a recession in 1990. In fact, the airlines collectively achieved record profit levels in 1988. This allowed the industry to place massive orders for new aircraft desired by the traveling public. The new aircraft will be more comfortable, quieter and more reliable. In addition, airline industry growth results in record levels of employment.	7. Deregulation is leading to ever greater industry concentration as major airlines fail and are not replaced. This development is detrimental to competition and creates skepticism that a competitive industry will survive over the long term. As weak remaining airlines fail in the 1990s, concentration is likely to increase even further with only a few carriers controlling the vast majority of traffic. Real competition amongst the few survivors is viewed as doubtful.

TABLE 7-3 Aspects of Deregulation

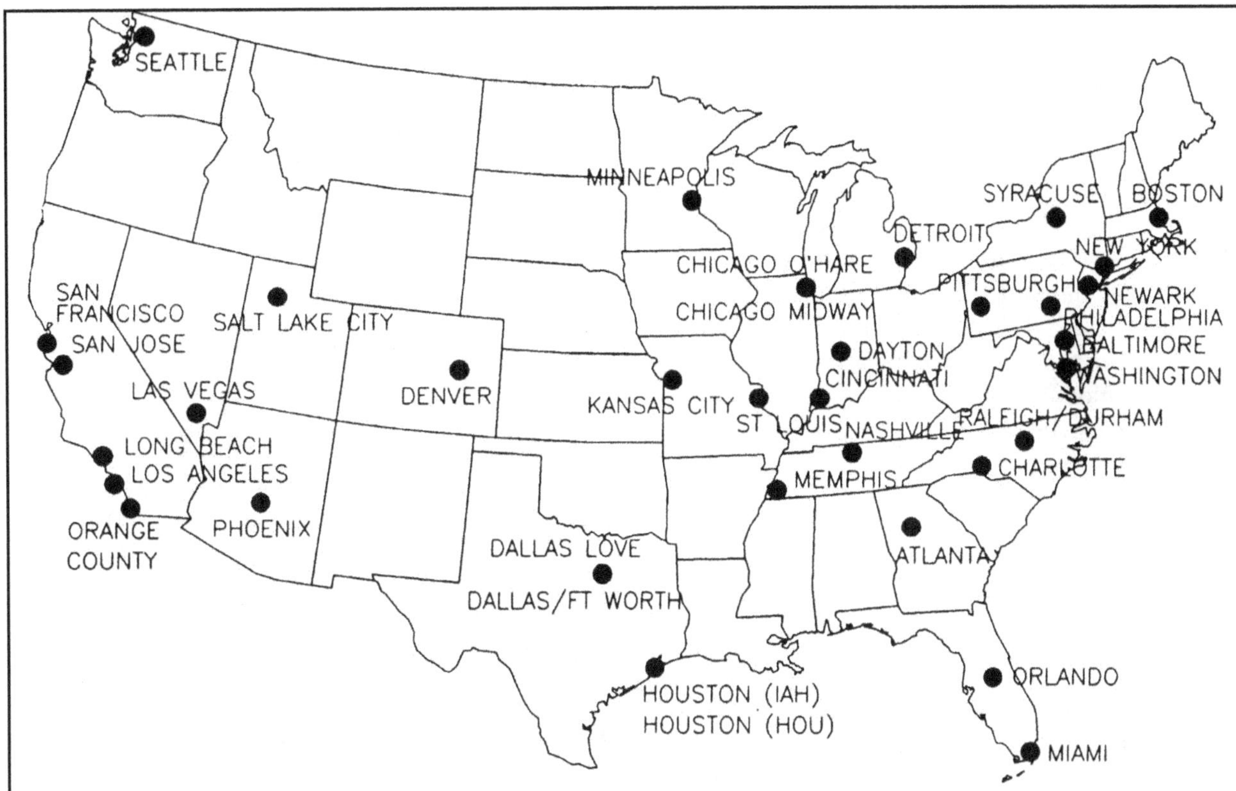

The HUB and SPOKE system is defined as anairport routing service pattern to feed traffic frommany cities to a central hub airport. It is designed to connect several flights to many other final destinations. The system flexibility connects many important destinations through the hub with fewer aircraft. This is more cost effective than would berequired if all such cities were to be connecteddirectly.

Source: Federal Aviation Administration

FIGURE 7-3 Major Hub Cities in the United States.

(defined as planes seating 60 or more passengers). By 1984, there were 123 such carriers.

The number declined in the late 1980s as weaker airlines went out of business or were taken over by stronger companies. Since 1993, however, the number again was on the rise as new airlines offering direct, low cost, no-frills service began to emerge. The new airlines were a result of several factors, most notably low prices for used aircraft and the availability of pilots, mechanics, and other airline professionals from some of the bankrupt carriers.

Increased Competition

The appearance of new airlines, combined with the rapid expansion into new markets by many of the established airlines, resulted in unprecedented competition in the airline industry. A study by the Department of Transportation a decade after deregulation found that well over 90% of airline passengers had a choice of carriers compared with only two-thirds in 1978. There have been some well-publicized airline bankruptcies and mergers, but the big airlines still in business compete with one another in virtually all major markets, as well as with the numerous new airlines on the scene today.

Between 1990 and 1996 the Department of Transportation certified about fifty new air carriers, 39 of them by 1993. By the end of 1996 only 25 of the first 39 were still flying and only 13 of those were carrying passengers.

Small airlines can offer significantly lower fares to passengers because they have much lower costs than the big airlines. Usually they start with older used aircraft models that were leased or bought inexpensively. They are not burdened by the costly labor contracts that the large air carriers must maintain. High costs have forced the failure of many smaller air carriers because of their limited financial support. Their financial reserves soon dissipated during economic slowdown or fare wars with the larger air carriers that could match or undercut the low fare of the smaller start-up air carrier. Some of these small new air carriers failed because of poor management and bad decision making.

Several major air carriers began price wars to discourage the new entrants. In some markets the larger air carriers engaged in predatory practices and engaged in below cost pricing.

The future of small air carriers is in doubt for several reasons. In addition to those discussed above there is the problem of acquiring gates and slots at air terminals to board and deplane their passengers. The major air carriers are having difficulty competing for terminal gates at the present. The chance of a new small air carrier obtaining a gate is indeed very slim. Their alternative is to schedule at off-hours. There are many in the air carrier industry, and particularly those of the small air carrier group, who are petitioning Congress to provide government help. This is a difficult matter because the local airports are governed by the political organization in which they are located.

Discount Fares

Increased competition spawned discount fares, and from the traveler's perspective, the discounts are the most important result of airline deregulation. Fares have risen half as fast as the rate of inflation since 1978. They have become so low, in fact, that interstate bus and rail service has been hard pressed to compete with the airlines, which today provide the primary means of public transportation between cities in the United States.

The Brookings Institute in 1993 estimated that the traveling public was saving $17.7 billion a year as a result of deregulation, measured in 1993 dollars. Fifty-five percent of the savings resulted from lower fares and 45% from increased service frequency which helps reduce the number of nights travelers must spend on the road. More than 90% of air travel today involves a discount, with discounts averaging two-thirds off full fare.

Growth in Air Travel

With greater competition on the vast majority of routes, extensive discounting, and more available flights, air travel has grown rapidly since deregulation, recessionary periods excluded. In 1977, the last full year of government regulation of the airline industry, U.S. airlines carried 240 million passengers. By 2005 they carried 739 million. A Gallup survey revealed that 77% of the U.S. adult population had flown at least once, more than 30% of them in the past 12 months.

Frequent Flyer Programs

Deregulation also sparked marketing innovations that equate to fare discounts, the most noteworthy being frequent flyer programs, which reward repeat customers with free tickets and other benefits. Most major airlines have such a program, and many small carriers have tie-ins to those programs. While the programs vary, the essential elements are the same. Once a customer enrolls, he or she is credited with points for every mile flown with the sponsoring carrier or with other airlines tied into the sponsor's program. The rewards (free tickets and upgrades that convert coach tickets to first class or business class tickets) are pegged to certain point totals.

A more recent development has been the marriage of frequent flyer programs with promotions in other industries, the credit card industry in particular. It is now possible to build up frequent flyer points by purchasing things other than airline tickets, and in some cases to exchange miles for other goods and services, like hotel rooms.

Computerized Reservations

Another important development following deregulation was the advent of computer reservation systems. These systems help airlines and travel agents keep track of fare and service changes, which occur much more rapidly today than they did when the government controlled such things. The systems also enable airlines and travel agents to efficiently process the millions of passengers who fly each day.

Several major airlines developed their own systems and later sold partnerships in their systems to other airlines. The systems list not only the schedules and fares of their airline owners, but those of any other airline willing to pay a fee to have their flights listed. Travel agents using the systems to check schedules and fares for clients, as well as to print tickets, also pay various fees for those conveniences.

Code Sharing

Computer reservation systems spurred the development of code sharing agreements in the mid-1980s between the major airlines and smaller, regional airlines. The agreements allow the small carriers to share the two-letter code used to identify their larger partners in the computer reservation systems. Their flights then are displayed as "through" flights rather than connecting flights, which appear lower in the computer listings.

The agreements also usually tie the regional carriers into their partners' marketing and frequent flyer programs, provide for schedule coordination for convenient connections between carriers, and in most cases permit the smaller airlines to paint their planes with markings similar to those used by their bigger partners.

All the major airlines have code sharing agreements with regional carriers, in most cases with several regionals. Some also own regional carriers outright, giving them greater control over these important services that feed traffic from outlying areas into the major hubs.

Code sharing also spread to international routes. Many U.S. and foreign airlines now have code sharing agreements that essentially enable those airlines to expand their global reach through the services operated by their partners.

Code sharing differs from "interlining," a much older industry practice where a carrier simply hands off a passenger to another carrier to get the passenger to a destination the first carrier does not serve directly. In such situations, the passenger buys a single ticket, and the airline issuing the ticket makes the arrangements for the traveler on the second carrier. However, schedules are not necessarily coordinated, there are no frequent flyer tie-ins, and there is no sharing of codes in computer reservation systems. The flights of each carrier appear independently in the computer reservation system.

Express Package Delivery

There was one other important development following deregulation—the rapid expansion of overnight delivery of documents and small packages.

As mentioned above, air cargo was deregulated a year before the passenger airline business, with dramatic results for all aspects of the cargo business but particularly express package delivery. Overnight delivery of high value and time sensitive packages and documents began in the early 1970s. However, it was deregulation that really opened the door to success for such services. Deregulation gave express carriers the operating freedom such high-quality services demand, and the result was outstanding growth for that end of the aviation industry over the next decade. In 1994, Congress further deregulated this part of the airline industry by preempting state efforts to regulate intrastate air/truck freight and express package shipments.

Endnotes

1. As amended by the Airline Deregulation Act of 1978.
2. Hearings on Oversight of Civil Aeronautics Board Practices and Procedures before the Subcommittee on Administrative Practice and Procedure of the Committee of the Judiciary, Senate, 94th Congress, 1st Session Vol. 1 at 1–3 (1975).

CHAPTER 8

The Federal Aviation Act

Responsibility for aviation safety had been lodged in the Civil Aeronautics Authority (CAA) by virtue of the Civil Aeronautics Act of 1938, as amended in 1940. Over the course of the ensuing twenty years, aviation safety had dramatically improved, largely due to the reliability of aircraft and engines and the development of instruments and navigation aids. The skies were much more crowded in 1958 than they had been in 1938, no doubt; but progress had been made in refining the airway (navigation) structure since the days of the lighted (beacon) airways. Beginning in 1947, VHF Omni-directional Radio transmitters (VOR) were installed across the country. An aircraft with a VOR receiver could track inbound from any point, directly to the station, by means of visual reference to the display shown in the onboard aircraft receiver. The direct routes between the VOR transmitters were established in 1950, designated as Victor airways, and were given numbers to distinguish one from another. Instead of flying from city to city as previously, or from one low-frequency radio transmitter to another (the radio range navigation system), aircraft that were equipped with these high-frequency radio receivers flew these routes as aerial highways.

The CAA was empowered to adopt rules and regulations pursuant to its mandate to administer safety concerns relating to aviation, and to establish through such rules and regulations standards for aircraft, engines, propellers, pilots, and mechanics, as well as for flight schools and the training of airmen. It was also charged with developing and administering the Air Traffic Control system (ATC), the system operated by the federal government that regulated the movement of aircraft throughout the United States. Control of aircraft by ATC could range all the way from taxi to takeoff, departure and enroute clearance, to arrival and landing clearance at destination. Most aircraft in the 1950s flew under Visual Flight Rules (VFR), which required little or no ATC control. Airline passenger operations, however, flew largely under Instrument Flight Rules (IFR), the procedure that was designed to ensure that an aircraft had airspace reserved specifically for it, so that no other airplane flying under IFR would occupy that same airspace.

These regulations were known as Civil Aeronautic Regulations (CARS) and were published in the Code of Federal Regulations. The CAA appears to have dutifully performed its administrative function regarding such matters. It can be argued with the aid of hindsight, however, that there was a lack of long-range vision within the CAA during its first twenty years of existence, between 1938 and 1958.

For one thing, the CAA was buried within the Department of Commerce along with the agencies that dealt with highways, maritime issues, textiles, the census, and myriad other matters. To make things worse, its appropriations had been slashed following World War II. Voice communication between pilots and ground controllers using high frequency radio had only been completely implemented in 1955. CAA air traffic control centers, which exercised control over all IFR traffic within large geographical areas throughout the country, now had direct voice contact with aircraft within their sectors. CAA controllers kept track of the location of each aircraft in the sector by means of position reports given by the pilots themselves. These position reports included time, altitude, last radio fix or location, next radio fix, and the estimated time of arrival at that fix. Position reporting was cumbersome and required extremely large blocks of airspace to be reserved for a single aircraft.

Radar, only recently invented (WW II), was slower to be adopted. Radar was first implemented by ATC only as an aid to making instrument approaches to airports during instrument or bad weather conditions. Gradually, as the number of large planes increased and began competing for the available airspace, it became more and more difficult for the air traffic control center personnel to keep track of aircraft. Distance Measuring Equipment (DME), the system that allowed an airplane to determine its distance from an equipped navigation facility, was incorporated into the VOR (Victor) airways system beginning in 1951. Now an aircraft could not only determine its azimuth location (bearing) from the station, it could determine its distance. DME was a big help in tracking aircraft in the "voice only" system, but even so, the control problem was becoming unmanageable.

In 1955 there were surveillance radars in place at thirty-two locations to service traffic arrival and departure at airports, but there were no long-range radars to control enroute traffic except in the mid-Atlantic region at Baltimore. Concerns were raised that ATC traffic congestion increased the likelihood of mid-air collision. Extension of long-range radars to provide positive radar control for enroute traffic was discussed, but not implemented. Then, on June 30, 1956, a United Airlines DC-7 collided with a TWA Super Constellation over the Grand Canyon, in Arizona, killing 128 people.

The lack of enroute radar was not actually a contributing cause, nor would it likely have prevented the collision. Both aircraft were flying under VFR, clear of clouds in good weather, and under visual flight rules their flight crews assumed the obligation to each other to "see and avoid." Flying under VFR was much preferred at the time due to the zigzag routes that aircraft were required to fly in order to fly the Victor airways, or to go from one VOR to the next VOR. In addition, nothing prevented other airliners, or any other aircraft for that matter, from flying VFR in your reserved airspace. Without area radar by which the controller could actually see both the controlled aircraft as well as any other aircraft that may constitute a collision hazard, flying under IFR in good weather was considered more trouble than it was worth.

Nevertheless, the collision focused public attention on the increasingly crowded skies developing over America as commercial aviation grew. The addition of jets, flying at almost twice the speed of the fastest piston aircraft, would only increase the hazard, perhaps exponentially. Another problem was the mix of military aircraft with civilian aircraft in common airspace. Bomber and fighter jet aircraft operated under one set of rules run by the military, and civilian aircraft operated under a different set of rules administered by the CAA.

On January 31, 1957, a second mid-air collision occurred, this one between a Douglas Aircraft owned DC-7 and an Air Force F-89 near Sunland, California. Head-on closure at high speed was deemed the probable cause. After a collision of another jet fighter with a United Airlines DC-7 in April 1958, unified authority for rule making for control of all aircraft, military as well as civil, in United States airspace was placed in the CAB. With this change, the CAB had rule-making authority, and the CAA was authorized to uniformly administer the rules.

With unified administrative authority in the CAA, positive control was mandated for all aircraft flying above 24,000 feet, as well as on three designated transcontinental routes at altitudes between 17,000 and 22,000 feet, so that it was no longer legal to fly under VFR in that airspace. This was the first step in an overall larger plan to gradually increase the airspace over which positive control would be exercised. Senator Mike Monroney of Oklahoma was instrumental in encouraging and coordinating these significant changes in the interest of safety, and, in the summer of 1958, he authored the legislation that became the Federal Aviation Act of 1958. He was later honored when the FAA's new national aeronautical center in Oklahoma City was named after him.

The Federal Aviation Act of 1958 incorporated virtually all provisions of the Civil Aeronautics Act of 1938 that related to economic regulation, and retained authority for such regulation within the CAB. The Act greatly expanded agency authority over aviation safety, however, and attempted to address the complaints voiced about the shortcomings of the CAA during the preceding twenty years. The CAA was abolished, and in its place was created the Federal Aviation Agency, which was organized as an independent agency answerable to Congress.

The Federal Aviation Agency was given authority to make long-range plans and to implement such plans without interference from competing government interests. All air safety research and development was consolidated and placed within the new agency; thus, the work of the Airways Modernization Board, the Air Coordinating Committee, and the National Advisory Committee for Aeronautics was assumed by the Federal Aviation Agency. An example of the work of the National Advisory Committee for Aeronautics includes the engine cowl research originally incorporated on the Boeing 247 and DC-1. Rule making was taken away from the CAB and placed in the new agency, as was the responsibility for recommendations regarding aviation legislation. Jurisdiction over suspension or revocation of airmen certificates was removed from the CAB and placed in the Federal Aviation Agency; the CAB was then designated as an appeals board to review the Administrator's certificate action with authority to reverse or modify the action taken by the agency against an airman. The CAB retained its responsibility for aircraft accident investigation as well as all aspects of economic regulation of the airlines.

In the fall of 1958, the first Federal Aviation Agency Administrator to be appointed was retired Air Force General Elwood R. "Pete" Quesada. He stepped up enforcement procedures in the airlines, assessing fines and issuing suspensions for rules violations. He led the way for the adoption of military-style radar for control of civilian aircraft. His tenure with the Federal Aviation Agency was marked by stormy relations with the airlines and its pilots, as well as with general aviation, but it set the country on a course of placing safety first for the flying public, a priority constant to this day.

The Department of Transportation

Prior to the creation of the Department of Transportation, the broad function of the administration of transportation fell to the Undersecretary of Commerce for Transportation. The Depart-ment of Commerce, a cabinet level executive department under the direction of a secretary and also a member of the president's cabinet, had been the catch-all repository for the various forms of transportation. The nation's regulation of transportation was administered by agencies, like the Interstate Commerce Commission and the Civil Aeronautics Administration, within the department created to deal with specific modes of transportation. Aviation matters had been removed from the Commerce Department by the Federal Aviation Act of 1958 so that, in 1966, both the CAB and the Federal Aviation Agency (formerly the CAA) were independent agencies. Others remained within the Commerce Department. Administration of the nation's transportation system was fragmented. Some transportation modes were over-funded and over-regulated, while others were under-funded and operated under a system of benign neglect.

The needs of the country from the earliest times were seen as including an efficient and accessible transportation infrastructure. But no overall plan had ever emerged to develop or administer transportation.

In 1965, then administrator of the Federal Aviation Agency, Najeeb Halaby, recommended to planners in the Johnson administration that a cabinet-level Department of Transportation be created based, in part, on his experience as head of that agency. For one thing, Halaby believed that the Federal Aviation Agency had been frozen out of the deliberations surrounding the administration's consideration of a supersonic aircraft transport program. To Halaby, this aviation endeavor was something that the FAA should be consulted on. He wrote to President Johnson that there existed ". . . no point of responsibility below the President capable of taking an evenhanded, comprehensive, authoritarian approach to the development of transportation policies . . ." and that no means existed ". . . to ensure reasonable coordination and balance among the various transportation programs of the government."

Others in the Johnson administration also saw the need for unification of transportation activities, legislation, and oversight. At the urgings of Joseph A. Califano, Jr., special assistant to the president, and Charles Schultze, director of the Bureau of the Budget, a special Task Force was created to explore the wisdom and feasibility of creating such a cabinet-level department. In October 1965, Alan S. Boyd, then Undersecretary of commerce for transportation and who had been appointed to head the Task Force, forwarded to the President recommendations that included the creation of a Department of Transportation. The Task Force report further recommended that all separate subagencies that dealt with transportation matters be included in the proposed department. Representative of these were the Federal Aviation Agency, the Bureau of Public Roads, the Saint Lawrence Seaway Development Corporation, the Interstate Commerce Commission, the Civil Aeronautics Board, and the Panama Canal Administration.

Legislation was forwarded to Congress on March 6, 1966, with a letter from Johnson in which he stated: "America today lacks a coordinated transportation system that permits travelers and goods to move conveniently and efficiently from one means of transportation to another, using the best characteristics of each." The thrust of the proposed legislation sought to create one venue for the coordination and management of government-funded transportation programs, and to provide a center for the development of a national transportation policy and its administration.

Debate on the bill was lively, given that many bureaucrats, with their supporters in Congress, had long staked out their turf with respect to their own agencies and authority. The maritime industry opposed the bill, and some in the Federal Aviation Agency voiced fears that its newly won independent status (by the Federal Aviation Act of 1958) would be lost. Nevertheless, by October 1966, a compromise had been reached, and President Johnson signed the bill into law.

The Department of Transportation began operations on April 1, 1967, becoming the fourth largest cabinet-level department within the United States government. It combined over thirty transportation agencies and functions, and their employees, who numbered some ninety-five thousand. During the organizational phase of setting up the DOT were born the Federal Aviation Administration, the Federal Highway Administration, and the Federal Railroad Administration. DOT absorbed functions that previously belonged to departments other than Commerce. Urban mass transit, for example, was removed from the Department of Housing and Urban Development, which in turn caused the creation of additional agencies (the Urban Mass Transportation Administration, later renamed the Federal Transit Administration). The National Transportation Safety Board (NTSB) was created, which assumed the investigative responsibilities formerly carried out by the CAB's Bureau of Aviation Safety. The administration of aviation was placed in the new department and named the Federal Aviation Administration.

The Federal Aviation Administration

When the Department of Transportation Act created the Federal Aviation Administration (FAA), the function of the government in promoting, regulating, and enforcing aviation safety standards finally found a permanent home. A quick review of the history of the administration of aviation safety shows the torturous path that it had taken.

The Air Commerce Act of 1926 first authorized safety regulation, the administration of which was placed within the Department of Commerce. The Aeronautics Branch was created as an agency in the Department of Commerce and became the first government agency to concern itself with aviation safety. This agency was renamed the Bureau of Air Commerce in 1934. Under the Civil Aeronautics Act of 1938 (as amended in 1940), these functions were transferred to the Civil Aeronautics Administration (CAA) and remained within the Department of Commerce.

The Federal Aviation Act of 1958 significantly reallocated existing authority in aviation regulatory matters. The CAA was renamed the Federal Aviation Agency, removed from the Department of Commerce, and organized as an independent agency that reported only to Congress and to the President. The Federal Aviation Agency was given the responsibility previously exercised by the CAB for proposing air safety legislation (statutory) and for rule making, designated under the CAB as Civil Aeronautic Rules (CARs), and now known as the Federal Aviation Regulations (FARs). All air safety research and development authority was consolidated within the Agency, including that previously carried out by the National Advisory Committee for Aeronautics, the Airways Modernization Board and the Air Coordinating Committee. The procedural responsibility in airman certificate actions was also transferred from the CAB to the Federal Aviation Agency. Under the Federal Aviation Act of 1958, the CAB retained its responsibility for the investigation of aircraft accidents as well as its economic regulation of the airlines, and it became an appeals review board for certificate action taken by the Federal Aviation Agency.

Under the provisions of the Department of Transportation Act, responsibility for aviation safety, and virtually all logical ramifications of safety issues, were placed within the authority of the FAA. Its basic mission is defined by its legislative mandate, particularly the Federal Aviation Act of 1958. In 1984, Congress authorized commercial space launches by the private (non-governmental) sector for the first time under the Commercial Space Launch Act. Regulatory authority was initially placed within the Department of Transportation in the Office of Commercial Space Transportation (AST), but in 1995 this function was moved over to the FAA under the same name (Office of Space Transportation (AST)). This office conducts the only space-related function within the FAA. FAA/AST regulates the commercial space transportation industry to ensure compliance with international obligations of the U.S. and to enhance safety and national security. It also licenses commercial space launches of both orbital and suborbital rockets and nonfederal launch sites, or spaceports.

The key functions of the FAA are outlined in the box opposite. A review of these functions will disclose how pervasive is the scope of FAA responsibility and authority. While safety has always been the mainstay of the FAA mandate, ongoing developments in aviation have caused new emphasis to be placed on related but separate concerns, such as security,[1] the environment, airport funding, international relations, and commercial space activities.

The functions of the FAA could logically be examined from several different perspectives, but for our purposes the following breakout of FAA responsibility should be the most instructive.

Key Functions of the FAA

1. Regulate and encourage aviation safety and security.
2. Develop, operate, and maintain a safe, secure, and efficient national air traffic management system.
3. Collaborate in developing a safe, secure, and efficient worldwide civil aviation system.
4. Regulate air commerce to fulfill the requirements of national defense.
5. Assist in the development of airports.
6. Help mitigate adverse environmental impacts of aviation.
7. Regulate the commercial space transportation industry.

Regulation

The FAA, as a government agency with rule-making authority, is required to follow certain procedures when originating or altering regulations that it issues. All federal agencies are required to issue Notices of Proposed Rule Making, published in the Federal Register, which are designed to allow those who may be affected by the proposed rule to be put on notice that a new regulation may be coming, and to allow input to the FAA on the impact of the proposed rule. Input from the aviation industry, or others who may be affected by the rule, often causes modification or abandonment of the proposed rule, and is an important and practical aspect of the regulatory function of the FAA. Industry groups, such as the Airline Transport Association, Aircraft Owners and Pilots Association, National Business Aviation Association, and National Business Aircraft Association, closely monitor the FAA for these Notices of Proposed Rule Making.

Regulations adopted by the FAA are published in the Code of Federal Regulations, Title 14, and referred to in the aviation community as Federal Aviation Regulations (FARs). These regulations have the force of law and are primarily concerned with safety, although environmental (noise) and funding issues are also addressed. The FAA is required by the provisions of the Aircraft Noise Abatement Act of 1968 to consult with the Environmental Protection Agency (EPA) to establish noise standards and to enforce those standards by regulation. In addition to the FARs, the FAA issues mandatory orders that have the force of law in the form of Airworthiness Directives (ADs). These directives generally require inspections or modifications to aircraft that are already certified and in use in the aviation community, and may be prompted by accidents, operating experience, or observations of pilots and mechanics.

Certification

The FAA enhances the safe operation of aviation by controlling, through certification, who may legally function in civil aviation, and by certification of certain equipment used in domestic civil aviation. Certification by the FAA applies to seven major categories:

1. Airmen
2. Aircraft
3. Air Carriers
4. Air Navigation Facilities
5. Air Agencies
6. Airports
7. Designees (representatives of the Administrator)

Airmen

Certification is required of pilots, flight engineers, navigators, air traffic controllers, aircraft dispatchers, mechanics, repairmen, and parachute riggers. Certification of airmen includes procedures not only for the written and oral testing of applicants, but also the requirement of practical demonstrations of required levels of proficiency.

Minimum physical and mental health standards are applied through periodic medical examinations of airmen. The FAA issues separate medical certificates to airmen through its network of Aviation Medical Examiners. The amorphous standard that airmen possess "good moral character" has been consistently required of all certificate holders since passage of the Air Commerce Act of 1926.

Aircraft

The FAA issues three types of certificates applicable to aircraft and their components—Type, Production, and Airworthiness. The aircraft components that must be certified include aircraft engines, propellers, and appliances. Every civil aircraft manufactured in the United States is subject to this inspection and certification regimen beginning with the design of the aircraft. A Type Certificate is issued to the aircraft manufacturer after flight and static testing confirms that the design conforms to the standards adopted by the FAA and published in the FARs.

The issuance of a Production Type Certificate follows on the manufacturer meeting all FAA standards designed to assure that all aircraft produced pursuant to the Type Certificate will faithfully conform to the approved design of the aircraft. This certificate is a sort of quality assurance requirement based on the manufacturer's production and inspection methods at its plant. The final requirement imposed by the FAA is the Airworthiness Certificate, which is awarded to each and every aircraft that comes off the assembly line and is required before the aircraft can be delivered to a purchaser. This certificate is valid only for a period of twelve months.

Air Carriers

There are two types of certificates issued by the federal government to air carriers in the United States. Air Carrier Operating Certificates are issued by the FAA under Part 121 of the FARs to carriers operating aircraft for hire with ten or more seats. The compliance requirements of Part 121 are numerous, but the two major areas addressed are the training of flight crews and aircraft maintenance programs. Air Carrier Fitness Certificates, also referred to as certificates of public convenience and necessity, are issued by DOT, not the FAA, and establish that the carrier has shown that it has the financial capacity and management expertise to carry on a scheduled airline operation. These are required only of carriers operating aircraft with 61 or more seats. While some Fitness Certificates authorize only cargo operations, the majority of such certificates apply to both passenger and cargo.

Air Navigation Facilities

Air Navigation Facilities include radio directional equipment and landing aids and are inspected and rated by the FAA to ensure compliance with safety operational standards. A certificate issued by the FAA as to such facility establishes such compliance.

Air Agencies

FAA regulations applying to aircraft repair stations and to maintenance technician facilities, pilot schools, and training centers include the requirement of meeting certifications standards relating to procedures, instructors, equipment, tools, and personnel.

Airports

Since 1982, the FAA has been charged with certificating operators of airports serving certificated air carriers (Airport and Airway Development Act). FAA regulations containing the requirements applicable to such airport operators are found in Part 139 and relate primarily to maintenance of minimum safety standards in airport operations.

Designees

The FAA designates individuals possessing certain skills, training, or education to assist it in carrying out its examination and inspection duties and through whom FAA certificates are issued. These individuals must first be certified, themselves, as representatives of the administrator and must hold that certificate issued by the FAA. These individuals include Aviation Medical Examiners who are issued a Certificate of Designation, and Pilot Examiners (Flight Standards Designated Examiner) who are issued a Certificate of Authority. In addition, Certificates of Authority are issued to Technical Personnel Examiners, Designated Aircraft Maintenance Inspectors, Designated Engineering Representatives, Designated Manufacturing Inspection Representatives, and Designated Airworthiness Representatives.

Investigation

The FAA conducts investigations of aircraft accidents subordinate to and in cooperation with the National Transportation Safety Board pursuant to an arrangement known as the Accident Investigation Selectivity Program. This program is formalized in an agreement between the NTSB and

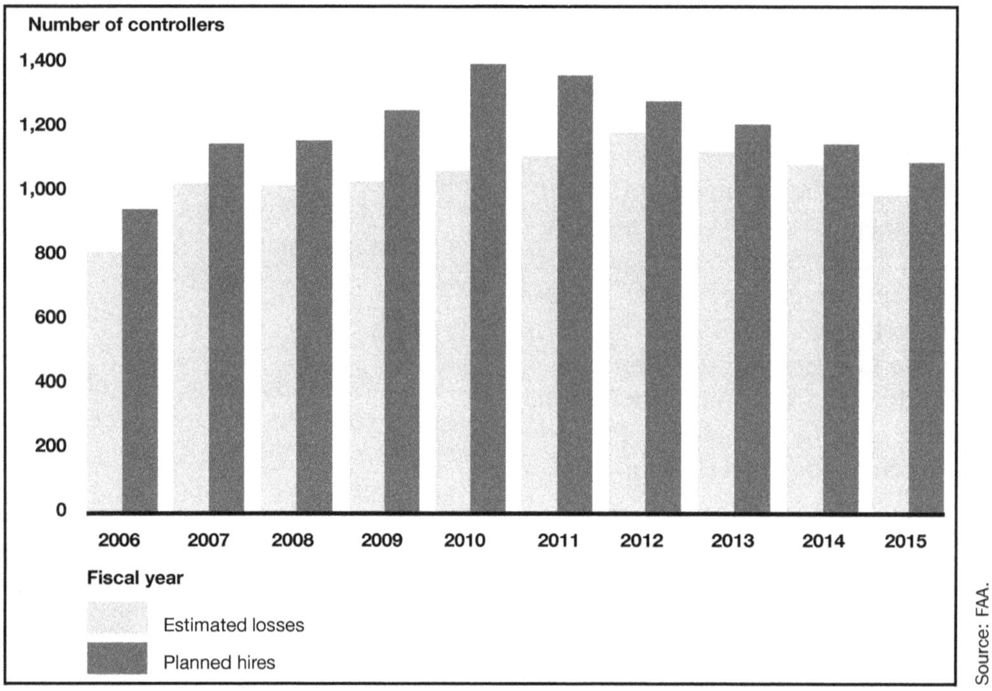

FIGURE 8-1 Estimated controller losses and planned hires, fiscal years 2006–2015.

the FAA and is designed to delineate responsibility in accident investigations and to avoid conflicts. Previously, separate investigations conducted by the NTSB and the FAA sometimes resulted in contrary findings and conclusions, and were the occasion for embarrassment to one or both.

The types of accidents investigated by the FAA are normally limited to general aviation accidents or those that, by comparison to airline accidents, are relatively limited in scope or impact in the aviation community. It should be noted that the objectives of an FAA investigation are different from those of the NTSB. In particular, the FAA is looking for violations of the FARs, and it scrutinizes whether the accident was a result of deviations from standards adopted by the FAA. FAA investigations seek to determine whether FAA facilities were a factor and whether the FARs were adequate. The FAA also investigates aircraft incidents that do not result in accidents, such as "near misses" by passing aircraft or other instances when aviation safety may have been jeopardized. The FAA is also mandated by Congress to investigate all reports of violations of the FARs.

Enforcement

Since passage of the Federal Aviation Act of 1958, responsibility for carrying out enforcement procedures for violations of the FARs has resided with the FAA. Enforcement options open to the FAA in any given case are normally dictated by considerations already well-established within the agency, and are generally handled either as administrative dispositions (warning notices and letters of correction) or by certificate action (suspension or revocation). Occasionally, civil penalties are assessed in lieu of certificate action (historically against corporations or against working pilots where certificate suspension is deemed too harsh).

Operations

The FAA is charged with the operation and maintenance of a vast array of facilities and equipment within the aviation system. We will briefly review the major categories of FAA responsibility.

Air Traffic Control

The ATC system includes airport control towers, air route traffic control centers (ARTCC), terminal radar approach control (TRACON), and flight service stations (FSS).

The FAA estimates that it will lose 10,291 controllers, or about 70 percent of the controller workforce, between 2006 and 2015 due to retirements. The large percentage loss is due to the unlawful PATCO strike in 1981, when President Reagan fired almost 11,000 controllers. From 1982 through 1991, the FAA hired an average of 2,655 controllers each year. These controllers will become eligible for retirement during the next decade.

In 1982, the FAA began a program of outsourcing operation at a limited number of VFR towers. As of 2006, 231 towers in 46 states participate in the FAA's Contract Tower Program.

In 2005, the FAA entered into a contract with Lockheed Martin to operate the 58 Flight Service Stations located in the contiguous United States.

Radio Aids to Navigation

These facilities include VORs, VORTACs, instrument landing systems (ILS), and microwave landing systems (MLS). The Global Positioning System (GPS) is operated by the Department of Defense, and Loran C is operated by the United States Coast Guard.

In 2003, the FAA inaugurated the Wide Area Augmentation System (WAAS) as a precursor for a new and extremely accurate navigation system. WAAS augments, or enhances, the Global Positioning System in order to provide the additional accuracy, integrity, and availability necessary for its use by the civilian aviation community. Previously, GPS data were unable to provide navigation capability for use in precision approaches. Through WAAS, precision approaches are conceivable for all 5,400 public use airports in the United States without local airport ground support facilities.

WAAS is an integral part of the FAA plan to replace ground-based Navaids entirely with satellite-based navigation capability, thus eliminating VORs, VORTACs, ILS, and MLS.

National Airports

The FAA no longer is responsible for the two major airports located in and near Washington, D.C., Reagan National and Dulles, since their operation has been assigned to the Washington Metropolitan Airport Authority.

Monroney Aeronautical Center

The Center was named for Oklahoma Senator Mike Monroney, who was instrumental in securing passage of the Federal Aviation Act of 1958. The Center is the repository for all aircraft registration, documents of title to aircraft, and lien recordations on United States aircraft (the Aircraft Registry). The Airman Records Branch contains the records pertaining to every person issued a certificate by FAA. It is the home of the FAA Academy, the training center for air traffic controllers, air safety inspectors, and other personnel. The Center also houses the Civil Aeromedical Institute (CAMI), which conducts research on various aspects of aviation safety, with an emphasis on human factors. CAMI specialists conduct tests on smoke toxicity, aircraft seats and restraint systems, air traffic controller selection and training methods, and the effects of fatigue, age, work, and rest schedules for ATC personnel. Teaching activities at CAMI include the training of pilots in water and arctic survival techniques and the sharing of the latest research in aviation medicine with designated Aviation Medical Examiners.

William J. Hughes Technical Center

Research and development programs are conducted at the technical center located just outside Atlantic City, N.J., on a former Navy airfield. Activities conducted at the Center include test and evaluation in air traffic control, communications, navigation, airports, and aircraft safety and security. The Center strives to develop innovative systems and concepts, new equipment and software, and modifications of existing systems.

Education

The FAA supports a large effort in the aviation community directed toward education of the flying public and the public at large. Periodic publications, such as the Advisory Circulars and Service Bulletins, and safety seminars for pilots, instructors, mechanics, and others reach out to all certificated airmen in an effort to facilitate improvements in all aspects of aviation safety.

Funding

Responsibility for distribution of federal grants under the Airport Improvement Program is assumed by the FAA under the Airport and Airway Development Act.

Registration and Recordation

A central registry for U.S. civil aircraft (N-numbered aircraft) is located at the Aeronautical Center in Oklahoma City. All aircraft operated within the United States are required to be registered, and it is the responsibility of the owner of every aircraft to secure the registration. The FAA issues its Certificate of Registration in the name of the owner upon satisfactory completion of the registration process. The Aircraft Registry also functions as the recordation site for establishing or determining legal title to aircraft, and is the one place that contains the entire chain of title of any aircraft. All legal encumbrances, or perfected security interests in aircraft, must be filed with the FAA at the Aircraft Registry.

Commercial Space Transportation

The responsibilities of the FAA discussed above grew and were assumed over time as civilian aviation sector activities developed. All of these responsibilities relate to civil aviation operations occurring on the surface of the earth and within the earth's atmosphere. When the United States began operations beyond the earth's atmosphere with the first space launch in 1958, and for many years thereafter, all U.S. space activities were the exclusive province of either NASA or the military.

With the passage by Congress of the Commercial Space Launch Act of 1985, the Office of Commercial Space Transportation (referred to as FAA/AST) was created within the FAA. Under this statute, AST has the responsibility to:

- Regulate the commercial space transportation industry, only to the extent necessary to ensure compliance with international obligations of the United States and to protect the public health and safety, safety of property, and national security and foreign policy interests of the United States;
- Encourage, facilitate, and promote commercial space launches by the private sector;
- Recommend appropriate changes in federal statutes, treaties, regulations, policies, plans, and procedures;
- Facilitate the strengthening and expansion of the United States space transportation infrastructure

FAA/AST is organized into three divisions:

- Space Systems Development Division (AST-100)
- Licensing and Safety Division (AST-200)
- Systems Engineering and Training Division (AST-300)

For a more thorough discussion of commercial space launch activities in the United States and the role of FAA/AST.

Dual Mandate

The historic mission of the FAA has been to not only administer the requirements of safety in the aviation community, but to "promote" aviation in the overall national transportation scheme. The FAA has come under criticism from time to time that this dual role really amounts to a conflict of interest in promoting the airlines, on the one hand,

and enforcing its regulations applicable to them, on the other hand. The issue resurfaced in the high visibility aftermath of the Valuejet crash in the Florida Everglades in 1996. The FAA had determined that the discount carrier was not in significant violation of the FARs, that FAA oversight and inspection of the airline had been standard, and reported its conclusion that the airline was "safe." Within six weeks, the FAA had shut down the company based on additional findings of serious violations of regulations relating to the transportation of hazardous materials, which led to the conclusion that such violations were the direct cause of the catastrophic crash, with the loss of all lives on board.

In 1996, Congress revised the FAA's mission in the Federal Aviation Reauthorization Act, removing the "dual mandate" by repealing the duty of the administrator to "promote civil aeronautics." Instead, Congress directed the FAA to consider as its highest priority the "maintaining and enhancing of safety and security of air commerce."

The National Transportation Safety Board

The National Transportation Safety Board (NTSB) was created by the Department of Transportation Act of 1966 as an agency within DOT. The primary responsibilities given to NTSB were to investigate transportation accidents, to determine the "probable cause" of the accident, and to make recommendations based on its findings designed to assist in preventing similar accidents in the future. The range of transportation modes subject to the scrutiny of the NTSB was commensurate with the DOT itself, that is, railroad, highway, aviation, marine, and pipeline. It was assigned the additional role of acting as a review board for airman appeals from certificate actions or penalty assessments by the FAA.

In 1974, the NTSB was removed from the DOT and established as an independent agency answerable to Congress pursuant to the provisions

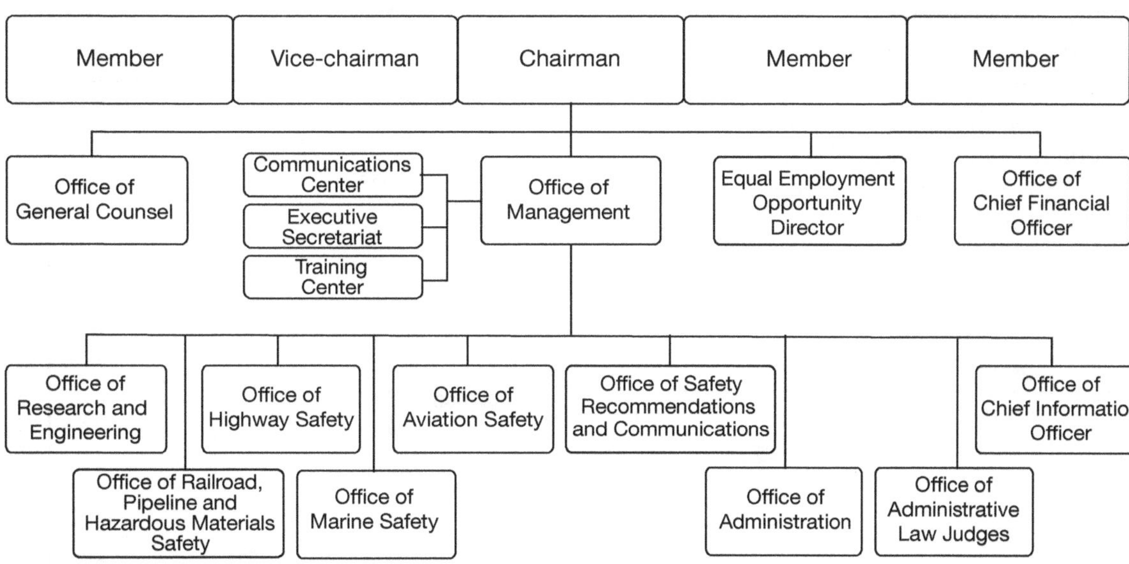

The agency is headquartered in Washington, D.C., and maintains 10 field offices nationwide and a training center in Ashburn, Virginia, in suburban Washington, D.C. In recent years, the agency has shrunk in size. In 2003, NTSB had 438 full-time employees compared with 386 in September 2006. During the same period, the number of full-time investigators and technical staff decreased from 234 to 203. (See fig. 3) NTSB's modal offices vary in size in relation to the number of investigators; as of September 2006, the aviation office had 102 investigators and technical staff; the rail, pipeline, and hazardous materials office had 31; the highway office had 22; and the marine office had 12 employees. An additional 36 technical staff worked in the Office of Research and Engineering, which provides technical, laboratory, analytical, and engineering support for the modal investigation offices. For example, it is responsible for interpreting data recorders, creating accident computer simulations, and publishing general safety studies.

Source: NTSB.

FIGURE 8-2 NTSB's Organization.

of the Independent Safety Board Act. This action was taken by Congress because it was determined that, given its unique role in investigation and recommendation, the agency should be completely independent of other agencies and departments to ensure that it could be direct, impartial, and uninfluenced in making assessments of fault and recommendations for changes.

The investigative role of the NTSB dates back to the Air Commerce Act of 1926 when Congress gave the Department of Commerce responsibility for investigation of air crashes. An Aeronautics Branch of the Commerce Department was created to carry out this responsibility and it did so until renamed the Bureau of Air Commerce in 1934. In 1938, the CAB took over the investigative role and performed this duty until the creation of the NTSB in 1966.

To facilitate its investigative and reporting responsibility, Part 830 of the FARs requires aircraft operators to provide notification to the NTSB of certain accidents and incidents, and in certain cases to follow up such notice by required reports. This notification and reporting regimen is important to the role of the NTSB is staying current with problem areas in aviation safety.

The role of the NTSB was extended to the investigation of nonmilitary public aircraft accidents under the provisions of the Independent Safety Board Act of 1994. Public aircraft, generally those aircraft owned or operated by various federal government agencies, were excluded from compliance with the airworthiness and maintenance requirements of the FARs by the Federal Aviation Act of 1958. In 1993, the Governor of South Dakota, George S. Mickelson, was killed in the crash of a government aircraft in which he was a passenger. In the wake of the investigation into that accident, Congress rewrote the law to bring most nonmilitary government-owned or operated aircraft within the authority of the FAA and the NTSB.

The Primary Functions of the NTSB

1. Aircraft accident investigation
2. Probable cause determinations
3. Accident report preparation
4. Safety recommendations
5. Receipt of Part 830 notifications and reports
6. Appellate review of FAA certificate actions and penalty assessments
7. Disaster assistance

Mode	2002	2003	2004	2005
Aviation[a]	1,949	1,997	1,870	1,937
Highway	52	45	45	33
Rail	11	9	12	8
Pipeline	1	2	2	1
Hazardous materials	2	1	2	1
Marine	6	6	7	4

[a]Aviation accidents include limited investigations in which NTSB delegates the gathering of on-scene information to FAA inspectors.

FIGURE 8-3 Number of accident investigations completed by NTSB by mode, fiscal years 2002–2005.
Source: GAO analysis of NTSB data.

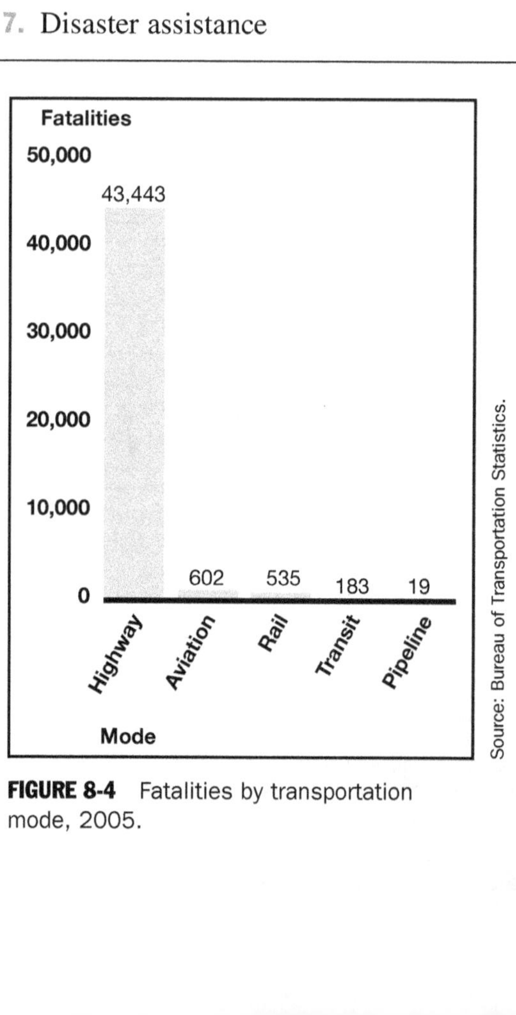

FIGURE 8-4 Fatalities by transportation mode, 2005.

In 1996, Congress further charged the NTSB, pursuant to the provisions of the Aviation Disaster Family Assistance Act, with the task of coordinating all federal assistance to survivors and families of victims of catastrophic transportation accidents. The NTSB strategic plan developed as a result of this mandate ensures that such people receive timely assistance from the carrier involved, and from all government agencies and community service organizations included in the program.

The primary functions of the NTSB are outlined in box shown on the previous page.

It should be noted that the NTSB has no authority over any other federal agency or any industry group. It has no regulatory or enforcement powers. Its effectiveness is enhanced by its resultant impartiality. The NTSB operates with a very small staff, historically fewer than 400 employees. Since 1967, the NTSB has investigated over 114,000 aviation accidents and issued more than 11,600 safety recommendations in all transportation modes. More than 80 percent of its safety recommendations have been adopted by those empowered to effect changes in the transportation system and in government agencies. At a cost of less than $.24 annually per citizen, it is said to be one of the best bargains in government.

Endnote

1. In 2001, responsibility for aviation security was transferred from the FAA to the Transportation Security Administration (TSA).

Prelude to Deregulation

The Airline Deregulation Act of 1978 (ADA) was signed into law by President Jimmy Carter on October 24, 1978, bringing to a close 40 years of government oversight, control, and protection of the airlines of the United States. Only in the fields of aviation safety and international relations would the government henceforth concern itself with the "public interest" implications of the nation's airlines and their operations. The ADA was a departure from established government tradition dating back almost 100 years to the Interstate Commerce Act of 1887, the first regulation of the railroads, and the beginning of a governmental philosophy of broad regulation of the transportation industry.

The Economic Nature of Transportation

The nature of transportation is such that, unlike most consumer services, it is an intermediate product, only a means to an end. Hardly anyone flies in the airline system just to go for an airplane ride. The reason people fly is to accomplish another goal, whether it is for a business purpose or a personal one.

This characteristic of the transportation industry likens it to a commodity, for example, wheat or petroleum. Wheat is not purchased because anyone wants a bushel of wheat, but to create an end product, perhaps a cake or a loaf of bread. Oil has no use except to facilitate a secondary purpose, like the lubrication of machines or as a means of propulsion.

There are no unique characteristics within a class of commodity; one bushel of wheat is like any other. One quart of oil is indistinguishable, and worth no more, than any other quart of oil. Economic theory teaches that the price of a commodity will seek the lowest possible level based on supply and demand. If the supply of a commodity is adequate to the demand, unit profit on any given quantity of a commodity will be very small.

If an airline seat is like a bushel of wheat, its price will be valued like any other airline seat absent some distinguishing characteristic, assuming an adequate supply of airline seats. If the market in any commodity is cornered, so that the supply can be controlled, the price of that commodity can also be controlled, not unlike the OPEC states' control of a large segment of the world supply of oil.

Under regulation of the airline industry, the Civil Aeronautics Board controlled both the supply and the price of airline seats. With deregulation that control was to end, and the supply and price of airline seats was to be opened to the free market. Under economic theory, a deregulated or free market air carrier industry should make airline seats adhere to commodity supply and demand principles, and to reduced prices. In a free market economy within the air carrier industry, the risk still exists that control of supply (the availability of seats) may be manipulated by a cornering of the market through economic power or anticompetitive practices. We will review these concepts later in this Part in connection with factors that limit competition.

The Nature of Regulated Transportation

Since early in their history airlines had been considered by government to be, to some degree, instruments of national policy.

Airlines were used:

- To carry the mail
- To facilitate commerce between the cities and states
- To establish a fast and efficient passenger transportation system
- To function in the national defense system in times of emergency
- To carry the flag internationally

To carry out these functions, the airlines had to be financially stable, and they had to be dependable in the long run and reliable at all times.

For 40 years, entry into the airline business had been controlled by the Civil Aeronautics Board. No major new entrants had been permitted, although some smaller feeder lines were allowed to enter the market on a limited basis. The number of trunk lines had decreased from sixteen in 1938 to ten in 1978. That year the five largest airlines accounted for two-thirds of all domestic revenue. Competition was seen as good by the CAB, but not destructive competition, so routes were awarded among the existing carriers based on the perceived needs of the communities and cities requiring service, and on the equitable allocation of routes to the airlines desiring and capable of delivering such service. Fares and rates were established mainly as a function of the airlines' cost of doing business, a system not particularly designed or administered to be cost efficient. But it was stable. Not one airline in 40 years had been allowed to go into bankruptcy.

Competition in the airline industry had been regulated, but it had not been eliminated. The more efficient the airline, the better its operating ratio and the more money there was at the bottom-line. The effect of airline economic regulation was to create both a financial ceiling and floor for the companies, guaranteeing that neither profits nor losses were excessive. While the support of the CAB limiting airline loss was comforting to management, the ceiling limiting innovation and profit was frustrating, particularly to the types of men who rose to run the airlines. The technological advance in aircraft and engine design kept airlines busy trying to stay ahead of one another in order to have the most appealing fleet available for the limited passenger market, most of which were business travelers. Aesthetics and service were high on the list of concerns, since these were two of the few discretionary operating decisions available to management. Marketing schemes were also highly competitive, the effort being limited to the best way to sell essentially the same product that every other airline sold. Each new innovation thought up by an airline, no matter how minor, was pushed as the reason to fly that airline. For instance, the introduction in 1965 of in-flight movies by TWA was highly advertised; it resulted in six to eight more passengers per flight. Airline management continually strived to bring some quality of uniqueness to their operations.

Until the 1960s and early 1970s, leadership of the major airlines had remained mostly in the hands of the young men who took over in 1934 after the Black investigation of the so-called "spoils conference" scandal. Newcomers like Robert Crandall at American and Richard Ferris at United were natural competitors and often went head to head on issues like market share, computer reservation system development, and travel agent loyalty. Crandall preferred to compete within the established regulatory framework, while Ferris, who came from the hotel industry, was open-minded about deregulation.

The market for cheaper air travel had been recognized since shortly after World War II, when charter operators began flying, using war surplus

DC-3s and later DC-4s. But CAB regulations governing charter operations were severely restrictive. Charterers could not sell individual tickets, nor could they fly published schedules. They were essentially relegated to selling the entire aircraft capacity to large, established groups by advance sales. The CAB monitored these operations carefully; ever watchful lest charter operations encroach on the CAB controlled scheduled airlines' established routes. The charter operators showed that a profit could be made with low fares, sometimes as much as fifty percent lower than CAB mandated fares, so long as the aircraft flew filled with passengers.

Charter operations failed to maintain market-share in a hostile environment controlled by the CAB, which often allowed the scheduled lines to meet the charter fares on a case-by-case basis. The trunk carriers in such cases would match the charter fares and, after the demise of the particular charter operation, the higher scheduled fares would be reinstated. Still, charter operators persisted and made money in specialty markets, particularly after the CAB began loosening the regulations applicable to charter operations in the 1970s.

Factors that contributed to deregulation of the airlines arose from several quarters, but came together at about the same time. By the 1970s, the railroad industry had been heavily regulated by the Interstate Commerce Commission for over eighty years, and the airline industry had been regulated by the Civil Aeronautics Board for almost forty years. The railroads were in serious financial trouble under regulation. Mergers entered into to stave off financial collapse, like the Pennsylvania Railroad and the New York Central, only succeeded in delaying the inevitable as the Penn Central entered bankruptcy in the 1970s. Railroad passenger service could no longer be sustained in the private sector, as Amtrak, subsidized by the government, was required to take over that service.

In the 1970s, the adversities realized by the railroad industry were perceived as a harbinger for the airline industry. It also seemed that the government was not trusted, nor was it respected as before, perhaps due to public disgust and unease created by the divisive issues of the 1960s and 1970s. It hardly mattered from which end of the political spectrum one viewed the situation. On the left, the Vietnam War and Watergate were examples of inept or corrupt leadership. On the right, the social experimentation of The Great Society programs of Lyndon Johnson, the rapid deterioration of inner cities, and civil disturbances seen nightly on the evening news were evidence of misguided governmental policy. Government seemed to be contributing to the problem, rather than offering rational solutions.

The anti-regulatory mood was also intellectual. Alfred Kahn, a professor of economics at Cornell University, produced a two-volume work, *The Economics of Regulation,* which basically postulated that the heavy hand of government regulation was inimical to the public interest, and that competition would naturally produce the best product for the best price for the public. Kahn had also become acquainted with a 1965 law review article in the *Yale Law Journal* written by Michael Levine, who as a law student had looked into the intrastate airline business in California. In comparing the intrastate airline fares with those mandated by the CAB, he concluded that CAB policies "fostered unnecessarily high fares, encouraged uneconomic practices, and limited the variety of services available to the public."

"I really don't know one plane from the other. To me they are just marginal costs with wings."

Alfred Kahn, 1977

The anti-regulatory mood was not only emotional and intellectual, but also political. In 1974, Sen. Edward Kennedy had aspirations concerning the office of Presidency of the United States, and he was searching for an issue. Kennedy had on his staff a Harvard law professor, Stephen Breyer, in residence in Washington for a summer sabbatical. Breyer, who was to become an Associate Justice of the United States Supreme Court twenty years later on, was apparently genuinely concerned about what he saw within the CAB as an accepted pattern of price fixing of fares, to the detriment of the consumer. Politically, Kennedy could expose the price-fixing pattern of the CAB as a pro-consumer advocate, and he could propose the elimination of the agency on the basis of a fiscal conservative. Hearings on the matter began in February 1975.

Dissatisfaction with the regulatory scheme at the CAB came even from within the agency itself. Careerists at the agency seemed to have run out of tolerance for much of the silliness associated with the matters that were required to be decided by them, from the price of drinks on airplanes to special fares for skiers. In 1975 a staff study within the CAB observed that restrictions imposed on the airline industry were not justified, and, heretically, that the industry was not even monopolistic, but naturally competitive.[1]

In 1975, an individual by the name of John Robson took over as chairman of the CAB under President Gerald Ford. He had been an undersecretary at the Department of Transportation, a career Washington bureaucrat, and he knew little about the airline industry. But he soon came to understand the mood of the place. He began the relaxation of the myriad regulations that applied to charter operations, allowing the charter operators to fly more routes, longer distances, and with fewer restrictions, thus creating immediate competition for the scheduled airlines. This, in turn, caused the scheduled airlines to make application to the CAB for permission to make a legitimate, long-range competitive response to the charterers. The CAB then began granting those applications, thereby setting up an incipient competition theretofore unknown under regulation. This would come to be known as "de facto" deregulation.

The idea of deregulation, a concept truly iconoclastic in the annals of government, was beginning to catch on. When Jimmy Carter took office as President in 1976, he was a political newcomer. Sen. Kennedy still loomed on the horizon as likely competition to him for the hearts and minds of Democrats, if not a large segment of the entire voting population itself. Carter got behind the deregulation issue and supported it. He advised that Congress pass the legislation in order to "reduce Federal regulation of the domestic commercial airline industry" in the name of a less burdensome federal government. Carter had earlier appointed professor Alfred Kahn head of the CAB. Kahn not only continued the course begun by John Robson before him, he actively applied his deregulation philosophy to the matters coming before the Board.

The CAB had no jurisdiction over intrastate carriers like Southwest Airlines, and although Southwest could charge what it pleased subject only to the rules of the Texas Public Utilities Commission, its fares were substantially less than those mandated by the CAB regimen. Southwest was stiff competition for any airline flying within the very expansive borders of Texas, and that competition included interstate carriers Texas International and American Airlines. When Texas International sought CAB authority for its "peanut fares," (its regular, CAB mandated fare discounted fifty percent) in order to compete with Southwest, the CAB obliged. Again, when American Airlines wanted to institute its "Supersaver" fare in March 1977, a flat out charter-like discount theretofore prohibited by CAB philosophy, the CAB approved. Significantly, the "Supersaver" fares applied to seats on regularly scheduled flights on which standard fare passengers had purchased tickets. Thus began the confusing and seemingly inequitable pattern of full-fare passengers seated beside someone who had paid a fraction of full fare. These fares also invited into the cabin leisure passengers, bringing with them their small children and babies, to occupy the center seat previously left unfilled. American Airline's coast-to-coast traffic soon increased by 61 percent. Nobody seemed to notice, but deregulation, for better or worse, had already begun.

De facto deregulation was limited to a revision of the "business as usual" regulation imposed by the CAB in only three major areas:

1. The loosening of the restrictions on charter operations
2. Approval of Texas International's "peanut fares" as a competitive response to Southwest and charter operations
3. Approval of American Airlines "Supersaver Fares" that allowed the first co-mingling of discounted fares with full fare passengers on the same flight

Still, in a regulated world the safety net was there. Carriers with financial problems could still rely on the CAB to make things right. The airlines were insulated from the Big Bad World of competition and all of the downside that unbridled competition can bring. But all of this was about to change.

The Airline Deregulation Act of 1978

The Airline Deregulation Act amends the Federal Aviation Act of 1958, stating as its purpose "to encourage, develop, and attain an air transportation system which relies on competitive market forces to determine the quality, variety, and price of air services." The Act completely changed the economic foundation for the domestic airline industry. It provided, among other things:

1. for the phase-out of the CAB and its authority over domestic routes and fares,
2. for the phase-out of existing economic regulations formerly constituting barriers to competition,
3. safeguards for the protection of air carrier service to small communities,
4. for the facilitation of entry of air carriers into new markets, and
5. for certain protection of airline employees who may be adversely affected by the results of the Act.

CAB route authority ended on December 31, 1981, and rate authority terminated on December 31, 1983. The CAB itself ceased to exist as of the close of business on December 31, 1984.

Prior law, as interpreted by the Supreme Court, had exempted the airlines from compliance with the antitrust laws governing commercial enterprise in the United States. After deregulation, upon the demise of the CAB, airlines became subject to these laws just the same as other corporations, with enforcement jurisdiction initially residing in the Department of Transportation.

With deregulation, there was no longer any requirement to secure from the CAB certificates of convenience and necessity before commencing service on a route. No longer were there artificial barriers to entry into the previously exclusive airline carrier club, nor was there any requirement to secure approval from the CAB for rate increases.

When deregulation became law, all of the pent-up competitive instincts of airline bosses were suddenly unleashed. Like adolescents let loose on a first unsupervised journey away from home, excesses might have been expected. The choices of how to proceed were practically unlimited. Unbridled optimism coupled with a fear of being left behind in the race to gain position on their competitors spurred frenzied activity of all sorts, and not a few miscalculations.

Braniff—A Case History Under Deregulation

Braniff's conclusion that deregulation would only be temporary, and that re-regulation was inevitable, was probably the biggest mistake of all. That conclusion prompted Braniff to believe that new routes should be established as quickly as possible, before the window of opportunity slammed shut, and that the equipment to serve these new routes should be immediately acquired before the aircraft manufacturers became backlogged with orders from all of the other airlines that were sure to come.

In 1978 Braniff International Airways was a successful, established carrier with a reliable business clientele responsible for about 70 percent of its traffic. When Postmaster General Farley ordered the re-bidding in 1934 for airmail routes after the so-called "Spoils Conference" affair of 1930, Braniff had acquired the coveted Dallas-Chicago route. Braniff was profitable and had been for much of its proud history as one of the 16 trunk carriers grandfathered under the Civil Aeronautics Act of 1938. After World War II, Braniff became the first international competitor to Pan American certificated by the United States government when it began service to South America along its East Coast. In December 1965, Braniff bought the 50 percent interest of W. R. Grace in Panagra, which served the West Coast of South America, and in March 1966 it completed this acquisition by buying out the remaining 50 percent interest from Pan American. Between 1975 and 1980, Braniff doubled in size.

Braniff's management decision after deregulation to rapidly expand its operations led to the purchase of many new aircraft, which in turn resulted in huge debt. The decision required expansion of its infrastructure at high cost to service the new, expanded route structure put in place after deregulation.

Beginning in 1980 and extending into the early years of the decade, fuel prices spiraled upward due to the OPEC oil crisis, interest rates shot up to 20 percent, and the attendant recession had a stifling effect on passenger traffic. When deregulation did not end, and upstart airlines continued to enter the field and pose significant competitive pressures on the Braniff's expanded routes, Braniff began suffering catastrophic losses. In order to maintain cash flow, Braniff began to sell off its newly acquired fleet of aircraft at distressed prices to its competitors, further weakening its position. It then turned to selling off its biggest prizes, its overseas routes, and the handwriting was on the wall.

By 1982, Braniff could no longer keep its doors open against the clamor of creditors, and it filed for protection under the bankruptcy act. It was the first United States airline to do so since regulation was begun in 1938. Ultimately, the assets of the company were liquidated and it ceased to exist.

The United States Bankruptcy Act[2]

The Constitution of the United States (Article 1, Section 8) specifically provides that Congress be empowered to establish "uniform laws on the subject of bankruptcies throughout the United States." Congress has done so on repeated occasions since 1801. Bankruptcy in the United States, therefore, is mainly a federal exercise, administered in the federal bankruptcy courts, which are an adjunct of the United States District Courts located in each state across the land.

The concept of bankruptcy first implies that one's debts exceed his assets. This is called "insolvency." Under the laws of the U.S., a petition in bankruptcy can be initiated either by creditors of the insolvent debtor, called "involuntary bankruptcy," or by the debtor himself, called "voluntary bankruptcy." As we saw in Part I of this book, the industrial revolution, and particularly the advent of the railroads, caused the rise of the corporate form of business entity. Under U.S. law, corporations are entitled to the same basic privileges as individuals, including the protection of the bankruptcy laws.

The bankruptcy code is sub-divided into "Chapters," each one dealing with a separate kind of bankruptcy. The most common form of bankruptcy, known as "straight bankruptcy" is found in Chapter 7 of the Code and results in the shutting down of the business. This procedure provides for the appointment of a Trustee to liquidate all of the debtor's assets and to distribute the proceeds to the creditors. It is a more complex procedure that allows the debtor to remain in business under the supervision of the bankruptcy court while it goes through a "reorganization" of its debt structure and contractual obligations.

The intent of this chapter, in allowing a company to remain in business under reorganization, is to provide a way to pay most if not all of the creditors, to save jobs, to preserve the engine of profitability (which is the corporation's operations in place, good will, experience, and hope of the future), and to allow the business to earn a "fresh start." One trade-off to accomplish this result is the cancellation or renegotiation of previously incurred debts and contracts, including labor contracts. This is accomplished either by compromise between the debtor and the creditors, or by rulings of the bankruptcy judge.

During the reorganization process, which may take months to years depending on the complexities of the reorganization, the debtor is considered "under the protection" of the bankruptcy court. This means that the debtor is shielded from lawsuits that could otherwise be brought by creditors, and from general harassment associated with its unpaid debts. At the same time, the operations of the debtor are subject to the scrutiny of the bankruptcy court and the creditors.

In the following chapters of this book, we will see how bankruptcy has become an integral part of the air transportation business in the deregulated world. Other sophisticated free market techniques, previously unheard of in commercial aviation, would be brought to bear as airlines attempted to cope with the new world of competition. Hostile corporate takeovers, leveraged buyouts, downsizing, outsourcing, and employee pay givebacks and salary cuts were only some of the new developments that loomed over the horizon. Chief practitioner of these ideas was a Harvard MBA by the name of Frank Lorenzo.

The Progression of Deregulation

When Congress passed the Airline Deregulation Act in 1978, there existed no empirical data, nor any experience, to reliably predict what would happen to the U.S. air carrier industry under deregulation. During the three years that Congress held hearings on the matter before the Act was actually passed, scores of proponents and opponents testified. The arguments made for deregulation were largely just that, arguments, and they were premised on anecdotal data. Economic theory was given much credence, but it was still just theory. The effect that deregulation would have on the nation and the airline industry was unknown, and there was little basis or experience from which to project any reliable conclusion.

Airline Fares

A study conducted by the Government Accounting Office (GAO) in the early 1970s concluded that airline fares under the CAB regulated system were anywhere from 22 percent to 50 percent higher than they would be under a deregulated system. The actual results of deregulation on median airline fares is depicted in Figure 9-1, and as of the end of the year 2005, airline fares were almost 40 percent less, in 2005 dollars, than in 1980. Fares have declined the most in long-distance and heavily trafficked markets, and the least in shorter-distance and less-traveled markets.

For many years after deregulation, most of the price benefit went to leisure travelers, who were not constrained to fly on short notice, like the business traveler. Due to computer reservation systems and the yield management programs available to the incumbent airlines, ticket prices were discounted most heavily for those who could purchase early. The airlines were able to charge much higher fares for those who could not plan very far ahead.

In the early years of deregulation, the system of yield management and other competitive

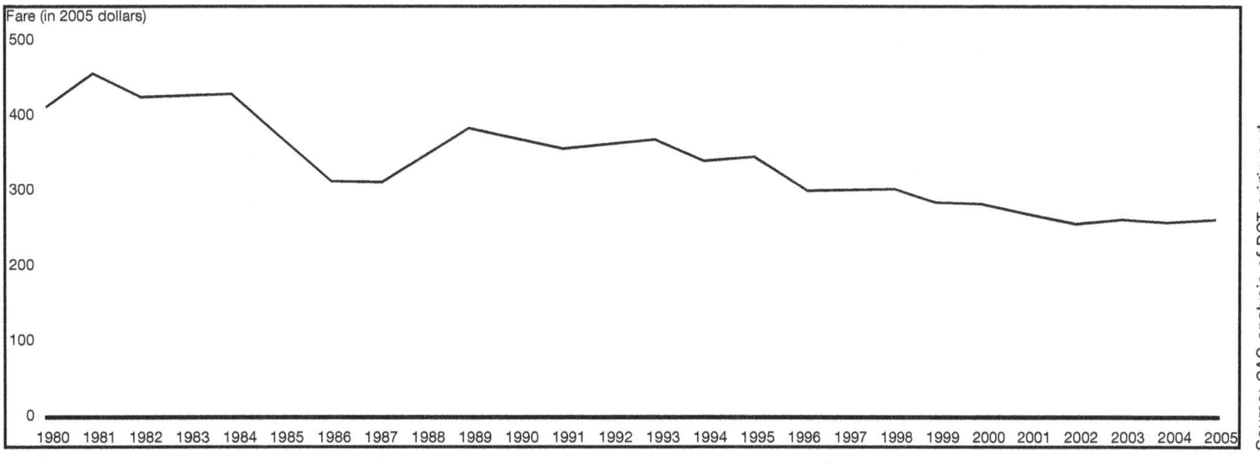

Source: GAO analysis of DOT origin and destination survey data.

FIGURE 9-1 Median fares have declined almost 40 percent since 1980 as measured in 2005 dollars.

pricing schemes resulted in an almost indecipherable pricing regimen, where the passenger in 20A, for example, may find that his fare is twice that of his seatmate in 20B, for no discernible reason. Business fares actually increased after deregulation, at one time by as much as 70 percent over similar fares charged during the CAB era.[3]

Recent evidence, however, discloses that businesses that were punished by the economic downturn beginning in 2000 began to resist such disparate pricing. Corporate travel offices were set up for the express purpose of economizing on costs of air travel. By 2002, airlines began to reduce fares across the board, and an increased corporate use of the Internet caused some tempering of fare disparity. Fares still remain higher at "fortress hubs," where one or two airlines have disproportionate market share, and at some small airports where there is little or no competition.

Passenger Travel

Passenger enplanements since 1978 have increased dramatically.

- In 1978, 275 million people flew on domestic airlines. By 1995, that figure had doubled, to 548 million. In 2000, the number had increased to 693 million. Because of 9-11, passenger travel dropped significantly, to a low of 641 million in 2002, but by the end of 2005, passenger enplanements were almost 780 million.
- Airline departures increased from 5 million in 1978 to 11.5 million in 2005.

Number of Carriers

The number of carriers has increased. In 1938 when the CAB took over regulation of the airlines, there were 16 trunk carriers; by 1978, that number had shrunk to 10 airlines, although small and feeder lines had brought the total number of airlines to 43. In 2000, there were a total of 90 airlines operating in the United States, of which 10 were classified as "major" airlines, a rough equivalent to the trunk airlines of regulation days.

The number of carriers operating in any given year since deregulation has fluctuated greatly. This variation has been referred to as the "ebb and flow" of entrants by former CAB chairman John E. Robson. A variety of reasons has been given for new entrant failures: inexperienced management, unrealistic business plans, lack of solid financial backing, public doubts about airlines' reliability, and poorly conceived pricing structures.

At the end of 2005, there were 138 domestic airlines operating in the U.S., including 20 Majors, 33 Nationals, 31 Regionals, and 55 Commuters.

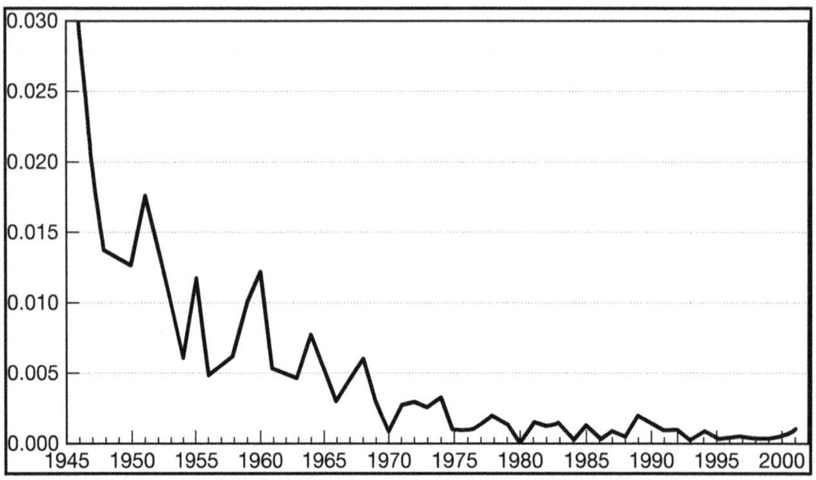

FIGURE 9-2 Safety record of U.S. airlines—fatal accidents per million aircraft miles.

Market Share

After deregulation, and for many years, at least through 2001, the incumbent airlines increased market share over that which existed during CAB regulation. In 1978, for instance, the five largest (incumbent) airlines took in 66 percent of domestic revenues. This control of market share was recited by proponents of deregulation during the Congressional hearings to demonstrate the anticompetitive impact of CAB regulation. In 2001, the five-carrier share had increased to 72 percent, revealing a central failure of deregulation theory. By the end of 2006, however, market share of all incumbent (legacy) airlines had fallen to 55 percent. This resulted, at least in part, due to the fact that after 9-11 four of the largest U.S. airlines went into bankruptcy reorganization and all legacy airlines cut capacity.

Safety

The airline accident rate has been steadily declining since the 1940s. With the introduction of jet aircraft into the civilian airline fleet, the rate of decline increased even faster, so that by the late 1980s, the total annual number of airline accidents had become historically miniscule. Commuter carriers, flying more turboprop equipment than their larger brethren, have had a proportionately higher accident rate, but over the last two decades even that rate has declined by 90 percent. Since deregulation, the overall fatal accident rate per million miles flown has averaged 0.0009, compared to 0.0135 during the forty years of regulation. It can be logically surmised that the difference in the accident rate before and after deregulation probably has more to do with the technological advance of aircraft and equipment than with regulation. It should be remembered that deregulation did not extend to safety issues. The airline industry remains today one of the most heavily regulated endeavors in the world.

Employment

Through the year 2000, the number of airline employees had increased by 50 percent, to almost 675,000. This increase was reflected across all related industries, like airport security, and construction projects at airports and access to airports. Beginning in 2001, however, airline jobs declined for the five consecutive years ending in 2005, to just over 550,000.

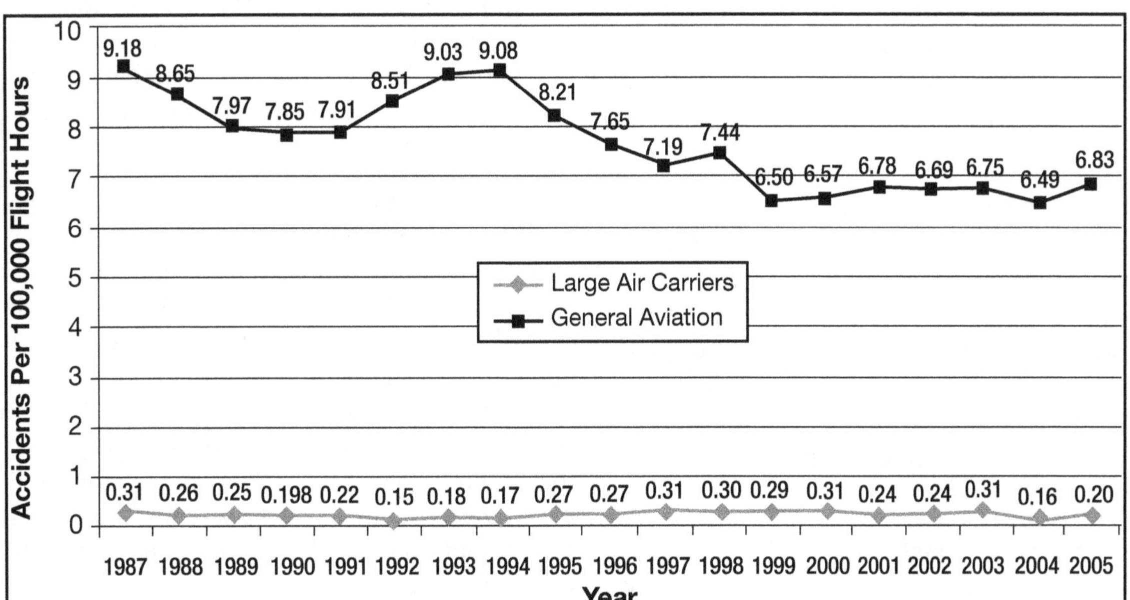

Source: National Transportation Safety Board Aviation Accident Statistics (Tables 5 and 10 available at: *http://www.ntsb.gov/aviation/Stats.htm*)

FIGURE 9-3 Accident rates.

The Evolution of Operating Practices

The air carrier industry has passed through a series of changes or "waves" since deregulation in 1978.[4] The initial wave was the creation of the hub and spoke system. The second wave was the inauguration of low-fare, point-to-point service, pioneered by Southwest Airlines. The third wave was the entry into the airline fleet of the regional jet. The fourth wave, now in process, is the abandonment of the financial and operational model of the legacy carriers from the period of CAB regulation.

Hub and Spoke

Before deregulation, it was said that if you traveled in the Southeastern part of the United States and you wanted to get to heaven, you would have to go through Atlanta and change planes. Delta Airlines is credited with creation of the hub and spoke concept that it centered in Atlanta, and, as discussed earlier, Delta began this service during the 1940s at the behest of the CAB in order to bring service to small outlying communities in the Southeast. The other trunk airlines that operated during regulation, however, were all point-to-point carriers.

After deregulation, the opportunities to serve when and where the airlines wanted, coupled with the economic necessity to fill their airplanes with as many passengers as possible, caused the adoption of the hub and spoke system nationwide. This system had two main advantages to the traveler:

- The passenger who lived in the hub city gained access to a greatly increased number of destinations directly from the hub airport.
- The passenger who lived in one of the smaller communities at the end of a spoke, who may not have had any service under regulation, was offered access to the same greatly increased number of destinations after one stop at the hub airport.

Hub and spoke brought to the airlines a much more efficient use of aircraft by allowing many more destinations to be served using far fewer airplanes. By way of example, if a carrier had 20 airplanes engaged in point-to-point service between city pairs, as was the case before deregulation, the number of origin and destination operations (O&D) would be limited to 20. In the hub and spoke system, the O&D number would suddenly jump to 400 (20×20).

The hub and spoke system has drawn complaints, too.

- First, passengers were said to be traveling "around their elbow," being required to stop at hub airports that were considerably distant from a direct line of travel, and losing the main advantage of jet aircraft, which is speed and the efficient use of time. According to this view, these passengers were traveling at the convenience of the airline, not themselves.
- Second, the system produced the natural result that the dominant airline gained tremendous market share at the hub city, a potential anticompetitive development. Dominated hubs include Atlanta (Delta), Denver (United), Detroit (Northwest), and Chicago (American and United).
- Third, the system required that all aircraft returning to the hub do so at or about the same time in order to make connections with aircraft departing from the hub to new destinations. This confluence of activity placed a huge strain on air traffic control and airport operations.

The introduction of these relatively short-haul operations altered the airlines' needs as to types of aircraft. Boeing, it is said, was in the process in the late 1970s of phasing out production of the 737. This decision was reversed after deregulation due to the adoption of the hub and spoke system, and production of 737 aircraft soared. The hub and spoke system also gave rise to an entirely new line of short-range aircraft, like the MD-80. Suddenly there was less need for the larger, fuel hungry 747s, and a general downsizing of aircraft began.

Year	Passenger Fatalities	Passenger Serious Injuries	Total Passenger Enplanements (millions)	Million Passenger Enplanements per Passenger Fatality
1987	213	39	458	2.2
1988	255	44	466	1.8
1989	259	55	468	1.8
1990	8	23	483	60.4
1991	40	19	468	11.7
1992	25	14	494	19.8
1993	0	7	505	No Fatalities
1994	228	15	545	2.4
1995	152	15	561	3.7
1996	319	19	592	1.9
1997	2	21	648	324.0
1998	0	12	650	No Fatalities
1999	10	46	676	67.6
2000	83	11	701	8.4
2001	483	7	629	1.3
2002	0	11	619	No Fatalities
2003	19	10	654	34.4
2004	11	3	711	64.6
2005	18	2	743	41.3
2006	47	4	750	16.0

Notes
- Injuries exclude flight crew and cabin crew.
- Since March 20, 1997, aircraft with 10 or more seats used in scheduled passenger service have been operated under 14 CFR 121.

TABLE 9-1 Passenger injuries and injury rates, 1987 through 2006, for U.S. air carriers operating under 14 CFR 121.

Low Fare, Point-to-Point Service

Southwest Airlines was a wholly intrastate carrier before deregulation, having been founded in 1971. It had developed a no-frills approach to air carrier service that it brought to the interstate market after deregulation, gradually expanding in a deliberate and cautious way. It used just one type of aircraft, the 737, thereby greatly reducing its parts inventory requirements and the training of its airframe and powerplant people, as well as its pilots. Southwest chose secondary airports, like Love Field in Dallas and Midway Airport in Chicago, where the turn-around time for its aircraft would be minimized. This practice avoided the congestion, the ATC delays, and the airport confusion that were becoming symptomatic of the hubs. Southwest's success moved it into the top ten airlines in the United States during the 1990s, and it continued to expand its service into the eastern United States market, opening up service to underserved or unserved airports. In 2006, Southwest had grown to be the second largest air carrier in the country as measured by passengers carried.

The success of Southwest caused startup airlines (increasingly known as "low cost carriers") to imitate its model of low cost and no frills, thereby creating even more service to even more destinations. The legacy airlines also began, in self-defense, to imitate the model of Southwest by creating subsidiaries that provided the same kind of no frills service and that used lower-paid crews operating a single type of airplane, just like Southwest.

The Regional Jet

The concept of the "regional aircraft" was born after World War II to describe the kind of airplanes used by "feeder airlines" authorized by the CAB to supplement the mainstay air carrier fleet. These airplanes were thus described to differentiate them from the long-haul aircraft flown by trunk carriers. At first these aircraft were older aircraft previously flown by the trunk lines, like the DC-3, Convair 240, 340, and 440, and the first commercial turboprop, the Vickers Viscount.

During the late 1950s and early 1960s, a new kind of turboprop was conceived to service the short-haul and feeder market. These were airplanes like the high-winged, 28-seat Fokker F27, delivered in 1958, and larger iterations of the same basic design. The F27 and its successor types would go on to become the most successful turboprop of all time. In 1963, the low-winged Avro 748 turboprop took to the skies, carrying over 20 passengers.

Turboprops worked well in this market, as their operating characteristics allowed them to service smaller airports, and their fuel economy was much better than turbojets. After deregulation, and during the 1980s, other manufacturers entered the 30 to 40 seat commuter market, like deHavilland with the Dash 8, also a high-wing turboprop. While these turboprops were well liked by passengers because of their relative roominess, they were slow compared to jets.

The regional jet (RJ) was introduced into the aviation community in 1992 by the Canadian aircraft manufacturer, Bombardier, with its 50-seat CRJ100 (Canadair Regional Jet), in part fashioned on its business jet, Challenger 604. In 1998, the company announced a stretched version holding 64 to 70 seats, designated the CRJ700, Series 701, and the 75-seat CRJ700, Series 705. A 90-seat version, CRJ900, joined the fleet in 2001. Canadair had some 55 percent of the regional jet market in 2002.

The Brazilian aircraft manufacturer, Embraer (Empresa Brasileira de Aeronautica, South America) entered the field in 1996 with the ERJ-145, with 50 seats. The 35-seat ERJ-135 was introduced into service in June 1999 to begin replacing the Brasilia, Embraer's turboprop workhorse. In 1999, Embraer launched a new family of twin-engined passenger aircraft consisting of the EMB-170, 175, 190, and 195 jets with seating in the 70 to 110 range. The first of this new family, the 170, flew on February 19, 2002. Embraer claimed about 40 percent market share of the regional jet market in 2002.

The Embraer 190 received FAA certification in September 2005. JetBlue Airways took the first delivery of this 98 seat RJ and has an order for 100 more. These new RJs are state-of-the-art airplanes, which rely on digital modeling and virtual reality concepts in its design. This airplane has an all digital cockpit and is equipped with fly-by-wire flight controls except for ailerons. Winglets at the wing tips are standard. The fuselage design features the "double bubble" idea, instead of the traditional circular cross section, which provides the look and feel of a larger cabin.

The Sukhoi Superjet 100, is a 75 to 95 seat RJ, developed by the Russian aerospace firm Sukhoi in collaboration with Ilyushin and Boeing, and is set to be launched in September 2007. The Chinese are in the developmental stage of an RJ called the ARJ 21, with 80 seats for the first phase production and 100 seats for its next phase. This airplane is expected to be certified by the Chinese Civil Aviation Administration in 2008.

There have been few other entrants into the RJ production market. (See Figure 9-4.) Fairchild Dornier, a subsidiary of the U.S.-German partnership, Fairchild Aerospace Corporation, marketed the 329Jet. Production stopped with Fairchild's financial reverses in the 1990s. The only other manufacturer of regional jets was Aero International, a consortium composed of Aerospatiale, Alenia, and British Aerospace. The Bae 146 series became the Avro RJ series (RJ 70/85/100) and 160 of these were produced before BAE Systems announced their discontinuation in the last quarter of 2001.

During the 1990s, RJs began to replace the turboprops used by commuter airlines. (Figure 9-5.) As airlines reconfigured and modified their hub and spoke concepts to utilize RJs, these small jets became commonplace and relatively popular in that

Producer	Aircraft	Capacity (class)	Mfg. Year	Airline
McDonnell Douglas	DC-9-10	60/78 seats (1/2 cl)	1965–67	NW/TW/YX
	DC-9-30	84/100 seats (1/2 cl)	1967–82	NW/TW/YX/US/CO
	DC-9-40	112 seats (2 cl)	1967–79	NW/TW
	DC-9-50	122 seats (2 cl)	1975–81	NW/TW
British Aerospace/ Aero International (Regional)	BAe 146-100 (Avro RJ70)	86 seats (1 cl)	1982–	UA
	BAe 146-200 (Avro RJ85)	86/89 seats (1/2 cl)	1983–	UA/NW
	BAe 146-300 (Avro RJ100)	100 seats (1 cl)	1988–	UA
Boeing	717	106 seats (2 cl)	1999	n/a
Bombardier	CRJ-200	50 seats	1992–	DL
	CRJ-700	70 seats	1998–	n/a
Embraer	EMB-145	50 seats	1996–	CO/AA
	EMB-135	37 seats	1998–	n/a
Fairchild Aerospace	328 Jet	32–34 seats	1999	n/a
	728 Jet	70 seats	2001	n/a
	428 Jet	44 seats	2000	n/a
Fokker Aircraft	F28-4000	25/68 seats (1/2 cl)	1976–87	US/AS
	F-100	96/107 seats (1/2 cl)	1986–96	AA/JI
	70	78 seats (1 cl)	1994–96	Mesa

FIGURE 9-4 A wide array of "regional jets."

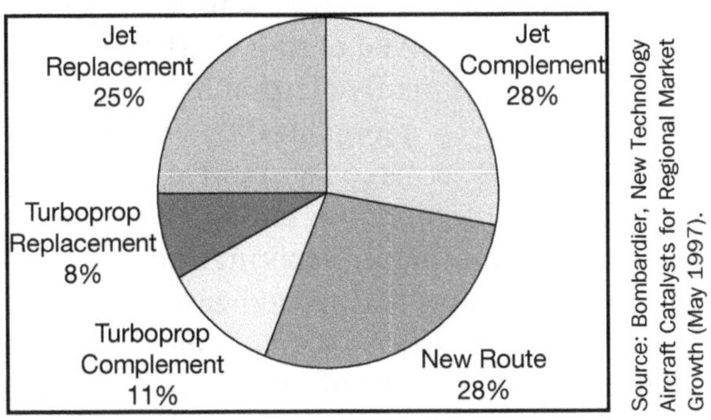

FIGURE 9-5 Uses of regional jets.

service. (See Figure 9-6.) The regional jets are faster, the engines are more reliable, and engine maintenance costs are lower. But compared to turboprops, the original small RJs were much more expensive to operate on a per seat basis. They were also more cramped than the larger turboprops, had less carry-on storage space, had lavatory issues, and minimal flight attendant service. As seen above, the trend in RJ size has consistently been toward larger and larger aircraft.

As the new RJ designs have increased their seating capacity, the line between between a medium-sized jet and a so-called RJ is being blurred. The Airbus 319, for instance, is normally configured for 124 seats, not much larger than the latest RJs. The original RJ concept that emerged during the 1990s of producing jets to replace similar sized turboprops is apparently being abandoned. Bombardier, for example, stopped production of its 50-seater in January 2006.

Because experience has shown that operating costs of RJs can make sense only on longer routes (400 miles seems to be the minimum), and as per seat operating costs have caused so-called "regional jets" to become larger, a market is appearing for a new era of turboprop aircraft to fill that niche. Most short-haul routes are less than 350 miles. Rising fuel prices have only reinforced this idea. Turboprops use about 30 percent less fuel than RJs.

There are only two companies producing turboprops in the 40-seat-plus capacity range as of 2007: Bombardier and ATR. The economic factors

	1978	1996	Average Annual Growth Rate
Passengers Enplaned (millions)	11.3	61.9	9.9%
Revenue Passenger Miles (billions)	1.28	14.22	14.3%
Aircraft in Service	1,047	2,127	4.0%
Average Seating Capacity	11.9	25.1	4.2%
Average Trip Length (seat miles)	117	230	3.8%
Operating Airlines	228	109	−4.0%

Source: RAA Annual Report, AvStat Associates.

FIGURE 9-6 U.S. regional airline growth, 1978–1996.

discussed above have caused increased orders for these companies' turboprop aircraft.

The old deHavilland Dash 8 production unit, which delivered the first Dash 8 in 1984, was sold first to Boeing and then to Bombardier in 1992. Bombardier turboprops are the Q100, first delivered in 1984 (33–37 seats), the Q200, first delivered in 1989 (33–37 seats), the Q300, a stretched version of the 100 (48–50 seats) and the Q400, first delivered in 2000 (68–78 seats). These airplanes have been fitted with a computer controlled noise and vibration suppression system since 1996 (the "Q" denotes "Quiet"), and produce a cabin decibel level equivalent to the CRJ regional jet. The Q400 has an impressive maximum cruise speed of 360 knots.

The European consortium ATR is a joint venture between EADS and Alenia Aeronautica. It produces the ATR 42-500 (48–50 seats) with a maximum cruise speed of 300 knots, and the ATR 72-500 (68–74 seats) with a maximum cruise speed of 276 knots.

As of January 2003, 90 percent of all regional jets were still used to service hub airports. As we have seen, this may be in the process of change, and JetBlue has announced that it intends to put its new fleet of EMB 190 aircraft (98 seats) to work overflying hubs on point to point service between city pairs.

Opening up smaller airports in point to point service by the use of RJs could also bring access to airline travel closer to home for the average traveler. Ninety percent of the country's population lives within 30 miles of an airport, yet only 64 airports (1 percent of all airports) serve 80 percent of passengers enplaned in the United States.[5]

The Fourth Wave—Demise of the Legacy Airlines Business Model

Pior to 9-11, the legacy carriers copied the Southwest model only by creating subsidiary operations patterned along the lines of Southwest, while maintaining their primary business models that evolved during CAB regulation. After 9-11, four of the five largest legacy carriers entered bankruptcy reorganization, where they secured approval to make substantive changes to their basic structure, including eliminating employee pension plans, securing reductions in wages, cutting capacity, and downsizing generally. It is yet too early to see how these changes will affect the operations of legacy carriers, but it is not too heretical to suppose that the mainline operations of the legacy carriers may be gravitating to the Southwest model.

This conceivable development will be augmented by the NextGen policy now in the planning stages in the Department of Transportation. Simply put, this complete transformation of the Air Traffic Control system from one of land-based navaids to a satellite-based navigation system, requiring on-board performance avionics and navigation capabilities, will likely drive new point-to-point operational models.

Endnotes

1. For a review of monopolies and the basis of regulation, refer to pages 15 and 16.
2. The bankruptcy law is codified at Title 11 of the United States Code. The U.S. Code is a series of books containing all of the laws of the United States arranged sequentially from Title 1 through Title 50A. Each Title is devoted to a particular subject matter. Title 49, for example, contains the federal statutes in the field of transportation.
3. Kahn, Alfred E., The State of Competition in the Airline Industry, Statement before the U.S. House of Representatives Commission on the Judiciary, June 14, 2000.
4. The first three waves are suggested by Robert W. Poole, Jr. and Viggo Butler, "Airline Deregulation: The Unfinished Revolution," Reason Public Policy Institute.
5. Miles, Richard B., Testimony before the Aviation Subcommittee, Committee on Transportation and Infrastructure, U.S. House of Representatives, *Competition in the U.S. Aircraft Manufacturing Industry*, June 26, 2001.

CHAPTER 10

Airports

Before Herbert Hoover was President, he was Secretary of Commerce in the Coolidge administration. He was called on to testify before the Morrow Board in 1925. The Morrow Board had been created for the purpose of studying the state of aviation and recommending to the President an aviation policy for the nation. Aviation, as the newest form of commerce joining maritime and land-based transportation, naturally followed some of the paths previously established by the older forms. It was also recognized that the promotion of aviation was in the national interest, much the same as it had been acknowledged that the nation needed the Post Office, a merchant marine, and the railroads.

Secretary Hoover drew an almost complete parallel between the needs of the fledgling aviation industry and the government's policy toward maritime commerce in the United States. He pointed out that government had accepted the responsibility of providing aids to navigation in the nation's ports and waterways by establishing markers, buoys, and lighthouses, and by providing surveys and geodetic charting. The government had provided land grants to the railroads in order to open up the West, in the name of the national interest. Roadways, too, were within the realm of government responsibility in part to facilitate motor commerce. The analogy was complete. Aviation needed and deserved federal assistance and direction if it was to develop in an organized manner. Otherwise, a fragmented and chaotic system of air transportation could be expected.

Included within the analogy was the need for airports to serve the various cities of the country. In the 1920s airmail service was being provided to some cities but not to others, sometimes largely or completely based on the fact that no landing fields were available to receive the planes. Airports then were truly landing "fields," sometimes referred to as "all-way airfields" since landings and takeoff could be made in any direction. Runways were the exception. Notwithstanding the favorable national policy toward aviation, there was no authorization for the direct participation of the federal government in the construction of airports. Municipalities, counties, and state governments recognized that their participation in air commerce was going to be dependent, in large part, on their own financial contributions. The Air Commerce Act of 1926 authorized the Commerce Department to survey and rate airports, and by 1929 some 181 airports had been catalogued. Only half of the airports had some kind of "prepared" runway, ranging from an oil-treated surface to cinders and concrete. Major cities, including Cleveland, Detroit, Buffalo, Milwaukee, Denver, and Boston, had fields acquired and improved with local money. Indeed, prior to World War II, most of the airports of the country were financed, developed, and operated by local or regional government, since no federal airport program had ever existed.

New York's LaGuardia airport was a local project. It started out in 1929 as North Beach Airport and, when Mayor Fiorello La Guardia began his expansive program of municipal works during the 1930s, including the city's famous bridges, tunnels, and highways, the airport was included. New York had been a central maritime port for over a century and had developed into a major transatlantic passenger seaport by the 1930s. Its piers, visibly surrounding the island of Manhattan, provided a gateway to the world. They also provided an aviation analogy for the advanced planners of New York.

LaGuardia airport was only eight miles from the center of Manhattan, and Pan American had built its Marine Terminal there. There was a concrete apron for the parking of the new DC-2s and DC-3s, and its runways were a mile long. La Guardia took advantage of a Depression era program known as Works Progress Administration (WPA), which was begun to provide work for the millions of unemployed men during the 1930s. At a time when practically all construction of any kind was stopped by the rigors of the Great Depression, and with commercial aviation just beginning to emerge as a new and viable transportation medium, federal monies expended through the WPA program greatly enhanced the progress of commercial aviation in the 1930s.

The Civil Aeronautics Act of 1938 lifted the ban on direct federal contributions for airports. One of the first cities to be benefited was Washington, D.C., whose airport, Washington-Hoover Airport, was described at the time by historian John R. M. Wilson:

> Bordered on the east by Highway One, with its accompanying high-tension electrical wires, and obstructed by a smokestack on one approach and a smoky dump nearby, the field was masterpiece of inept siting. Incredibly, the airport was intersected by a busy thoroughfare, Military Road, which had guards posted to flag down traffic during takeoffs and landings. In spite of such hazards, Washington-Hoover had a perfect safety record—for the simple reason that whenever even a slight breeze was blowing, planes refused to land there.[1]

By 1941, Washington National Airport had taken the place of Washington-Hoover, having been literally dredged up out of the swampy ground next to the Potomac. It immediately became the second busiest airport in the country.

Civil airport construction languished, largely because of World War II. During the war, the federal government had created many airfields for military use under a program known as Development of Landing Areas for National Defense, spending $3.25 billion. After the war, pursuant to the Surplus Property Act of 1944, about half of these bases were turned over to local and state governments. Still, airports of the size and quality for use by growing commercial aviation were few, and those few were abysmal. As reported in *Fortune* in 1946.[2]

> The half-dozen largest city airports handle millions of people a year. LaGuardia airport with 2,100,000 people, Washington with 757,000, Chicago with 1,300,000, and Los Angeles' Lockheed Air Terminal with 760,000 give clear indication of the size of the new air traffic. By standards of the huge railroad terminals, such as New York's Grand Central, which handles 65 million people a year, a million passengers is not so much. But a million passengers jamming through one small room, such as Chicago's filthy little air terminal, instantly creates a problem solvable only by a fresh start in new surroundings, by new design on functional lines.
>
> Chicago is the worst; its airport is a slum. Chewing gum, orange peel, papers, and cigar butts strew the floor around the stacks of baggage. Porters can't keep the floor clean if people are standing on it day and night. At almost all hours every telephone booth is filled, with people lined up outside; the dingy airport cafe is filled with standees. To rest the thousands there are exactly twenty-eight broken-down seats. One must line up even for the rest rooms. The weary travelers sit or even lie on the floor. The drooping grandmothers, the crying babies, the continuous, raucous, unintelligible squawk of the loudspeaker, the constant push and jostle of new arrivals and new baggage tangling inextricably with their predecessors, make bus terminals look like luxury.

To say that the airports at San Francisco or Los Angeles are less squalid than Chicago is faint praise, for the difference is so slight that anyone passing hastily through would notice no real improvement. Almost all U.S. airports are utterly barren of things to do. The dirty little lunch counters are always choked with permanent sitters staring at their indigestible food; even a good cup of coffee is a thing unknown. The traveler consigned to hours of tedious waiting can only clear a spot on the floor and sit on his baggage and, while oversmoking, drearily contemplate his sins.

Federal Airport Act of 1946

In 1946, Congress authorized the expenditure of funds for use by the political subdivisions of state, county, and city governments in building or improving the airport infrastructure. Pursuant to the provisions of the Federal Airport Act of 1946, the government would contribute fifty percent of these improvements, and the local government would fund the rest. Small communities found it difficult to come up with even fifty percent of such large expenditures. The program worked better in large cities that could float bond issues to finance their portion of the cost. A series of amendments to the Act extended the duration and funding of the program and added provisions designed to guarantee proper standards of construction and operation, open and non-discriminatory access by the public, and appropriate zoning of lands adjacent to the airport.

New York utilized the program to construct a third major airport, named Idlewild (now JFK), and then expanded the original construction with $60 million of terminal buildings and parking areas. Los Angeles raised a bond issue in the amount of $60 million for its international airport. St. Louis expanded its airport and built a new terminal. The busiest airport in the country was Midway, in Chicago, with an arrival or departure every 80 seconds. The activity at Midway led to the building of O'Hare, which opened in 1955. Dulles airport, conceived to serve Washington, D.C., finally found a home in Chantilly, Virginia, in 1958. Still, it was generally realized and noted at the time that United States airports were not ready for jets, and that they were not keeping up with the growth of commercial air travel. The number of passengers enplaning on domestic flights would almost double between 1954 and 1959, from thirty-two million to fifty-five million. And it was reckoned that less than ten percent of the American population had ever set foot on a commercial airplane.

Airports faced new problems as a direct result of the larger piston airplanes, and these problems were compounded by the arrival of jets. Fuel storage facilities were inadequate, both in size and in type, since jets burned a form of kerosene, not the gasoline of piston engines. Taxiways had to be redesigned, relocated, and widened, due both to the size of the aircraft and to allow for the new, low positioning of jet engines on the pylons hanging below the wing. FOD, the new acronym for foreign object damage, was a major concern to the health and integrity of jet engines because they could not tolerate ingesting small rocks, gravel, and debris of any kind lying about the taxiway. Runways were too short for jet aircraft, particularly those needing a full load of fuel for transatlantic or transcontinental routes. These requirements were exceedingly costly to the airport owners, usually a governmental entity, whether city, country, or regional political subdivision.

Not only was it costly and time-consuming to enlarge and upgrade existing airports, new construction was becoming a contentious issue in the communities in which they were located. Noise associated with jet operations was becoming a serious environmental issue, an issue that had constitutional and legal ramifications that extended far beyond the boundaries of the airports. And if improvement of existing airport facilities was a problem, then the building of new airports was next to impossible to pull off. In 1952 when authorities began looking for a site near Washington, D.C., to relieve the already overcrowded National Airport, an exercised public in Fairfax County, VA, rose in protest. It would require six years of effort spent in negotiation, in cajoling, in conducting studies, and in spending lots of money before the eventual site for Dulles International Airport was selected and approved. It was worse in the metropolitan New York area. Attempts to locate an agreeable site anywhere in northern New Jersey were defeated; twenty-two other proposed sites in New York and New Jersey went down to defeat. A group comprised of the New York Port Authority, the FAA, and some airlines finally settled on a site forty-eight miles from the city, but opposition cancelled that one too. Plans to expand and upgrade the ancient White Plains Airport were defeated by local opposition. To this day, New York has the same three airports that it had in 1950.

The resistance was not limited to the Northeast. Attempts in Dade County, Florida, to build a new airport to serve Miami failed. Miami International Airport, like most metropolitan airports, has been situated in the same place since bi-wing, open cockpit airplanes first began using it. One exception was the Dallas-Fort Worth Airport, in Texas, which only came about because of the cooperation of the citizens and local governments of the two cities that it serves. Located midway between Dallas and Fort Worth, DFW was completed in 1974. It became a symbol of pride and progress for the area, and it was appreciated by its business and leisure travel communities. One motivating factor for success was the appallingly inadequate Love Field in Dallas, which it replaced, as well as the fact that Fort Worth had no commercial airport. After completion of DFW, no new airport would be built in the United States for another twenty years.[3]

The launch of the Airbus 380 in December 2000 created another challenge for U.S. airports, as well as airports all over the world. The size of the A380 will require that modifications be made at airports which will serve the new gigantic aircraft. Although the A380 can land on runways that will accept the Boeing 747, modifications to taxiways, terminal gates, and aprons will have to be made. Airports will also have to fund the purchase of new servicing vehicles for the A380. It is estimated that some 20 airports worldwide will be ready to receive the A380 by 2007, with that number increasing to 40 airports by 2010.

Airport Security

Exploding passenger traffic after World War II and the advance in aircraft technology were only part of the problem confronting airports. Aircraft hijacking began as a means for oppressed citizens from communist regimes to escape to freedom in the West. As long as the hijackings were of communist-controlled airlines and the destination was freedom, the public generally applauded this audacity. But when the hijacking traffic started moving in the other direction, from the United States to Cuba, people began to view hijacking a bit differently. The first such U.S. to Cuba effrontery occurred in 1961 when a National Airlines airliner was commandeered to Havana, and then more followed. It was unthinkable. No federal laws adequately covered the activity, so Congress hurriedly passed appropriate legislation making the hijacking of an aircraft a federal crime. Things then cooled off and it began to look as though the few hijackings from the United States had been an aberration. But in 1968, there were seventeen hijacking attempts; in 1969, there were thirty-three more.

Hijacking commercial airliners soon became a worldwide phenomenon when two Arabs grabbed a TWA flight bound for Tel Aviv. Hijacking was gradually evolving into terrorism; that is, hijacking for a political purpose. Palestinians seized a Pan American 747 in September 1970 and forced it to Cairo, where it was blown up. This kind of terrorism was repeated on several occasions, resulting in destruction of the aircraft amid full-blown television coverage. Hostages were taken; ransoms and the release of imprisoned terrorists were demanded.

The United States initiated a program of air marshals, G-men who rode anonymously aboard selected airliners for the purpose of foiling would-be hijackers. Eastern Air Lines began using metal detectors for boarding passengers. Then the infamous D. B. Cooper, a thief with no particular social or religious philosophy, took over a Northwest Airlines 727. He directed the aircraft to a designated landing site where he released the passengers but demanded and got $200,000 and four parachutes. After the aircraft departed, and while airborne over the vast forests of the northwestern United States, he lowered the 727's unique rear stairwell and disappeared, with his cash, into the night.

But hijackings worldwide were taking on a deadly and tragic caste. In October 1972, an Eastern 727 was hijacked to Havana by wanted murderers, and in November, an escaped convict and two accomplices seized a Southern Airways DC-9 and proceeded to take the airplane and crew on an extended twenty-nine hour odyssey, making eight landings. The airplane and crew finally wound up safely in Havana, even after agents in Miami shot out the tires during takeoff. The FAA responded by ordering the installation of metal detectors at all gates. In December 1972 the FAA changed the passenger airline business forever by ordering the airlines to carry out electronic screening of all boarding passengers, as well as the inspection of carry-on luggage.

In the years since the first electronic screening began, scrutiny of airline passengers has steadily intensified. The list of prohibited carry-on items has steadily lengthened. Surveillance methods and sources have increased. Many more security personnel have been employed. Since the terrorist attacks on September 11, 2001, security procedures have been greatly amplified, as well as modified to address the threat of international terrorism. Responsibility for airport security has been removed from the FAA and given to the new Transportation Security Administration (TSA).

Evolution of a New National Airport Policy

The operation of airports changed in many other ways. The physical size of new airports serving modern jet airline traffic, the noise considerations inherent in airport operations, the large facilities necessary to accommodate the millions of passengers passing through the airports, and the new-found safety concerns resulting from the criminal and social developments of the 1960s called for a new and aggressive national airport policy.

Under the Federal Airport Act of 1946, the Federal-Aid Airport Program (FAAP) had been the first peacetime program of financial aid aimed exclusively at promoting development of the nation's civil airports. It endured for 24 years, but the growing demands of modern commercial aviation rendered that program obsolete.

In 1970, Congress had passed the Airport and Airway Development Act in order to address the obvious shortcomings of the nation's airports and the airway system. The policy statement for this law recognized the inadequacy of the nation's airport and airway system, and committed the government to its substantial expansion and improvement in order to meet the demands of interstate commerce, the national defense, and the postal service. Congress created the Airport and Airway Trust Fund, from which grants would be made for the purposes of the Act. The statute provided for the collection of certain user fees to further fund the program.

In 1982, after deregulation, Congress amended the existing statute with the Airport and Airway Development Act of 1982, reestablishing the FAA's airport grants program (which had been inactive since 1981) and renaming the Trust Fund program the Airport Improvement Program (AIP). The Trust Fund was originally administered by the CAA in 1946, and sequentially thereafter by the Federal Aviation Agency and then the Federal Aviation Administration.

This Act also amended the Federal Aviation Act of 1958 by requiring, for the first time, that operators of airports serving certificated air carriers secure "Airport Operating Certificates" by application to the FAA, demonstrating the ability to conduct safe and properly equipped airport operations. These requirements are set forth in Part 139 of the Federal Aviation Regulations.

Airports and Deregulation

Throughout the period of Civil Aeronautics Board regulation of air carrier routes and rates, airports competed with each other for the limited amount of traffic that was available. Traffic was limited

Aviation Taxes	Comment	Tax Rate
		PASSENGERS
Domestic Passenger Ticket Tax	Ad valorem tax	7.5% of ticket price (10/1/99 through 9/30/2007)
Domestic Flight Segment Tax	"Domestic Segment" = a flight leg consisting of one takeoff and one landing by a flight	Rate is indexed by the Consumer Price Index starting 1/1/02 \$3.00 per passenger per segment during calendar year (CY) 2003 \$3.10 per passenger per segment during CY2004. \$3.20 per passenger per segment during CY2005. \$3.30 per passenger per segment during CY2006. \$3.40 per passenger per segment during CY2007.
Passenger Ticket Tax for Rural Airports	Assessed on tickets on flights that begin/end at a rural airport.	7.5% of ticket price (same as passenger ticket tax) Flight segment fee does not apply.
	Rural airport: <100K enplanements during 2nd preceding CY, and either 1) not located within 75 miles of another airport with 100K+ enplanements, 2) is receiving essential air service subsidies, or 3) is not connected by paved roads to another airport	
International Arrival & Departure Tax	Head tax assessed on pax arriving or departing for foreign destinations (& U.S. territories) that are not subject to pax ticket tax.	Rate is indexed by the Consumer Price Index starting 1/1/99 Rate during CY2003 = \$13.40 Rate during CY2004 = \$13.70 Rate during CY2005 = \$14.10 Rate during CY2006 = \$14.50 Rate during CY2007 = \$15.10
Flights between continental U.S. and Alaska or Hawaii		Rate is indexed by the Consumer Price Index starting 1/1/99 \$6.70 international facilities tax + applicable domestic tax rate (during CY03) \$6.90 international facilities tax + applicable domestic tax rate (during CY04) \$7.00 international facilities tax + applicable domestic tax rate (during CY05) \$7.30 international facilities tax + applicable domestic tax rate (during CY06) \$7.50 international facilities tax + applicable domestic tax rate (during CY07)
Frequent Flyer Tax	Ad valorem tax assessed on mileage awards (e.g., credit cards).	7.5% of value of miles
		FREIGHT/MAIL
Domestic Cargo/Mail		6.25% of amount paid for the transportation of property by air
		AVIATION FUEL
General Aviation Fuel Tax		Aviation gasoline: \$0.193/gallon Jet fuel: \$0.218/gallon
Commercial Fuel Tax		\$0.043/gallon

TABLE 10-1 Current aviation excise tax structure (Taxpayer Relief Act of 1997, Public Law 105-35)

not only because CAB policy restricted the number of available routes, but because it also mandated high airfares. During the 1950s, 1960s, and most of the 1970s, airports maintained marketing departments whose functions included lobbying the airlines and the CAB for service.

After deregulation, by the year 2000, some 380 million more passengers a year were passing through America's airports than before deregulation. The problem was no longer one of how to secure more service; rather it was one of how to service the existing, and increasing, traffic.

Commercial-service airports were suddenly faced with the need for more gates and runways (airside expansion) and for more groundside support facilities (parking, restaurants, rental car counters, ticket counters, and shops). Methods of funding for renovation and expansion had to be addressed anew. The developing operational practice of the airlines known as hub and spoke, with its concentration of passenger arrivals and departures all at the same time, put severe pressure on airport operations. Control of existing gates had to be reevaluated in view of the increasing number of airlines requiring access to the limited number of gates. Environmental concerns, both from increased aircraft operations and from ground traffic to and from the airports, brought additional pressures on airport management practices.

The commercial air service industry consists of three essential components:

1. Air carriers
2. Air traffic control
3. Airports

While all three of these components must combine to function in an orderly and cohesive manner, the provisions of the Airline Deregulation Act directly controlled only the air carriers. In other words, only the air carrier portion of this combination had been deregulated. The federal government is solely responsible for the operation of the ATC system, and it is significantly involved in the operation of airports. Airports operate under a matrix of federal regulations, and commercial-service airports are all owned by local, regional, or state governments. While deregulation had unleashed the power and innovative potential of private enterprise in air carrier operations, the inherent limitations of government control in the ATC and airport sectors still overlay, and constrained, that potential. We will review how these factors have impacted modern airport operations.

Airport Ownership

There are around 18,300 airports in the United States, of which 5,300 are for public use. The majority of these airports (4,166) are publicly owned, and over 500 of these are commercial-service airports. Commercial-service airports are publicly owned airports that enplane more than 2,500 passengers annually in regularly scheduled passenger service. All of these airports are included in the FAA's National Plan of Integrated Airport Systems. While all such airports are currently publicly owned, a limited number have contracted out some of their operations to private, commercial management, primarily at medium hub airports.[4]

Airports are further classified by the FAA as

- Large hub airports (at least 1.0 percent of total national enplanements)
- Medium hub airports (less than 1.0 percent but more than .25 percent of total national enplanements)
- Small hub airports (less than .25 percent but more than .05 percent of total national enplanements)
- Non-hub airports

As of 1999, over 75 percent of all passenger enplanements occurred at large hub airports, 16 percent at medium hub airports, and 6 percent at small hubs. Less than 3 percent of all passenger traffic passed through non-hub airports.[5]

Airport Funding for Operations and Capital Improvements

Each airport is operated under a master plan, the primary purpose of which is to provide safe and efficient air carrier service and to enhance airport capacity. Capital improvements include construction of new facilities or renovations of existing facilities. Sources of funding include:

- Airport Revenue Bonds
- The Airport Improvement Program (AIP)
- Airport user charges
- Passenger facility charges (PFCs)
- State and local funding programs

Airport Revenue Bonds

Bonds are the primary means of financing airport capital development projects. Bonds are debt instruments (like I.O.U.s) issued by the airport owner to raise the money necessary for new construction or renovations. These bonds are tax-exempt and are known as General Airport Revenue Bonds (GARBs). GARBs accounted for 36 percent to 70 percent of airport capital development expenditures during the 1990s, a sum in excess of $3 billion per year. Projections for the time period 2007 to 2011 are for bond funds to account for 44 percent of planned capital projects.

A secondary source of revenue is derived from bonds known as Special Facility Bonds. These bonds are secured directly by the owner of the facility, such as a fixed base operation or aircraft maintenance facility, and the owner of the facility is usually responsible for paying off the principal amount of the bond and all interest charges.

Airport Improvement Program Funds (AIP)[6]

AIP funds are federal monies administered by the FAA. AIP grants are distributed both according to statutory formulas and FAA discretion. During the 1990s, AIP funds constituted 21 percent to 40 percent of the total airport capital development expenditures. The majority of AIP funding went to Landing Area Construction (52.9 percent) and Noise Control (11.1 percent). Airport owners or sponsors must provide a minimum of 10 percent share of any project funded by AIP grants.

Airport User Charges

Airport user charges are either (1) aeronautical user charges or (2) non-aeronautical user charges.

1. Aeronautical user charges include landing fees, apron, gate-use or parking fees, fuel-flow fees and terminal charges for rent or use of ticket counters, baggage claims areas, administrative support quarters, hangars, and cargo buildings.
2. Non-aeronautical user charges include rentals to terminal concessionaires, automobile parking, rental car fees, and rents and utilities for hotel, gas station, and related facilities.

Passenger Facility Charges (PFCs)

In 1990, Congress authorized airports to charge a per-passenger enplanement fee to be used for the financing of airport capital improvements and the expansion and repair of airport infrastructure. A fee of up to $3 (fees over $3 require special justification pursuant to AIR 21, but as of April 1, 2007, only two applications out of 1,517 had been denied) per passenger may be imposed for the funding of three specific purposes: (1) to preserve or enhance safety, security or capacity, (2) to reduce noise or mitigate noise impacts, and (3) to enhance air carrier competition. Pending legislation (the 2007 FAA Reauthorization Act) would raise the PFC ceiling to $6.00. In 1996, PFCs accounted for 15 percent of total airport capital development expenditures. (See Figure 10-1.) Projections for the time period 2007 to 2011 are for PFCs to account for 20 percent of planned capital airport projects.

State and Local Government Programs

State and local funding is most often used as matching funds in order to receive federal support, although direct funding of maintenance projects is sometimes provided. During the 1990s, state and local funds accounted for 7 percent to 11 percent of airport capital development expenditures.

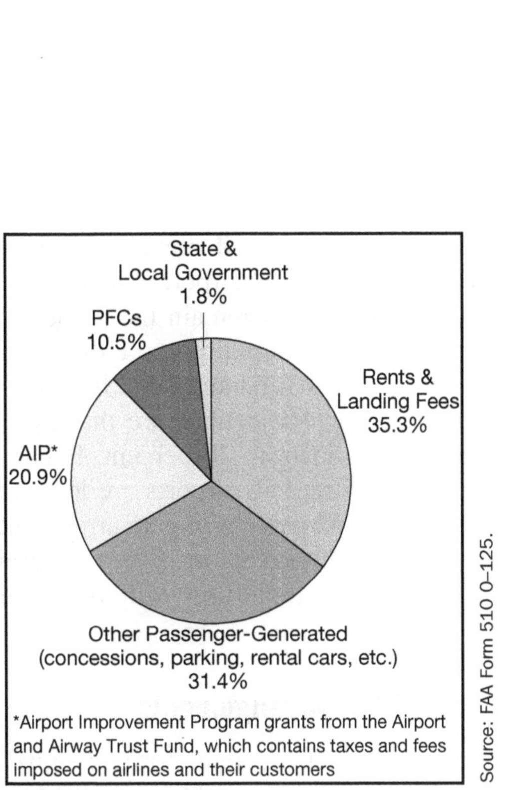

FIGURE 10-1 98% of airport revenue comes from airport users: U.S. Airport Sources of Revenue, 2001.

Airport Facilities Use Arrangements

Use of airport facilities, buildings, and land is normally arranged under use and lease agreements between the airport management and the user. Most of these agreements are between the airport owner and the airlines. These agreements generally fall into three separate categories: residual, compensatory, and hybrid agreements.

Residual Use and Lease Agreements

Under this type of agreement, airlines agree to assume the financial risk of running the airport. Airlines guarantee that the airport will not lose money by agreeing to make up the difference between total cost of operations and the total of non-airline revenue received by the airport. These types of agreements originated during the period of regulation when traffic volumes were low and the airlines were much more powerful than after deregulation. The trade-off an airport makes when entering into this kind of agreement is two-fold:

1. The airport gives up the opportunity to make any profit in its operations since any surpluses are credited to the airlines.
2. The airport gives up the right to make autonomous decisions over capital expenditure programs because of provisions in the agreements, called "majority-in-interest" (MII) clauses.

These MII clauses allow the airline to approve or disapprove, or at least delay, specific capital projects, the costs of which would be included in the future charges to the airline.

Some of these capital projects (like construction of new gates) could benefit new entrant airlines into the airport, which would be adverse to the interests of the signatory airline. The right to disapprove the project, of course, is anticompetitive, so that the airline has been placed in a position of unfair advantage over its would-be competitor. These clauses are also against the public interest in having better, bigger, and more efficient airport facilities for use by passengers. MII clauses, however, do not give the airline approval authority over projects funded by AIP, PFCs, or special facility bonds, and airlines are legally barred from exercising veto rights over PFC-funded projects. Eighty-four percent of residual use and lease agreements have MII clauses, and the average length of the agreement at large hub airports is 28 years.

Compensatory Use and Lease Agreements

Under these types of agreements the airline pays only for the facilities and services actually used. The airport assumes the responsibility for meeting the costs of operation like any other owner. One survey reported that 20 percent of these types of agreements have MII clauses. Compensatory agreements, without MII clauses, allow the airport full latitude in the use of its funds and allow the incentive to add additional service in the interest of the public and the airport instead of the interests of the incumbent airlines at the airport.

Hybrid Use and Lease Agreements

These agreements typically exclude non-aeronautical uses (restaurants, rental car operations, etc.) from the residual pool, so that an airline's guarantee is limited to the cost of aeronautical operations only. The airport retains earned revenues from its non-aeronautical operations while being guaranteed a break-even on airfield activities. Seventy-four percent of these agreements contain MII clauses, the average length of which is 20 years. While these agreements are anticompetitive with respect to capital improvement and expansion projects, they do give to the airport the incentive to expand its non-aeronautical sources of income.

Gate Leasing Arrangements

Airport gates obviously are a finite commodity and an essential element of both airline service and competition in the airline industry. In combination with overriding use agreements, or as separate undertakings with airlines, airports have normally utilized three methods of allocating gate use to airlines serving that airport:

1. Exclusive-use contracts
2. Preferential-use contracts
3. Airport controlled gates

Exclusive-Use Arrangements

Exclusive-use arrangements typically assign to one airline the right to use and occupy gates and facilities for a specified duration, as well as the right to sublet or assign those gates and facilities conditioned on prior consent of the airport. These types of arrangements are recognized to constitute potential barriers to entry based on complaints by new entrant carriers that incumbent airlines hoard gates, require substantial sublease premiums, offer access at less preferable times, force the new entrant to use the incumbent's ground personnel, or refuse to sublease altogether. These arrangements also have the effect of hindering airport management from properly exercising its legislative mandate of providing equal, nondiscriminatory access to its federally-funded facilities. Moreover, since gate leases are considered assets in law, secured creditors of bankrupt carriers, like banks, can wind up possessing the proprietary rights to airport gates, and preventing recovery and use of those gates by airport management.

Exclusive-use gates remain the predominant arrangement at large hub airports, although their number fell between 1968 and 1992 from 63 percent to 55 percent. Projections are that by 2004 their use will be down to 40 percent. Exclusive-use gates at medium hub airports are less prevalent, constituting 35 percent of gate arrangements in 1992, down to 29 percent in 1998, with projections that by 2004 their uses will be down to 30 percent.

Preferential-Use Arrangements

Preferential-use arrangements normally give the tenant airline the primary right to use the facility only when it has operations scheduled. These arrangements constitute a shared control between the airport and the airline with an explicit contractual right remaining in the airport authority to allow use of the gate by other airlines. Importantly, this type of arrangement preserves the airports' ability to honor the legislative mandate of providing nondiscriminatory access to its facilities and to use the tenant's gates for new entrants.

Preferential-use agreements differ in their specifics, some containing "use-it-or-share-it" or "use-it-or-lose-it" requirements, as well as other types of recapture provisions. These types of gate arrangements have become more prevalent at large hub airports, increasing from 28 percent in 1992 to 32 percent in 1998, with projections that by 2004 their use will be up to over 46 percent. Reagan Washington National has reported a gate composition of all preferential-use gates, and Boston Logan expects to have all preferential-use gates beginning in 2002. Detroit Metropolitan Airport is in the process of transitioning to all preferential-use gates.

Airport-Controlled or Common-Use Arrangements

Airport-controlled or common-use arrangements are completely under the control of the airport authority. The airport may assign gate and facility usage on a temporary, per-turn basis or for a short-term duration. These types of arrangements have been popular in Europe and other foreign regions for some time. The concept has gained popularity to the point that it has acquired the acronym C.U.T.E., or Common Use Terminal Equipment, to describe what is being increasingly seen as the best way for an airport to organize its gates and check-in counter facilities.

The International Air Transport Association (IATA) has even issued a recommendation (No. 1797) favoring common use systems as a means of efficient and cooperative use of available terminal facilities worldwide.

Various proprietary contractors have developed expertise in assisting airports in setting up these common-use arrangements so that, rather than being blocked off, the available facilities can be distributed as needed to different airlines. These facilities include check-in counters, gates, holding rooms, and electronic equipment. The systems control and integrate all components necessary to the carrier, including computers, displays, and boarding pass printers and readers.

Two airports have all airport-controlled gates—Miami International with 121 gates and Phoenix Sky Harbor with 84.

DOT Interest in Airport Practices—Unfair Competition

Prior to 2001, the Department of Transportation's authority over the gate practices of commercial-service airports was severely limited.[7] While DOT had jurisdiction over the gate practices of airlines, constitutional principles prevented interference in policies and practices of state-owned airports.

With the passage of the Wendell H. Ford Aviation Investment and Reform Act for 21st Century (AIR-21), DOT was given authority to require certain large and medium hub airports to submit competition plans as a condition of receiving federal grant monies and as a condition for authority to impose PFCs at their airports. These airports (including those at which competition among the airlines was threatened by airline domination, gate control, and other anticompetitive practices) were required to provide DOT detailed information concerning their gate practices. DOT has used its authority to approve or disapprove these competition plans as a means to insure that gate practices at those airports are fair. This includes insuring broader access to gates by new entrant airlines.

Airline gate practices continue to be monitored by DOT to insure that airline control of gates does not unduly impede competition. Such authority could compel an airline to surrender control of airport gates, or prevent tying arrangements involving subleases by one airline to another, where, for example, the lease requires the use of the lessor airlines' ground forces by the lessee airline. So too may DOT apply its authority to situations where an airline, with market power, exercises it contractual rights under a MII clause to block the construction of facilities for competitors merely to maintain its own monopoly power.

Airport Privatization

It would be fair to assert that the promise of airline deregulation has, to some extent, been compromised by continued government operation of the remaining two components of the air service industry: airports and air traffic control. Thirty years after deregulation of the airlines, airports are still run under principles similar to those applicable to the airlines prior to deregulation. It would also be legitimate to ask whether privatization of airports might not relieve some of the constraints now imposed on the air service industry, and whether many of the consumer complaints heard as a result of airline operations since deregulation might not be alleviated.

The international community is well out in front of the United States on this issue. Over one hundred foreign airports are now privatized, including Auckland, Buenos Aires, Dusseldorf, Johannesburg, London, Melbourne, and Rome. Others, such as the huge Frankfurt airport and Amsterdam's Schiphol, are partially privatized, and still others are in the process of privatization, including Hong Kong and Tokyo. Most of these airports are operated under a lease of the entire airport facility to a private operator on a long-term basis, usually 30 years or longer. The governmental owner of the airport normally retains ownership of the land and airport facilities.

Government-owned and managed airports are characterized as risk-averse, passive, and noninnovative. Privatized airports are more willing to take new risks, like expansion of facilities and gates. The lack of gates is cited as one of the constraints on new entrant airlines, which is a competitive disincentive. As we have seen, long-term leases with incumbent airlines allow a large degree of control to be lodged with those airlines, another competitive disincentive. Oxford University has conducted recent research into airport management strategies and found that the management approach at privatized airports is significantly more "passenger-friendly" than at government-operated airports.[8]

At least one effort has been made toward privatization in the United States. In 1996, Congress passed the 1996 Airport Privatization Pilot Program. Under this statute, up to five airports, called "pilot program slots," were eligible for privatization, The owners of these airports were to be exempted from provisions of federal statutes that required repayment of prior federal grant monies to those airports and from restrictions on the use of proceeds of sale or lease of the airport to private persons or corporations.

As a result of airline lobbying, however, the statute provided that approval of the privatization would require consent of at least 65 percent, in landing weight, of the airlines serving that airport.

Only one airport has been privatized under this statute. Stewart Airport in Newberg, New York, received FAA approval in April 2000 of its 99-year lease to a British company. It was unable, however, to secure consent of the airlines serving the airport. As a result, the state of New York is required to use revenues derived from the lease of the airport for improvements to the airport or other state-owned airports.

In recent years there have been several additional applications for privatization, including New Orleans Lakefront Airport, a large general aviation field, and the State of Illinois for approval of its new Abraham Lincoln National Airport to be situated in Peotone, Ill. There is also activity by the City of Chicago toward potential privatization of Midway Airport in Chicago.

Under the 2007 FAA Reauthorization proposal, the number of "pilot program slots" would be increased to fifteen, and would eliminate the restrictions on the classification (sizes) of the applicant airports. It would also eliminate the airline approval requirement.

Wendell H. Ford Aviation Investment and Reform Act for the 21st Century (AIR-21)

On April 5, 2000, the far ranging Wendell H. Ford Aviation Investment and Reform Act for the 21st Century became law. This statute, designed to alleviate many of the intransigent problems faced by the aviation infrastructure arising from deregulation, will increase funding for civil aviation in the United States by $10 billion over levels current at the time of its passage. Total authorized funding for federal aviation programs for the years 2001 to 2003 total $40 billion. The majority of the increased funding was earmarked for airport construction and improvement and for radar modernization. The major components of the law, broken out into the general categories that are affected, are outlined in the box on page 261.

It is recognized that deregulation was directed only at the economic regulation of the airlines, and that the aviation infrastructure, comprising the airports and the air traffic control system, were left in their pre-deregulation posture in 1978. Airports were unprepared—physically, financially, and legally—to deal with the developments in civil aviation wrought by deregulation. The ramifications of deregulation of the airlines were not fully appreciated at first, as they realistically could not have been, and only after some years' experience following deregulation has that experience now defined the major problems confronting the aviation infrastructure. An example of how the federal government is now addressing these concerns is discussed

Box 10-1 The Major Components of the Wendell H. Ford Aviation Investment and Reform Act

1. Safety
   a. Increases FAA facilities and equipment budget by almost 50 percent for ATC system modernization
   b. Increases funds for runways and airport equipment
   c. Provides funding for FAA hiring and retention of controllers and inspectors
   d. Makes runway incursion prevention devices eligible for AIP funding
2. Competition
   a. Funding for new terminals, gates, and taxiways
   b. Abolishes slots at O'Hare in 2002; modified by FAA in 2004
   c. Abolishes slots at LaGuardia and Kennedy in 2007
   d. Creates 24 new slots at Reagan National, 12 of which are to be used for flights within the 1,250 perimeter and 12 to be used outside of the perimeter
   e. Requires certain and medium hub airports to submit a competition plan
3. Environment
   a. Increases funding for noise abatement
   b. Establishes guidelines for air tours over national parks
4. Small Communities
   a. Increases funding for non-hub airports
   b. Guarantees funding for general aviation airports
   c. Doubles the small airport fund
   d. Creates an incentive program to help airlines buy regional jets to be used to serve small airports
   e. Creates a new funding program to assist small airports to promote their air service
5. Large Airports
   a. Doubles the amount of annual passenger funding for primary airports (those with 10,000 or more passengers per year)
   b. Raises the cap on annual funding for large airports from $22 million to $26 million
   c. Doubles the funding for cargo airports
   d. Raises the cap on Passenger Facility Charges (PFC) by $1.50 (to $4.50) to facilitate airport improvements that cannot be funded through the Airport Improvement Plan (AIP). PFCs can only be used to fund airport projects that increase safety and competition or are used for noise abatement
6. FAA Reforms
   a. Creates an oversight board (similar to IRS reform legislation)
   b. Makes changes in FAA management structure to ensure spending integrity
   c. Creates a management board to oversee the ATC modernization program (the DOT is to consult with Congress in board appointments)

above as a result of the passage of AIR-21. It should be particularly noted that AIR-21 requires certain large and medium hub airports, as a condition of imposing PFCs at those airports, to submit a competition plan for DOT approval designed to assure a competitive environment. Central to this effort is gate use monitoring, gate sublease oversight, and procedures for gate assignment.

Airports had in place legally enforceable contracts with airlines and with others that could not, and in some cases, still cannot be abrogated or modified. The anticompetitive effects of these financial, legal, and physical constraints are only now being addressed, but at least they are now recognized to exist. Enlightened policies at DOT concerning competition and a proactive Congress concerning the funding needs of airports point to a remedying of the worst shortcomings present today in the aviation infrastructure.

Airports and the Environment

As a result of the industrialization of society and the growth of heavy industry during the late 19th and early 20th centuries, pollution of air and water became a significant by-product of progress. In the 1950s, prevailing wisdom held that such pollution was the inevitable price of such progress.

Visible air pollution was seen from stationary sources like plants and factories and in "smog" accumulations in places like the Los Angeles Air Basin due to automobile emissions. Industrial and municipality discharges into waterways caused wide-spread prohibitions against swimming and fishing due to health risks. Catastrophes like the Cayuga River fire (the river caught fire) and the Love Canal scandal (toxic waste seepage caused a declaration of a federal emergency) were high-profile examples of pollution. When jet transport aircraft were introduced into the air carrier fleet in 1958, the dense, black exhaust emissions created on takeoff at ground level and during climbout were vivid evidence of yet another encroachment on air quality levels.

Prior acceptance of pollution as an inevitable by-product of progress has now been roundly rejected. Current environmental policy is concerned with almost all aspects of the quality of life on earth. In this chapter we will briefly review the evolution of current policy generally, and we will look at how current policy has attempted to address the specific environmental impacts of aviation.

The Air Pollution Act and The Clean Air Act

The first attempt by Congress to address the problem came in the Air Pollution Act of 1955. Subsequent efforts to strengthen controls on pollution occurred in amendments beginning in 1963 in The Clean Air Act, which was amended almost yearly until 1970. That year proved to be a watershed year for environmentalists, with the passage of The Clean Air Act of 1970 (an amendment to the 1963 Act), which created the Environmental Protection Agency as an independent agency reporting directly to the President. Its broad authority is over control of pollution, noise, and radiation.

Although The Clean Air Act mandates a national policy, the statute gives state and local governments primary responsibility for regulation of pollution from power plants, factories, and other stationary sources. The EPA retains primary responsibility for "mobile source" pollution.

Aviation Impacts

Aviation activities impact the environment primarily in two ways: (1) engine-generated emissions into the air and (2) noise. Noise legislation has been addressed by the Congress in separate, specific legislation, and we will discuss those statutes later in this chapter. Engine emissions result from both aircraft operations and airport ground operations. Aircraft emissions and airport operations were not, at first, on the radar scope of the EPA. As it became more evident that airport operations were contributing to environmental concerns, however,

Pollutant	Representative Health Effects
Ozone	Lung function impairment, effects on exercise performance, increased airway responsiveness, increased susceptibility to respiratory infection, increased hospital admissions and emergency room visits, and pulmonary inflammation, lung structure damage
Carbon Monoxide	Cardiovascular effects, especially in those persons with heart conditions (e.g., decreased time to onset of exercise-induced angina)
Nitrogen Oxides	Lung irritation and lower resistance to respiratory infections
Particulate Matter	Premature mortality, aggravation of respiratory and cardiovascular disease, changes in lung function and increased respiratory symptoms, changes to lung tissues and structure, and altered respiratory defense mechanisms
Volatile Organic Compounds	Eye and respiratory tract irritation, headaches, dizziness, visual disorders, and memory impairment

TABLE 10-2 Representative Health Effects of Air Pollutants

Pollutant	Representative Environmental Effects
Ozone	Crop damage, damage to trees and decreased resistance to disease for both crops and other plants
Carbon Monoxide	Similar health effects on animals as on humans
Nitrogen Oxides	Acid rain, visibility degradation, particle formation, contribution toward ozone formation
Particulate Matter	Visibility degradation and monument and building soiling, safety effects for aircraft from reduced visibility
Volatile Organic Compounds	Contribution toward ozone formation, odors and some direct effect on buildings and plants

TABLE 10-3 Representative Environmental Effects of Air Pollutants

the EPA initiated a study in 1999 to assess these impacts and has followed up with increased monitoring of aviation generated air pollution.

Aviation-Generated Air Pollution

Emissions associated with aviation are the result of burning fossil fuels that power aircraft, the equipment that services them, and the vehicles that transport passengers to and from airports. These pollutants include volatile organic compounds (VOCs), carbon monoxide, nitrogen oxides (NOx), particulate matter (PMs), and carbon dioxide (CO_2). As a group, these are considered deleterious to human health, to plant and animal life, and to the future of the planet. Comparatively, aviation-related activities are a minor contributor to total air pollution in the United States, comprising only 0.5 percent as of 2003. Projections of significant increases in commercial air traffic, however, have raised concerns about aviations's long-term effect on the environment.

Success in reducing particulate matter in the exhaust of jet engines was realized fairly early. This was due in part to the fact that this type of emission was the most noticeable and a solution was, therefore, quickly sought. In addition, jet engine-produced black smoke consists of fine carbon particles, partially burned fuel, and raw fuel. These PMs contain carcinogen content and descend to earth, creating health hazards. The reduction of exhaust smoke was accomplished by a more complete combustion process whereby the hydrocarbons in jet fuel were converted to carbon dioxide and water.

Take off and landing operations produce the highest rate of PMs and NOx. Nitrogen oxides are a primary contributor to the formation of ozone, which is the most significant air pollutant in urban areas and which is a greenhouse gas in the upper atmosphere. Paradoxically, as innovative jet engine designs have produced more power while using less fuel and with lower carbon monoxide emissions, NOx emissions have increased due to their higher operating temperatures. Research is ongoing to find ways to reduce NOx.

Aircraft idle and taxi operations produce the highest rate of VOCs and carbon monoxide. Ground operated support equipment is mostly powered by gasoline or diesel engines, which produce VOCs, carbon monoxide, NOx, and PMs. Carbon dioxide, while not considered a pollutant in the lower atmosphere, can form ozone.

Airport operators have no direct control over aircraft emissions, although two foreign countries[9] have imposed landing fees based on the amount of aircraft emissions.[10] Primary regulation of aircraft emissions is through EPA standards that apply to new engine designs. The EPA has elected to adopt international emission standards as set by ICAO. The FAA has been appointed by the EPA to function as the U.S. representative in formulation of these standards within ICAO. Engine manufacturers must now meet these standards in order to have new engines certified.

The imposition of emissions standards for new engines is a complex undertaking, since consideration must be given to the feasibility of mandating standards in view of the many other considerations inherent in engine manufacture, such as fuel efficiency, safety, and cost. Realizing this, the FAA and the EPA have established a consultation process known as "the stakeholders group," which includes representatives from government, airports, air carriers, and the aerospace industry. This group focuses on achieving lower aircraft emissions through a voluntary program because this strategy offers the potential for achieving desired goals with less effort and less time than a regulatory approach.

Airports are theoretically subject to nationally supervised state control of emissions through State Implementation Plans (SIPs). These plans must be submitted by the states to the EPA for reducing emissions in areas that fail to meet the National Ambient Air Quality Standards set by by the EPA under the Clean Air Act. The power of states in controlling pollution at airports is limited, however, since the EPA retains control of regulating mobile sources of emissions and because states are preempted from regulating aircraft operations generally. This is a federal responsibility (in order to maintain a consistent national policy) and the FAA is responsible for enforcing emission standards. For these reasons only three states have even attempted to target airports for emission reductions.[11] Operators of some of the busiest airports in the country have initiated voluntary programs to reduce emissions from sources over which they have control under the Voluntary Airport Low Emission Program (VALE). The goal of this program is to reduce emissions caused by ground service vehicles by investing in proven low-emission technologies, including alternative fuel vehicles.

	Average Emission per Landing/Takeoff		
Emission	**Older Boeing 737 (pounds)**	**Newest Boeing 737 (pounds)**	**Changes**
Nitrogen oxides	12.1	17.8	47% increase
Carbon monoxide	16.8	10.7	37% decrease
Hydrocarbons	1.2	1.1	10% decrease

TABLE 10-4 Additional Information on Comparison of Older and Newest Model Boeing 737 Landing/Takeoff Emissions

Source: GAO.

Notes: Landing and takeoff data for U.S. aircraft in 2001 obtained from AvSoft. Emissions were calculated using FAA's Emissions and Dispersion Modeling System, version 4.01. The following variables were assumed to be the same for all aircraft: (1) taxi-time: 15 minutes, (2) auxiliary power unit time: 26 minutes, and (3) ceiling height for emissions mixing with local air: 3,000 feet. The model's default was used for takeoff weight.

Emission	Emission per Aircraft During Landing/Takeoff		Changes
	Boeing 747-400 (pounds)	Boeing B777-200ER (pounds)	
Nitrogen oxides	103.5	124.2	20 percent increase
Carbon monoxide	47.7	30.4	36 percent decrease
Hydrocarbon	4.1	2.4	41 percent decrease

TABLE 10-5 Additional Information on Comparison of Boeing 747 and 777 Emissions on a Per Aircraft Basis

Source: GAO.

Notes: Landing and takeoff data for U.S. aircraft in 2001 obtained from AvSoft. Emissions were calculated using FAA's Emissions and Dispersion Modeling System, version 4.01. The following variables were assumed to be the same for all aircraft: (1) taxi-time: 15 minutes, (2) auxiliary power unit time: 26 minutes, and (3) ceiling height for emissions mixing with local air: 3,000 feet. The model's default was used for takeoff weight.

The Boeing B77-200ER data is the weighted average (based on 2001 landings and takeoffs) for three different engines. The nitrogen oxides and other emission characteristics of these engines vary significantly.

The 58 Boeing 747-400s in the 2001 U.S. fleet have PW4056 engines and average 361 seats per aircraft. The 101 Boeing 777-200ERs in the 2001 U.S. fleet have the following engines: PW4090 (37 aircraft averaging 302 seats), GE90-90B (16 aircraft averaging 283 seats), and TRENT 892B-17 (48 aircraft averaging 249 seats). The three engine types for the Boeing 777-200ERs emit 138.6, 123.6, and 112.3 pounds of nitrogen oxides emissions per landing/takeoff, respectively.

Aircraft Noise

While air pollution considerations have become increasingly important in the years since jet transport aircraft were introduced into the civil aviation fleet in 1958, noise levels associated with the operation of jet transport aircraft at the nation's airports were an immediate concern to anyone within audible range. Jet noise had previously been limited to military operations, and most people were familiar with jet operations only as they were usually conducted at higher altitudes.

Citizens' groups became quite vocal about this intrusion into what had theretofore been a relatively peaceful coexistence with airport operations. Jet noise brought a new challenge to this attempt at coexistence. It took regulators until 1968 to accumulate enough anecdotal and empirical evidence to bring the matter successfully before the Congress. In that year, the first aircraft noise legislation was passed.

Aircraft Noise Abatement Act of 1968

That year the initial step was taken to confront what was coming to be recognized as not only an environmental issue, but also a health issue. The Aircraft Noise Abatement Act of 1968 required the FAA, in consultation with the new Environmental Protection Agency, to establish noise standards for aircraft and to apply them through issuance of civil aircraft certificates.

Noise Control Act of 1972 and Aviation Safety and Noise Abatement Act of 1979

In 1972, Congress passed the Noise Control Act, which amended the Federal Aviation Act of 1958 to specifically involve EPA in the regulation of airport noise. This was followed in 1979 by the Aviation Safety and Noise Abatement Act, which authorized the Secretary of Transportation to formulate a national aviation noise policy and authorized the FAA to promulgate regulations pursuant thereto, including "air noise compatibility planning." These regulations are contained in 14 Code of Federal Regulations, Part 150.

The FAA established a program under the 1979 statute to help airport operators develop comprehensive noise reduction programs. Known as the Part 150 program (derived from CFR Part 150), this voluntary program encourages airport operators to develop Noise Exposure Maps (NEM) and Noise Compatibility Programs (NCP). NEMs identify noise contours and land use incompatibilities. NEMS are used to evaluate existing noise impacts and to discourage future development not compatible with the airport plan. If FAA approves the NEM, the airport operator can submit an NCP, which describes measures that will improve noise and land use compatibility.

In 2005, 266 airports were participating in the Part 150 program and 226 airports had NCPs approved by the FAA. An FAA approved NCP allows an airport to obtain federal aid for noise mitigation projects. Since 1982, 247 airports have received a total of $4.3 billion for this purpose in addition to AIP noise grants (discussed below).

Airport Noise and Capacity Act of 1990

In 1990, *the first comprehensive airport* noise regulation statute, the Airport Noise and Capacity Act (ANCA), became law.

ANCA recognized that a national aviation noise policy was vital to the fitness of the country's air transportation system. Former Secretary of Transportation Skinner is on record as asserting that ANCA is "the most significant piece of aviation legislation since the deregulation act." ANCA effectively altered the landscape in matters of aviation noise.

Federal noise regulations in 1990 classified aircraft as Stage 1, Stage 2, or Stage 3 aircraft, with Stage 1 being the loudest. All Stage 1 aircraft have been phased out of service. ANCA mandated that no Stage 2 aircraft could be added to the fleet or imported into the United States after November 5, 1990, and that all unmodified Stage 2 aircraft be phased out of service by December 31, 1999. Stage 2 aircraft include the 727, DC-9, and early versions of the 737 and 747. These airplanes were developed in the 1960s and 1970s.

Stage 3 aircraft must meet separate standards for takeoff, landing, and sideline measurements, depending on the aircraft's weight and number of engines. Stage 3 aircraft are the newer and quieter 757, 767, and MD-80 series, later versions of the 737 and 747, and aircraft that have been retrofitted with quieter engines by the noise reducing "hush kits."

Under the provisions of ANCA, which apply to aircraft of at least 75,000 pounds certificated weight, airport operators were specifically regulated as to when and how they could restrict Stage 2 and Stage 3 aircraft operations at their local airports, reaffirming the supremacy of the federal government over aviation policy in the United States. Airport operators were prohibited from issuing unilateral restrictions on Stage 3 aircraft, since such aircraft comprise the state-of-the-art in aircraft noise. To paraphrase, the federal government in effect said, "This is the best we can do in engine noise, these are the airplanes that are necessary to be used in air transportation, and they will be allowed to fly no matter what the locals say." Any attempted local regulation of Stage 3 aircraft would thus amount to an unlawful usurpation of the federal prerogatives regarding aviation. Subject to due process safeguards, such as notice and opportunity to be heard, airport proprietors were allowed to apply certain reasonable restrictions on the operation of Stage 2 aircraft as long as such local authorities did not impair the national policy of "phase out" articulated in the statute.

In June 2001, the International Civil Aviation Organization (ICAO) adopted new, more stringent noise standards that went into effect on January 1, 2006. These standards mandate a noise reduction level of 10 dB below the standards previously required.[12]

Much of the work for the development of a Stage 4 noise standard has taken place in the ICAO Committee on Aviation Environmental Protection (CAEP), which was established by ICAO in 1983. The United States, as a member of ICAO, participates in the work of CAEP.

In 2005, the FAA, following the lead of ICAO, issued a rule that adopted a new noise standard for aircraft designs submitted after January 1, 2006. The new FAA rule requires new airplane type designs to comply with Stage 4 standards. This noise standard is intended to provide uniform noise certification standards for Stage 4 airplanes certificated in the United States and those airplanes that meet the new ICAO Annex 16 noise standard. There is no weight limitation or exclusion for airplane type designs submitted after January 1, 2006; thus, all aircraft types will be subject to the Stage 4 noise standard.

Care should be taken to note the difference between the requirements of ANCA and the new FAA rule. ANCA, which requires compliance with Stage 3 standards, only applies to aircraft with certificated weight of 75,000 pounds and above. The new FAA rule, which requires new type designs to comply with Stage 4 standards, applies to all new airplane type design submissions, regardless of weight.

Although noise control of aircraft is exclusively a federal function, airport authorities and local governments do have the option to mitigate noise effects through land use controls, such as zoning and land acquisition, which the FAA agrees is the exclusive domain of state and local governments. Indeed, federal policy respecting Airport Improvement Program (AIP) funding favors the use of such funds for that purpose. Airport operators applying for funds for these purposes must design noise exposure maps and develop mitigation programs consistent with federal requirements to insure that noise levels are compatible with adjacent land uses. Noise compatibility projects include residential and public building sound insulation. They include land acquisition and relocating residents from noise sensitive areas. Airports have also installed noise monitoring equipment and noise barriers to reduce ground run-up noise.

ANCA also provides for additional funding sources by permitting the use of passenger facility charges (PFCs) for land use control. Airports have collected and used PFCs funds for noise studies and mitigation totaling $15 billion as of 2005.

Overall, ANCA provides a framework for the implementation of a national policy of aircraft noise control, and reaffirms that local governments have the continuing obligation to adhere to that policy and to cooperate with federal authorities to secure the achievement of such national interests. The policy is working. According to statistics supplied by FAA, exposure to airline noise has decreased significantly and consistently from 1975 to 2001. Airline noise levels are calculated using the number of persons exposed to 65 dbA, in millions. In 1975, some seven million people were subject to noise levels in excess of that number, while in 2001, the number of persons exposed to 65 dbA had declined to just 0.4 million. (See Figure 10-2.)

In 2005, FAA evaluation of the AIP set-aside program caused the FAA to set goals that will reduce residential and school populations exposed to aircraft noise at 65 dbA by 100,000 over the five year period from 2006 to 2010.

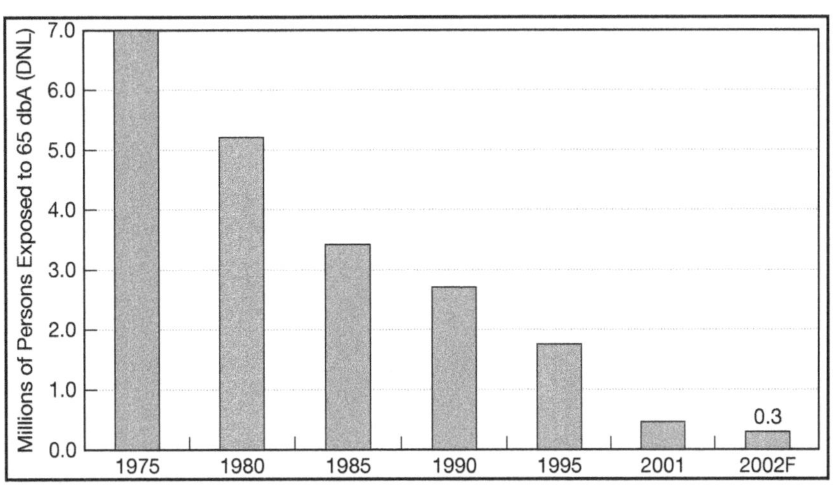

FIGURE 10-2 Significant progress in reduction of aircraft noise.

Endnotes

1. Wilson, John R. M., *Turbulence Aloft,* 34–35.
2. *Fortune* magazine, August 1946, 78.
3. Denver International Airport opened in 1995.
4. U.S. Department of Transportation, FAA/OST Task Force, Airport Business Practices and Their Impact on Airline Competition, October 1999.
5. U.S. Department of Transportation, Bureau of Transportation Statistics, Airport Activity Statistics of Certificated Air Carriers, Summary Table ending December 31, 1999, BTSO 1-03, Washington, D.C.
6. See pp 185–186 for the history of the Airport Improvement Program.
7. This was because of long-standing constitutional principles which limited federal interference with the rights of the states and their subdivisions, as confirmed by the United States Supreme Court case of ***Parker v. Brown,*** 317 U.S. 341 (1943).
8. Cited in Poole, Robert W., Jr., "More Airline Competition—Yet Another Reason for Airport Privatization, Reason Public Policy Institute
9. Switzerland and Sweden.
10. The FAA has indicated that it does not believe that this program will be effective in reducing emissions.
11. California, Massachusetts, and Texas.
12. Aircraft must be built to meet noise certification standards established by ICAO, found in Annex 16, Environmental Protection, V. 1. The first generation of jet aircraft (707, DC-8) preceded the Annex 16 standards. These are Stage 1 aircraft. Chapter 2 of the Annex applied to aircraft built before 1977, and these are referred to as Stage 2 aircraft. Chapter 3 of the Annex covered the latest production aircraft, and are referred to by the FAA as Stage 3.

CHAPTER 11

Human Factors in Aviation

Basic Concepts

Human Factors

The term Human Factors has been clearly defined by the International Civil Aviation Organization (ICAO):

> "Human Factors is about people: it is about people in their working and living environments, and it is about their relationship with equipment, procedures, and the environment. Just as importantly, it is about their relationships with other people.... Its two objectives can be seen as safety and efficiency'" (ICAO Circular 227).

The development of human factors as an area of study is derived from a series of questions set by NASA and the subsequent research these questions initiated. The study of human factors involves:

- Physiology
- Psychology
- Sociology
- Anthropometry
- Biomechanics
- Technology

The human element in aviation includes not just flight crew, but all others in the aviation community such as cabin crew, mechanics, air traffic control officers, etc. The human element in aviation is the most reliable, adaptable, and valuable part of the aviation system; however, it is also the most vulnerable to influences that can adversely affect its performance.

Aircraft Accidents

With the improvement in aircraft technology and reliability over the past few decades, aircraft accidents due to mechanical failure have declined significantly. However, human error has been cited as causal factors in the majority of aircraft accidents and incidents. Sixty to seventy percent of accidents and incidents related to the human element occurred over the past 40 years (See Figure 11-1), but through the proper application of human factors principles, they would have been largely preventable.

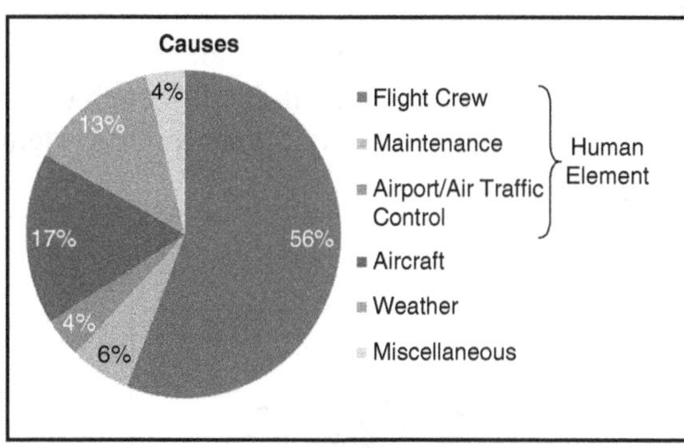

FIGURE 11-1 Leading causes of aircraft accidents over past 40 years

If the accident or incident rate is to be decreased, human factors must be better understood and the knowledge more broadly applied. The expansion of human factors awareness presents the international aviation community with the single most significant opportunity to make aviation both safer and more efficient.

SHELL Model

The SHELL model (See Figure 11-2.), first developed by Edwards (1972) and later modified into a *building block* structure by Hawkins (1984), is a conceptual model of human factors. The SHELL model clarifies the scope of aviation human factors and assists in understanding the human factor relationships between aviation system resources/environment (the flying subsystem) and the human component in the aviation system (the human subsystem) (Hawkins & Orlady, 1993; Keightley, 2004).

The name is derived from the initial letters of its components: Software, Hardware, Environment, and Liveware, and the following are the accepted interpretations of these components: Liveware—the human element; Hardware—the actual machine, controls, and switches, etc.; Software—all the procedures, symbology, manuals, etc.; and Environment—the situation in which the L-H-S system must function.

Any mismatch between the blocks (interface) is just as important as the characteristics of the blocks themselves since a mismatch can be a source of human error. The edges of the block are not drawn straight to show that all the components of the system must be aligned to each other if weak points and eventually breaks in the system are to be avoided.

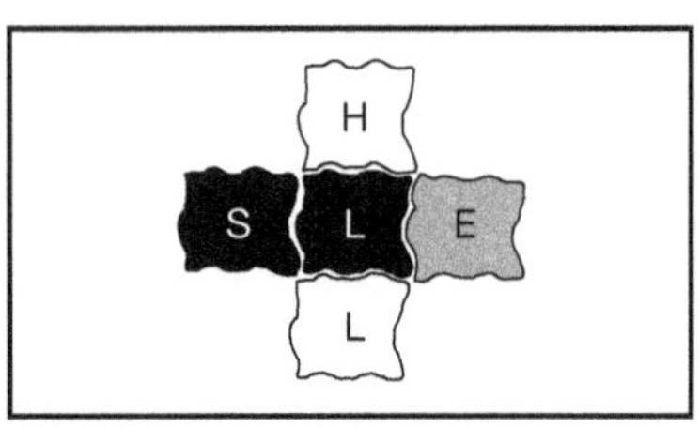

FIGURE 11-2 The hawkins SHELL.

Liveware-Software

Liveware-software is the interface between the central human element and the nonphysical aspects of the system. The nonphysical aspects of the system include such things as standard operating procedures, manuals, checklists, symbology, and computer programs. The problems in this interface are often not as noticeable as problems in other interfaces and are consequently more difficult to resolve, for example, misinterpretation of procedures.

Liveware-Hardware

Liveware-hardware is the interface between the central human element and the machine systems, the equipment the human element requires to complete the task. Although the human body is very adaptable, there is still the requirement to design the equipment, seats, displays, and controls to meet the specific needs of the human body. Problems in this interface are very difficult to identify because the adaptability of the human element tends to mask them.

Liveware-Environment

The liveware-environment is the interface between the human element and the environment in which the human is working. For a pilot this would include the environment both inside and outside the cockpit. This interface was recognized very early in aviation, and clothing/equipment was designed to adapt the human to the environment, for example, flying suits, helmet and goggles, oxygen masks, etc. Later, technology was developed that allowed the environment to be adapted to human needs, i.e., pressurization, air conditioning, soundproofing, etc. However, a pilot still faces problems associated with the environment outside the cockpit, and these need to be addressed. These include such things as ozone and radiation hazards when flying at high levels, disturbed biorhythms when crossing multiple time zones, and illusions/disorientation brought on by certain environmental conditions.

Liveware-Liveware

The liveware-liveware interface concerns the interaction between the human elements. Although working in aviation generally involves working within a team, early training and proficiency testing were done on individuals. The belief being, that if individuals were properly trained and were proficient in their jobs, then teams of these individuals would also be proficient. History has shown that this assumption is not always true and, many aviation accidents/incidents have been attributed to poor teamwork, not just by flight crew, but also by other teams within the aviation infrastructure.

This interface, therefore, is concerned with leadership, crew cooperation, teamwork, and personality interactions at all levels and in all areas of the aviation industry.

At the center of the model is the human element; this is the person who is the focus of interest within the system, i.e., the particular human subsystem. As stated previously, the human element in aviation is the most reliable, adaptable, and valuable part of the aviation system; however, it is also the most vulnerable to influences that can adversely affect its performance. The Liveware is the center of the SHELL model, and the other components must be adapted and matched to this central component. The objective of human factors is also about increasing the awareness of the central human element to the reasons why they are so susceptible to these influences and, by increasing their knowledge, making them less susceptible to these influences. As stated previously, human factors involve the study of many areas, including physiology and psychology.

Basic Physiology

The Sensory System

The different senses that humans have and use are sight, smell, touch, hearing, and taste.

Human perception is multi-sensory, and individuals use all of the five senses to acquire information about the world around them. This information is then combined with *stored* information, gained from past experiences, to arrive at a perception of the current situation/event. Everybody's perception of the same situation/event will be different, since our *arrived at* perception is unique to ourselves.

Central and Peripheral Nervous System

The main parts of the central nervous system (CNS) are the spinal cord and the brain. The spinal cord enables communication between the brain and any particular region of the body. If the spinal cord is cut, the brain loses the ability to communicate with all parts of the body below the cut. Hence, there will be no sensation from, and control over, that part of the body.

The brain has the basic function of *central processor* of different sensory inputs and is the origin of cognitive processes and the site of memory. The human brain is made up of three major components: the cerebrum, cerebellum, and brain stem.

The cerebrum is the largest part of the brain, about 85% of the total weight, according to the National Institute on Aging. The cerebrum is divided into two hemispheres—the left and right hemispheres. The Centre for Neuro Skills states that each hemisphere contains four lobes: the frontal lobe, parietal lobe, occipital lobe, and temporal lobe. The frontal lobe, located at the forehead, is responsible for emotions, consciousness, and word associations. The parietal lobe, located behind the frontal lobe toward the top of the head, controls visual attention, touch perception, and voluntary movements. Located at the back of the head toward the top, the occipital lobe is the region responsible for vision. The temporal lobe, located near the ears, deals with hearing and memory acquisition.

The National Institute on Aging states that the cerebellum is the second largest part of the brain, taking up slightly more than 10% of the brain's weight. Located in the back of the brain, below the occipital lobe, the cerebellum controls balance and coordination.

The last major component of the human brain, the brain stem, as stated by the National Institute on Aging, is located at the base of the brain and connects the brain to the spinal cord. While it is the smallest part of the brain, the brain stem controls essential bodily functions, such as breathing, heart rate, and blood pressure. If damage occurs to the brain stem, the effects can be fatal or put the patient into a vegetative state.

The peripheral nervous system (PNS) has two divisions: the somatic nervous system, which consists of the nerves that convey messages from the sense organs to the CNS and from the CNS to the muscles and glands, and the autonomic nervous system, which is a set of neurons that control the heart, the intestines, and other organs.

This whole system senses and controls the human body and therefore sometimes is called a biological control system. In a biological control system, the neuro-hormonal processes are highly self-regulated in the normal environment.

Human Information Processing

Selection and Execution

The senses are used to provide information (data), from which a mental model of our perception of what is actually happening is built. From this perception, decisions are made and action taken. The simplest functional model is displayed at Figure 11-4.

The two-way arrows represent a feedback mechanism, which is a complex process whereby confirmation of the actions taken is made by the new sensations that are received through the senses. Figure 11-5 is a more complete model of Human Information Processing (HIP) and shows the processes from receipt of information (stimuli) to the implementing of a response (action).

Attention

The two major limiting factors to processing information are the number of items that can be

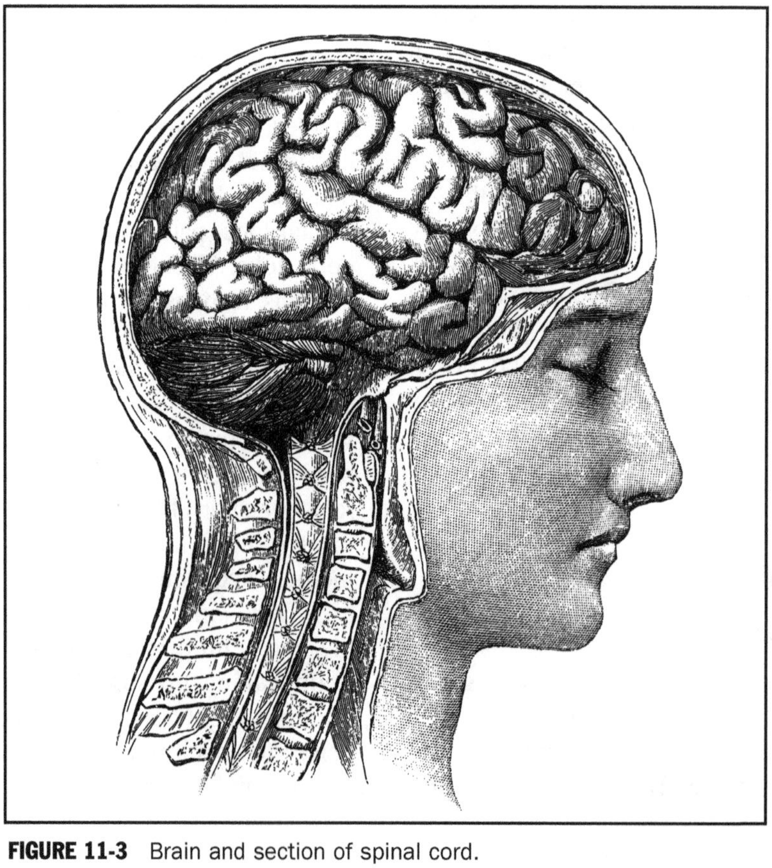

Source: Image © Hein Houwens, 2012. Used under license from Shutterstock, Inc.

FIGURE 11-3 Brain and section of spinal cord.

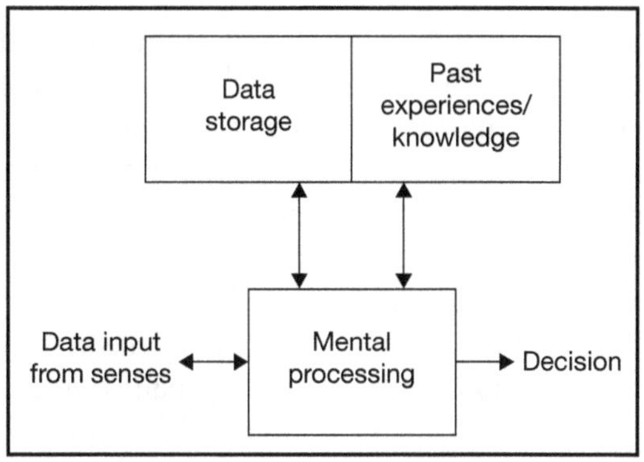

FIGURE 11-4 Simple human information processing.

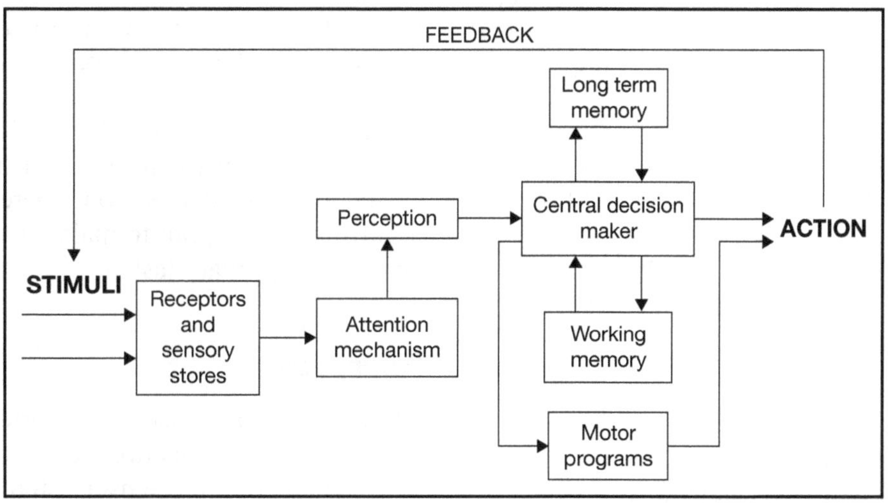

FIGURE 11-5 Human information processing.

held in working memory and attention. Attention is a term referring to the selective aspects of perception, which functions so that at any instant, an organism focuses on certain features of the environment to the (relative) exclusion of other features. Attention may be conscious, in that some stimulus elements are actively selected out of the total input, or it may be unconscious.

Attention is needed throughout several of the HIP stages, for example, during the perception stage, to let information into the mental process. Attention Mechanisms select the stimuli (sound, vision, muscular) that we can detect and then pass them on as a basis for thought and (perhaps) decision making. During that progression, attention is required to transfer information in and out of working memory and to keep information in the working memory, if required.

There are two basic types of attention that should be explored:

a. *Selective Attention.* Attention is like a spotlight at night in that it will not illuminate all the stimuli, but it can be directed to those stimuli that are found attractive. (The advertising industry is built on this phenomenon!) It determines the way inputs are sampled at the sensory store (e.g., how relevant is the information?) and competing demands on attention can lead to important information being missed/omitted. An example of this is that if pilots are dealing with one malfunction, then they can miss another important action, perhaps an altimeter check during the descent. However, if certain stimuli impinge on the senses (e.g., our name is spoken), attention may be diverted to the new stimulus. This is known as the cocktail party effect.

b. *Divided Attention.* Within certain personal limits, attention can be divided. This is seen as being able to do two things at once, like driving a car and talking to the passenger. However, if the demands of the task exceed the mental processing capacity, any further stimulus is going to cause one or both of the tasks to have an error.

Which of the following do you think you are most likely to be able to do at the same time?

Fly "straight & level" + solve a problem.

Read a book + listen to a radio message.

Write notes + carry on a conversation.

Listen to lecturer + solve a problem.

When thinking about the concentration and mental effort in each individual action, the least demanding combination is to fly straight and level and solve a problem. However, although the use of motor skills will free up the attention mechanism for the problem at hand, caution should be taken to avoid being consumed by the problem such that the motor skill goes unmonitored.

Vigilance

Vigilance is defined as maintaining overt attention to a perceptual task, often with infrequent stimulus events, for prolonged time periods. For example, imagine driving your car down a long, straight, and featureless road. Every now and then, another vehicle may grab your attention, but in between, the tendency will be for your attention to wander away from the task of driving safely. When maintaining vigilance, the level of arousal has a significant effect on performance. It does not follow what might be thought of as common sense, because performance does not always improve with increased levels of arousal. In fact, performance peaks at an intermediate level of arousal.

Hypo-vigilance is a lowering of attention to a perceptual task, something of particular concern to long-haul pilots, especially those on overnight flights. The key to maintaining vigilance during such a flight is in maintaining a moderate workload or level of stimulation. In the *automated cockpit* at night, the workload is lower because it is quiet up in the front, it is dark outside, and the body is at its lowest part of the temperature cycle, all of which conspire to lower the pilot's vigilance.

If, however, the workload is high, then vigilance may be lowered for the same reasons as discussed in divided attention (there is too much occurring for the pilot to maintain complete vigilance to the primary tasks).

Perception

Perception is the conscious experience of objects and object relationships, converting physical information from the sensory stores into an individual's conception of reality. It is, collectively, those processes that give coherence and unity to sensory input, for example, patterns of sounds converted into a message. This is an error-prone process because the brain uses past experience stored in memory and will rely on its expectations to recognize the stimulus. To this is added the wishes and desires of the individual, and the three elements apply pressure to sensory inputs, to create a perception.

Memory

Memory can be split into three separate forms, namely, working memory, long-term memory, and motor (skills) memory. All of these have three elements in common, which are encoding, storage, and retrieval.

Working Memory (Short-term Memory)

Information is retained for a short period in working memory, for example, looking up a telephone number or remembering a frequency you have just been passed. Generally speaking, however, the numbers will be forgotten unless they are rehearsed, and so you would probably repeat them in your head until you had dialed them. Acoustic information is easier to remember than visual because it is easier to rehearse sound than written data.

The capacity of the working memory is limited and, on average, people can retain seven (+/– 2) unrelated items. Exceed this number, and one or more items are likely to be lost or transposed. Chunking, a process where items are either artificially related or are conveniently grouped together and therefore only take up one station, seems to expand memory. However, unless actively rehearsed, items will be lost in 10–20 seconds by interference from previously stored information (pro-active interference), or by new information (retroactive interference).

Long-term Memory

Long-term memory (LTM) stores a great deal of information about one's past. One of the elements stored in long-term memory is rule-based behavior, drawn from LTM to a conscious level and actioned in conjunction with the central decision-maker. Also of relevance to LTM is knowledge-based behavior, where the central decision-maker compares the current situation with memories (events stored in LTM), to derive a plan of action.

Long-term memory is further subdivided into two components called the semantic and the episodic memory:

a. *Semantic Memory.* Semantic memory is the memory of facts, for example, the meaning of a word or where a place is on the map. It is generally believed that information is never lost from the semantic memory, although at times, an individual may have great difficulty remembering. In fact, the information is still there, but sometimes, one simply forgets exactly where the information is filed.
b. *Episodic Memory.* Episodic memory is *personal* knowledge relating to a person or, more specifically, events in that person's life. An interesting feature of this memory is that, although stored, it is not static, but influenced by desires and expectations, and thus, subject to modification. This is a great problem when reconstructing an accident from eyewitnesses, and often demonstrates the difference in evidence gathered from an expert and a layman. Most amnesia affects episodic memory, but not semantic memory. One may not remember who one is, but still know where New York City is!

Behavior

The Skills, Rules, Knowledge (SRK) framework defines three types of behavior or psychological processes present in operator information processing (Vicente, 1999a). The SRK framework was developed by Rasmussen (1983) to help designers combine information requirements for a system and aspects of human cognition. The three categories essentially describe the possible ways in which information, for example, from a human-machine interface is extracted and understood:

Rule-based Behavior

Rule-based behaviors are those that use a learned routine or procedure, such as following a checklist. The execution of this behavior may require the use of a set of skills and, because this behavior is based on rules, it is stored in the LTM. These behaviors are actioned at a conscious level and require the utilization of the central decision-maker and working memory.

Knowledge-Based Behavior

Whereas rule-based behavior relies on the application of a set of rules, knowledge-based behavior depends upon memory and recall. Since the knowledge upon which the behavior will be based is stored in LTM, the LTM is very important.

Skill-based Behavior

Humans practice chains of events to form skills, and the more practice done, the less attention needs to be given to the task. As the task becomes more familiar, it forms a behavioral sub-routine called a motor program, such as driving a car for example. An experienced driver drives without much effort because humans do not need to rely on frequent feedback telling them that they are successfully performing the activity. One advantage of a skill is that it frees attention, which is a limited resource. Skills, however, are prone to errors called action slips, when the motor program mixes up the sequence of events (pouring tea in the sugar bowl instead of the cup, for example). Action slips tend to occur at the initiation phase, rather than the execution phase, of an action. Since there is no feedback, because no attention is being paid to the skill, it can take some time for the error to be recognized. For example, pilots have landed wheels-up as a result of inadvertently lowering the flap rather than the undercarriage.

Phases of Motor Memory

Three phases are generally associated with motor memory: the cognitive phase, associative phase, and automatic phase:

a. *Cognitive Phase.* As the name implies, this is a thinking phase, during which the subject must consciously think about the actions within the task and the task as a whole.
b. *Associative Phase.* During this phase, the subject must practice each element of the task, until it can be completed successfully, and then each step is then combined to form the total task. Anderson calls this stage knowledge compilation.
c. *Automatic Phase.* In the automatic phase, the total task is capable of being executed as a smooth entity, without conscious control. All that is usually required is a conscious decision to initiate the total task. Anderson calls this phase procedural knowledge and at this stage, it is difficult for individuals to describe their skill to others. The skill knowledge is described as nondeclarative knowledge.

Receptors and Sensory Stores

Stimuli (light, sound, motion, etc.) are detected by the sense organs (retina, cochlea, semi-circular canals, etc.) and then transduced into electrical representations. The representations are stored briefly in the Short-term Sensory Store (peripheral or sensory memory).

For the information to progress further, attention must be paid to it, and we have already discussed the nature and limitations of the attention mechanism. The senses are continually bombarded by stimuli, possibly more than the brain can cope with, and therefore, the brain needs to let some of this information dissipate while retaining that which is believed to be important, so the short-term sensory store acts as a filter. Information will only be held in the sensory stores for a limited period, listed below, and if attention is not paid to a stimulus within the quoted period, it will be filtered out (lost).

Visual (iconic) store = 0.5–1 second.

Auditory (echoic) store = 2.0–8.0 seconds.

Kinesthetic (motion) store = 2.0–8.0 seconds (muscular/movement perception, e.g. touch/press on something—you can still feel the pressing after lifting your hand).

Learning

Response Selection

Learning is said to involve the acquisition of new skills or information, and the main methods by which this is achieved include the following:

a. Behavioristic approach:
   i. Classical conditioning.
   ii. Operant conditioning.
b. Cognitive approach.
   i. Latent learning.
   ii. Cognitive maps.
c. Observational learning.
   i. Skill learning.
   ii. Forming associations.

Behavioristic Approach

Classical Conditioning

This process involves the pairing of a neutral stimulus with a stimulus that already triggers the desired reflexive response until the point is reached where the neutral stimulus elicits the response.

Operant Conditioning

Operant conditioning (or instrumental conditioning) requires the use of some kind of reinforcer to encourage the required response. Experiments using animals have been used to illustrate this concept, such as laboratory rats having to operate a lever to get food. In this case, pulling the lever is the required action, and the food reward is the reinforcer. With training, the speed of response increases as the subject learns. Take away the reinforcer, however, and the speed of response will drop markedly.

Cognitive Approach

Conditioning should not be mistaken as a method of cognitive learning, which is more a process of gaining and remembering knowledge. Cognitive processes are evident in the way that different people represent, store, and use information, elements that are important in the learning process.

Latent Learning

Latent learning is that form of learning that is not evident when it first occurs. At some later time, when a reward is offered, the subject is very quick to perform the desired task. An example of this is the child in school, who appears either to not understand or not care about a particular type of problem. When the reward of an early finish to class is offered, the student is quick to show mastery of the problem.

Cognitive Maps

Cognitive maps are mental models of the world. When you first drive to your new place of work, it will take quite some time to become familiar with the various routes to it. When a cognitive map has been developed, the task of driving to work will become almost automatic.

Observational Learning

The process of learning by watching others is called observational learning, or, in some cases, simply mimicking or social learning. In this process, one learns by watching how others behave. We can identify four basic requirements for observational learning, which are:

a. *Attention:* One must pay close attention to what is happening, in order to learn.
b. *Retention*: One must be able to remember that which is being observed.
c. *Ability to Reproduce the Behavior*: One must be capable of performing the act.
d. *Motivation*: Generally, an individual will only perform an act if there is some reason for doing so. Likewise, one is most likely to imitate those whom they see rewarded for their actions.

Skill Learning

Learning a skill depends mainly on imitation or following instructions combined with a lot of practice, and also involving automatic actions and feedback. However, not all people have the co-ordination skills required to be a good pilot, and therefore, an element of natural ability still is involved.

Learning Techniques

Several techniques are used to enhance learning and memory. The following paragraphs give an overview of these techniques.

Memory Techniques

Memory techniques are broadly divided into internal and external strategies.

a. *Internal Memory Aids:* The main internal memory aids are mnemonics, methods for remembering information that is otherwise quite difficult to recall. A very simple example of a mnemonic is the "30 days hath September" rhyme. The basic principle of mnemonics is to use as many of the best functions of the human brain as possible to code information.
b. *External Memory Aids:* These are strategies for improving memory that are outside the body, in other words, mental training. There are many different forms of aids to mental training available.
c. *Noncommercial Memory Aids:* These are common aids that many people use, such as lists, notes, and calendars—checklists for example. Another noncommercial aid is the symbolic reminder, like a rubber band around the wrist, although this type of aid does not remind the person of the exact message.
d. *Reminding Service:* This type of aid serves to remind people about important obligations. Many healthcare professionals use reminder services to reduce patient "no-shows."
e. *External Knowledge Sources:* These sources, such as the textbooks you use for studying, contain information that a person has forgotten or information that a person never knew.

Skill Development

We have already said that skill learning involves automatic actions, practice, and feedback. Further development of the skill involves the same elements, which means that one should repeatedly practice the skill until its actions become automatic. Feedback should be obtained after each attempt to provide information on corrective action, so that the next attempt should be closer to perfect. Feedback also provides positive reinforcement of a task completed successfully, an extremely good tool for achieving desired results.

Motivation

Motivation is the catalyst that provides the desire to achieve a goal. The goal of every airline pilot is, or should be, to consistently demonstrate a high performance standard. To perform well, it is necessary to gain skills and knowledge, both of which require some kind of learning process. To learn anything effectively, the pupil must be able to pay attention to the teacher. If motivation is not present, then attention is greatly reduced. Thus, no motivation leads to lack of attention, leading to a lack of skills and knowledge, which will result in a poor performance.

Response Execution (Actions)

Sooner or later, intention has to be translated into a response or action, which may be in the form of a sequence of motor commands. The individual takes conscious actions or gives conscious oral commands, unless there is already a motor program in the memory.

Factors affecting response execution (actions) are:

a. If the result of a delay could be serious, then stress may become a major factor.
b. To respond too quickly can lead to error and, in most cases, there is a human tendency to trade speed for accuracy.
c. Arousal leads to a faster but less accurate response. Since auditory stimuli are more arousing than visual, there is a greater chance of error.
d. An expected stimulus and response is quick and accurate, e.g., ATC instructions during a normal approach.

e. Age is a factor in that, from age 20 to 60, the response becomes slower but is generally more accurate.

Response Times

Response or reaction time is the interval between the onset of a given signal and the execution taking place. For example, pressing a single button to illuminate a single light takes about 0.2 sec to respond to. Having a choice of two buttons to press, depending upon which of two lights you wish to illuminate, takes longer because you need to decide which button to press. It would be a good design factor to ensure that the light was adjacent to the button that illuminated it, but removed from the other button/light combination. This view is supported by the accident report recommendations of the B737 accident at Kegworth, UK.

Feedback

All the time the brain is processing the information, there is a self-monitoring system in action. This process, known as feedback, is a self-check on each of the stages discussed so far; the feedback loop extends back to the input (the sensory stores). The flying environment is one of many environments where a continuous monitoring process must be in action, to keep the pilot aware of the consequences of his/her actions and thus avoid unsafe situations.

Influence of Stress and Arousal

Stress increases the sense of arousal, influencing for example, the scanning pattern employed at different phases of flight. During the cruise, there is low arousal and a low scan frequency, whereas during the approach and landing, arousal increases and the scan rate becomes high. When stressed, the sampling rate increases, but may become focused or fixated, restricting the attention to the primary task. This effect is also called fragmented performance and involves a narrowing of the attention spotlight.

Integration of Sensory Inputs

This section deals with some of the problems that may occur in the human version of information processing. As discussed, mental models are based upon perception (i.e., past experience and expectations) and an illusion occurs whenever the mental model differs from the real world. Some examples of visual illusions are at Figure 11-6. An important situation is when large, distant objects are confused for smaller, closer ones. An expectation of seeing large distant cliffs on breaking cloud was satisfied when the pilot of the Air New Zealand DC-10 approached Mount Erebus in the Antarctic. What he actually saw were smaller cliffs much closer, but he did not realize it because his mental model was confirmed.

Visual Perception

Five important visual cues give the mental model an impression of depth, or 3-D, and these are:

a. *Convergence:* the amount our eyes converge to focus a visual target onto each fovea.

b. *Stereopsis:* an object produces an image on each retina. These images differ more from one another when an object is close than when the object is at a distance. This is sometimes called Binocular Vision.

c. *Obscuration:* near objects occlude (or blur) distant objects.

d. *Atmospheric Perspective:* objects become less colored and less distinct the further away they are.

e. *Retinal Size/Angle:* objects become smaller/subtend a reduced angle on the retina with increased range.

FIGURE 11-6c&d Some common visual Illusions.

Visual Perception in Flight

Approach and Landing

Final approach to a landing is the phase of flight that requires the most critical judgment. Visual cues are needed for distance and height assessment, for judgment and maintenance of the glide path and, finally, ground proximity before touchdown. VASIs (Visual Approach Slope Indicators) and PAPIs (Precision Approach Path Indicators) help initial judgment of the glide path, as does positioning the aircraft at predetermined heights above known ground features (local knowledge). Retinal size is vital in the final approach where height is likely to be judged by the retinal size of the runway. To use retinal size as a cue, the pilot does need to have sufficient local knowledge of the likely size of the runway (expectation). To maintain a constant glide path, the visual angle between the horizon and the impact point must also be constant.

In conditions of little or no horizon, i.e., poor visibility or at night, it is still possible to judge where the horizon is. You can use the extended sides of the runway as they converge at the horizon (but only if the runway is level). Texture gradients (if visible and also level) in the surrounding terrain also help to establish the horizon. However, if the runway or terrain slopes, then it causes incorrect horizon placement and poor approaches.

To maintain the visual glide slope:

a. Aim for the visual impact point (texture flows away from the impact point).
b. Keep the visual angle constant (at the angle of approach).

Height Judgment

The two principal visual cues in height assessment are the apparent speed of ground texture passing beneath the aircraft (ground proximity), and the retinal size of known objects on the ground. The cues can be removed if the approach is made over water, if snow covers the ground, if the terrain is featureless, or at night. Some trickier situations can arise if the runway width is wider or narrower than that normally anticipated. The pilot might think he/she is lower/higher (respectively) than reality, leading to overshooting the threshold or impacting in the undershoot!

An up-sloping runway can also induce a low approach if the pilot is unaware of upslope. He/she may use a visual assessment to position for a 3° glide path, and perspective says that all is well. However, the glide path will be a shallower angle and the distance between aiming and touchdown points increases accordingly. The obvious danger is that he/she may well land in the undershoot. Good approach lighting can help by giving visual cues in the undershoot to aid the judgment of ground proximity.

Constant Relative Bearing

If there is a line of constant bearing between two aircraft and the distance between them is decreasing, then there is a collision risk, even if the aircraft is climbing/descending. What this means is that the other aircraft appears to be stationary (no relative movement) to you, and if there is something like a windscreen pillar between your eye and the other aircraft, you are not going to see it until very late. This is due to the fact that the retinal size of the hazard aircraft increases at a nonlinear rate.

Figure 11-7 shows the rate of change of apparent size of another aircraft closing head-on. This is believed to be a contributory cause of a DC-9/Trident mid-air over Zagreb.

Visual search for other aircraft is constrained by the lack of relative motion between aircraft on a collision course. The other aircraft must first be detected, generally at the point of the highest acuity, i.e., at the fovea, as the rods will not pick it up (in controlled conditions, only 51% of other aircraft were seen, even when the pilot was alerted to their presence). Eye movement is a factor that can be utilized to improve the chances of detection. However, the eye does not move smoothly but moves in jerks with a rest in between, each movement/rest cycle being known as a saccade. Each saccade takes about a third of a second,

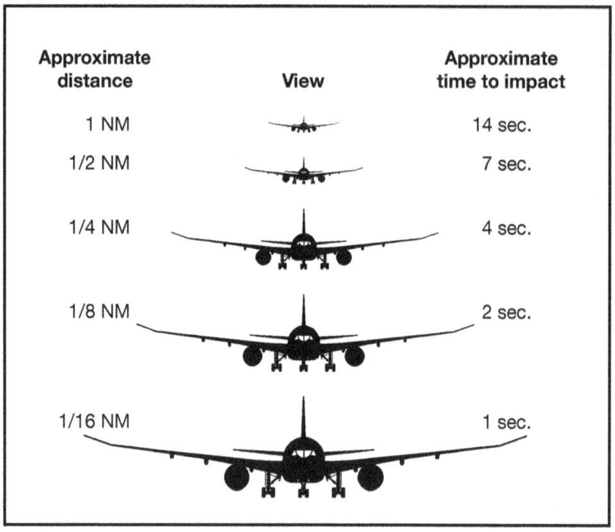

FIGURE 11-7 Rate of change of apparent size.

and during the rests, the brain samples what the eye sees. Keeping the eye moving and thus reducing the rest cycles can increase the sampling rate. Thus, the best scan technique is to make as many eye movements as possible with the head still, or moving slowly.

Illusions

Illusion is defined as a false impression formed when the brain misinterprets sensory information. Illusions in aviation are mainly of three types:

a. *Visual:* relating to the input from the eyes.
b. *Vestibular:* relating to the input from the vestibular system (semicircular canals and otoliths).
c. *Central:* Specific to the higher mental processes (imagination).

Visual Illusions in Flight

The visual sense is the most important contributor to perception of orientation. Misinterpretation of visual information is a visual illusion and will lead to spatial disorientation. Some of the more common visual illusions are:

a. *Sloping Cloud Bank:* accepted as the horizon.
b. *Ground/Sky Confusion:* well away from large light sources, it may be difficult to differentiate between fishing boat lights and stars.
c. *Autokinetic Illusion:* tiny motions of the neck and eye muscles can induce motion into stationary single light sources like stars.

Black Hole: Another danger is the black hole phenomenon. This is particularly relevant when approaching airports at night over the sea, jungle, or desert. When all around is dark except the distant runway or airport lights, there is an illusion of distance. The pilot will believe that he/she is closer to the airport than reality, possibly resulting in an early descent. This phenomenon was cited in the report of a 707 accident in Pago Pago, Samoa, where all on board were killed as the plane struck the sea some distance short of the runway. A combination of the black hole and up-sloping lights phenomena led to the destruction of four Boeing 727s in 1966. Each approach was made visually, at night, over water, into runways with up-sloping city lights behind.

Empty Field Myopia: An effect called the Mandelbaum Effect draws focus closer to the observer because of things like windscreen pillars and dirt on the windscreen. This makes detection of distant objects less likely. If there is nothing to be seen, day or night, focus tends to settle one to two meters ahead of the observer rather than at infinity. Imagine flying at high level by day over a calm blue sea with no clouds.

Vestibular Illusions

The vestibular system is a fairly sluggish and approximate system that can cause two types of illusions: somatogravic and somatogyral illusions—the body's response to gravity and rotation. When these are combined with poor visual cues, two more types of illusions can be generated: oculogravic and oculogyral illusions—eyes' response to gravity and rotation.

Gravic Illusions

Gravic illusions relate to the inability of the pilot to correctly sense (other than by good visual cues) the vertical, i.e., that defined by gravity. For example, during a prolonged turn, the semi-circular canals will sense the initial application of bank. This input will dissipate, however, if the rate of the turn is constant. The cues are similar to level flight, i.e., gravity appears to be acting vertically through the body, even though the body is angled to the horizon. During a flat turn, with no bank, 'g' is thrown outward causing the otoliths to create the illusion of an 'outside' turn.

The otoliths can also produce illusory effects during acceleration and deceleration. During acceleration in level flight, the resultant of 'g' and the inertial force due to the mass of the otoliths causes them to bend backward, giving the illusion of a pitch up. If the pilot 'corrects' according to this illusion, i.e., bunts the aircraft, the lessening of 'g' and the increased speed caused by the dive combine to throw the resultant even further backward, i.e., greater pitch up. The more the pilot bunts, the greater the pitch up sensation. As an example, many years ago during a missed approach, a Vanguard aircraft nose-dived into Gatwick airport. The cause was attributed to such a somatogravic illusion in pitch. Unfortunately, for the pilot, the effect of the acceleration on his air-driven artificial horizon would have 'confirmed' his illusory perception of pitch up.

If poor visual cues are present, then the brain will accept the vestibular sensations and provoke reaction to its own perception of motion. This is the same phenomenon as causes the room to continue spinning briefly even after open the eyes when intoxicated. If the mental model formed by the brain says 'pitch up' and it detects a stationary light source on the retina, then it will assume that the light has pitched up as well; this is known as an oculogravic illusion.

Gyral Illusions

Gyral illusions are caused by the decaying input from the semi-circular canals during prolonged constant speed turns. During the turn, the input will decay (over 15–30 sec), giving the illusion of straight and level. When the pilot rolls out of the turn, then the illusion is that he/she has rolled into a turn in the opposite direction—a somatogyral illusion. This feeling will decay over time, but again, it can lead to the movement of poor visual cues to produce revolving lights—the oculogyral illusion.

Coriolis Illusion

During maneuvering, particularly if this is associated with head movements, complex interactions between the cupola and otoliths (both forming the static organ) can produce a tumbling illusion—the Coriolis illusion. This is caused by cross-coupling in the semi-circular canals. Simply put, this means that rotation about two axes of the semi-circular canals is sensed as rotation around the other axis.

The Leans

One of the features of the semi-circular canals is their lack of sensitivity. Rotation at low rates of <0.5 deg/sec will not be sensed because they are "sub threshold." The leans will occur when a pilot allows a wing to drop at a sub-threshold rate, i.e., undetected. The pilot believes he/she is still straight and level, but if the pilot notices the drop (on the attitude indicator), he/she will pick the wing up. This will be done, however, at a rate that will be detected. The pilot's perception is, therefore, of having bank applied, even though the instruments say otherwise. To resolve these sensations, the pilot will lean over to align him/herself with the illusory vertical.

Get on Gauges

The obvious conclusion that can be drawn from the material above is that our vestibular system is not to be trusted. Experienced pilots have been handed control of an airplane in straight and level flight in cloud and asked to fly without reference to instruments. The longest any of them lasted, before a loss of control that required 10,000 feet of airspace for the safety pilot to recover, was about a minute and a half. Therefore, in these conditions, it is imperative that pilots fly by reference to the instruments and believe these rather than trust their senses.

Central Illusions

The brain, without major inputs from the visual or balance mechanisms creates central illusions. They are often called Disassociative Sensations or Break-Off Phenomena. Some flight crew have reported sensations with varying severity, from a break off with reality to out-of-body experiences. Typically, these are described as a feeling of detachment (and can be pleasant), or the feeling of being balanced on a knife-edge, where 5° angle of bank feels like 30°. They tend to affect flight crew on long, high-altitude, straight transits, when they are in little contact with other people, i.e., boring sectors when you may allow your imagination free reign. If you are affected, concentrating on the job, or human contact, will usually dispel the illusion.

Flicker Vertigo

Another source of discomfort to aviators can be flicker vertigo. This is most frequently encountered in rotary wing and single-engine propeller aircraft. Some of the conditions that can produce light flicker are:

a. Rotor blade passing over the cockpit, usually worse around the middle of the day, when the sun is overhead.
b. Slowly revolving propeller exaggerated by flight into a rising or setting sun.
c. Reflecting anti-collision/strobe/position lights, especially in restricted visibility, e.g., the gloom in cloud.
d. Windscreen wipers.
e. A vibrating instrument hood.
f. A searchlight deliberately arranged to flicker.
g. Polarized sunglasses used in aircraft with electrically heated windscreens. Light waves, generated by certain types of heated windscreens, flicker when seen through polarized sunglasses.
h. Looking through a moving propeller.

Light Flicker protection measures include:

a. Turn the aircraft to a point where sunlight does not flicker through the blades.
b. Turn off interfering lights.
c. Remove polarized sunglasses.
d. Vary RPM.
e. Make passengers keep their heads still and not look out the cabin windows.

Fatigue and frustration tend to increase the annoying quality of flicker vertigo and make the manifestations more pronounced. Low g tolerance, hypoxia, and hyperventilation are believed to increase susceptibility to the effects of light flicker.

Decision Making

Rational and Intuitive

The science of decision making can be divided into two basic parts: rational and intuitive. The rational part involves the use of reason, or the process of reasoning, and is primarily cognitive in nature as opposed to emotional. The intuitive part is characterized as direct and immediate, and occurring without conscious thought or judgment. A problem is a difference between an actual situation and the desired situation, and a decision is a choice among alternative courses of action for dealing with the problem. Thus, problem solving is the process of identifying a discrepancy between an actual and desired state of affairs and then taking action to resolve the discrepancy, which will involve decision making.

Decision-making Process

The decision-making process may be divided into five basic stages, which are:

a. *Analyze the Situation and Define the Problem:* To define the problem, it is necessary to gather input from all the available sources. This includes input from other people, as well as from any instrumentation.
b. *Evaluate Solution Alternatives:* There are usually several possible solutions to a single problem, so it is wise to try to evaluate each one and not to force oneself into only one choice. To do this, one must be objective and be sure that each solution solves the problem. Furthermore, care must be taken not to fall into the trap of merely solving the symptoms. British Midland 737 crash at Kegworth, UK, is a good example of this.
c. *Choose One Alternative:* Out of all the choices, one must be selected after looking at all the positive and negative consequences. At this stage, it is very important always to keep the ultimate goal in mind.
d. *Implement Decision:* Implementing the decision, in this business, means assigning tasks so that the necessary actions are carried out.
e. *Evaluate:* Evaluation, or review, is possibly the most important stage of the decision-making process. It is vital to ensure that the actions carried out produce the desired result, and if not then now is the time to implement changes as necessary.

Three decision-making methods are generally available to aircraft captains, and the key to using these methods is in knowing when, and how, to use them effectively. They are:

a. *Individual Decisions:* The captain acts alone and makes the decision alone. In this case, it must be presumed that the captain has sufficient information and understanding to make a good decision.
b. *Consultative Decisions:* The captain seeks advice, information, or opinions before making the decision, but still makes the decision himself.
c. *Group Decisions:* The captain consults with the crew and asks for assistance in making the decision. This method results in an increased acceptance of the final decision and a better understanding of the decision.

Decision-making Styles

Democratic

Democratic decision making allows the crew to vote, and the majority vote makes the decision. Advantages of this style include group participation and contribution, as well as a fairly fast decision. The major disadvantage is the lack of individual responsibility, which would be problematical in aviation, since the captain should remain ultimately responsible for decisions and actions.

Autocratic

Autocratic decision making allows the captain to maintain control and ownership of the decision, and thus, he/she is responsible for the outcome of the decision, whether good or bad. There are advantages in that there is personal responsibility and a fast decision, but a major disadvantage is a lack of desired effort to carry out the decisions by those who were not responsible for the choice made.

Consensus

Consensus decision making is when the captain gives up control of the decision and only the crew is involved in the process. The captain, therefore, is not responsible for the outcome, but rather the entire crew is responsible. Advantages include group commitment and responsibility for the outcome but, once again, the decision-making process is very slow and extremely time consuming.

Collective-Participative

Collective-Participative decision making involves the captain encouraging the members of the crew to take part in the decision-making process. The captain maintains control of the decision, and the crew is used for the information they can provide. The advantage of this approach is the positive nature of group involvement, but a disadvantage is that the decision process is very slow and time consuming.

Decision-making Concepts

In aviation, rule-based behaviors are of most importance in emergency situations, and thus, they tend to be committed to documentation. Provided that the use of such documentation would not inhibit the actions required by the emergency, there is no point committing such things to memory when a checklist would do the job. These procedures are regularly practiced in the simulator, and only the initial actions (e.g., calling up the checklist) are committed to memory. In this scenario, although errors are minimal because of the documentation, one must guard against misidentification of the emergency, possibly leading to the selection of the wrong checklist.

Error Sources and Limits in Decision Making

There are various sources of error and the respective limits in decision making, the most obvious of which is the limited human decision-making channel. A pilot receives information from a variety of sources simultaneously, but only one decision can be made at a time. The channel is therefore time-shared between the various input systems. The information remains either in short-term memory, or in the sensory stores, until capacity is available in the channel. Sometimes, this results in the loss of information from the sensory stores.

Although people may feel that they can do several things at once, this is really a result of rapid scanning of the input channels.

Possibly the biggest source of error in the decision-making process is the false hypothesis. Usually, a false hypothesis is formed on the basis of a faulty or incorrect piece of information and, once formed, the human mind sets about proving that it is correct. Unfortunately, this often means that the mind ignores all information that does not support the hypothesis. There are a number of situations in which a false hypothesis is likely to occur:

a. *High Expectancy:* If many experiences of a similar situation have resulted in a particular action, then when these conditions arise again, the expectancy is high that the outcome will be the same. For example, in a certain situation, the pilot gets a specific ATC clearance One day, however, the clearance is different and the tendency is to fill in the space with what is expected.
b. *Diverted Attention:* Pilots will often make assumptions about other aspects of the flight when their attention is focused on a problem.
c. *Defense:* Humans do not like to hear news that is bad or uncomfortable and, as a result, the incoming message sometimes is subconsciously modified to a form that is more acceptable.
d. *After High Concentration:* After a period of high stress or concentration, the mind tends to relax and adopt a less cautious approach. This relaxation may be too great and result in a bad decision being made.
e. *Motor Memory Efforts:* This is the result of selecting the wrong motor program. Motor programs are not monitored or challenged,

thus a pilot will always believe that the correct program was selected. A simple example is sitting at the breakfast table, reading the paper, and mistakenly pouring the tea from the teapot into the sugar bowl. It may be sometime before the mistake is realized—probably when taking the first sip of very sweet tea!

Finally, because the decision-making channel is limited in its capacity, it is also subject to overload. The usual reaction to overload is to carry out "load-shedding." Sometimes when this occurs, vital information may be discarded, resulting in decision-making errors due to incomplete information.

Decision-making Models

Decision-making models are designed as a tool to aid an individual in the making of a correct, or most appropriate, decision. The normative model is the one most commonly used, involving all the decision-making stages covered earlier. A good example of a decision-making model based around a mnemonic is one used by British Airways, DODAR, which stands for:

D Diagnose
O Options and consequences
D Decide
A Assign tasks
R Review

Decision-making Processes

Skill-based Decision Making

Stored routines, such as motor programs, are fully utilized during skill-based behavior. These skills have been learned with practice such that, although one may consciously initiate them, little or no conscious thought or attention is required thereafter.

There are two ways of acquiring skills, the first of which is by concentrating on individual aspects or components of single events and practicing, such that they become integrated into a smooth and automatic process. The second is by practicing the whole pattern or procedure until it produces consistently accurate results.

Where an individual has a particular skill, but cannot pass it on nor describe it, then it is known as nondeclarative knowledge. For example, if a golfer is asked to fully describe his swing, he may have difficulty describing it, but can demonstrate it with ease. This is why flying instructors need to be trained on how to instruct and not just how to fly.

If a skill is thought about, the execution is often spoiled. Certainly in civil commercial flying, it is usually better to trade off the slick but unmonitored skill for the monitored action (Action Slips, Environment Capture, etc.).

Rule-based Decision Making

Rule-based behavior involves the use of a learned routine or procedure that carries a set of rules (e.g., the order of the response to a fire warning). In aviation, most sets of events are anticipated and reduced to a set of rules. Rule-based decision making does not rely on motor programs, but is stored as sets of rules in the long-term memory. It requires a decision followed by conscious actions, thus the pilot is unlikely to suffer either an action slip, or environmental capture.

Training in flight simulators or procedural trainers is designed to encourage rule-based behavior by practicing the use of Standard Operating Procedures (SOPs). Emergency procedures (rule-based) that cannot be fully practiced when flying can be practiced in simulators, refreshing the long-term memory. Some procedures can be too involved to be committed to memory and therefore only vital actions are memorized, but supplemented by checklists. There is little virtue in unnecessary memorizing and rule-based decision making should rely on documentation, unless circumstances require immediate action.

Long-term memory failures are rare and SOPs generally robust, but errors can occur in rule-based behavior because of misidentification of the problem. Such misdiagnosis may cause an incorrect procedure to be used, for example, an audio warning signifying cabin pressure failure is misinterpreted as an over speed. Another route to error is if the pilot believes it is safe to depart from a procedure, for instance, making a visual approach into an illusion-prone airport, even though the SOP is instrument approaches only.

Knowledge-based Decision Making

A knowledge-based decision-making process is one that follows a situation where there is no established procedure. In this scenario, the pilot evaluates information, and then uses knowledge and experience to deal with it. The DC-10 that crash landed at Sioux City, USA, in 1989 following catastrophic tail-engine failure and subsequent loss of all flight controls is a classic example. The pilots of the DC-10 used their knowledge and skill to control the airplane by differential engine power instead the flight controls. This ability to apply knowledge-based decision making is the prime asset man has over the machine. An autopilot can exercise most of the skill-based tasks facing the pilot, and computerized management systems can apply the rules, but a pilot must evaluate unusual information and reach a conclusion.

When a situation requires the application of knowledge-based behavior before a decision is made, normal human behavior is to look for patterns or rules that fit the evidence. The natural tendency is to then accept evidence that confirms the hypothesis reached, but disregard that which does not fit—a condition called Confirmation Bias.

A crew, on the approach into Nairobi, thought that they had been cleared to descend to 5,000 ft (Nairobi is approximately 5,600 ft above mean sea level!). The crew disregarded the evidence from the ILS glideslope indicator, which showed them to be well below the glide path. Even when they broke cloud, they accepted the unusually large trees and bushes as a visual illusion. It was only the unusually large size of the giraffe and antelope that caused them to save their situation.

Risk Assessment and Options

As the captain of an aircraft, an individual is expected to assume a managerial role in the aircraft operation, and part of this task is the assessment of risk factors, both during normal and emergency operations. It is the responsibility of the captain, although often over-ridden by the company, to determine what is an acceptable level of risk for a particular operation.

Once the level of acceptable risk is determined, it is possible to look at various options that fit within that level. When all available options have been assessed, it is the captain who must decide which option to elect. The alternate path to this is to look at all possible options for a situation and, once they have been determined, narrow the field of selection by only further assessing those that are within a reasonable level of risk.

Decision and Response

The decision and response stage incorporates that part of the processing system called the Central Decision Channel, which acts as a channeling of information about which decisions must be made. Although generally only one decision can made at one time, if a motor program (a behavioral sub-routine) is being run, then it is possible to continue that program and make other decisions (e.g., drive and converse at the same time).

Three kinds of action could be carried out as a result of the decision process. First of all, the information could be held in memory for a brief period, while deciding what to do with it. Secondly, the information could be memorized and held in memory long term for use later. Finally, the information could be used to start a skill process, such as a motor program.

A suitable example would be a Ground Proximity Warning System (GPWS) alert. The information could be held briefly in memory (working memory) while checking the information against other sources (chart, approach plate, etc.). This aircraft is prone to spurious GPWS warnings, and this as an example of it (long-term memory). The third option is to execute a go-around and climb away from danger (motor memory). The first two options have resulted in accidents and incidents.

Principles of Navigation

Form of the Earth

The Solar System

The Solar System consists of the sun, eight major planets (of which the earth is one), and thousands of asteroids (sometimes called minor planets or planetoids) and comets, which orbit the sun. The earth is the third planet from the sun and has an elliptical orbit with an average distance of 93 million statute miles (81 million nautical miles) from the sun. The orbital path and orbital speed of the earth around the sun are governed by Kepler's laws of planetary motion.

The Earth

Besides orbiting the sun, the earth rotates about its own axis presenting a continuously changing location to the sun, hence, resulting in day and night. The earth's axis is inclined at an angle of 66½° to the orbital plane.

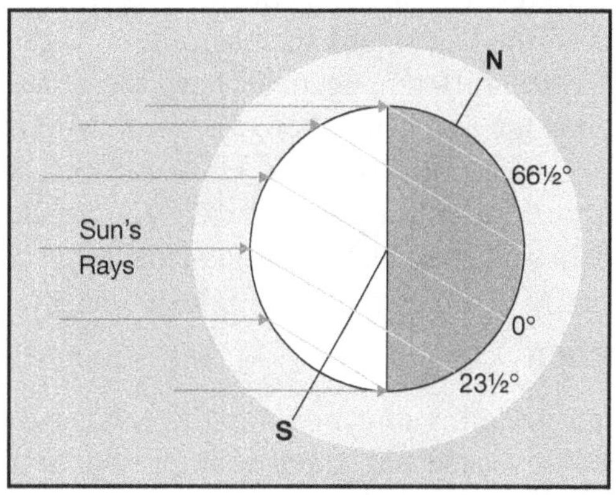

FIGURE 12-2 Earth's inclination.

The inclination of the earth's axis is the cause of the sun's movement between the earth's northern and southern hemispheres during the year, hence, resulting in the changing seasons. It is also the cause of the changing time interval between sunrise and sunset throughout the year.

The poles are defined as the ends of the axis about which the earth rotates. When viewed from above the North Pole, the earth appears to rotate in a counter-clockwise direction. When viewed from above the South Pole, the earth appears to rotate in a clockwise direction.

Although the earth is commonly represented by a sphere, this is not the earth's true shape. The earth's shape is a sphere slightly flattened at its poles, its polar diameter is 23 nautical miles (or 27

FIGURE 12-1 Solar system.

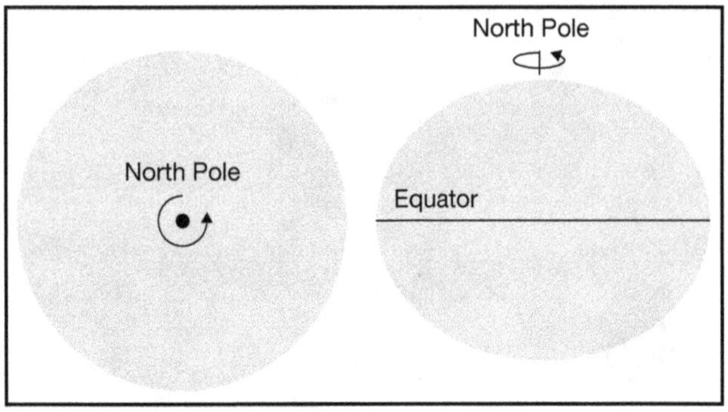

FIGURE 12-3 Earth's shape.

statute miles) less than its equatorial diameter, and it is described as an oblate spheroid.

Several different agencies have measured and modelled the earth, and there are small differences in the models. However, the USA uses the World Geodetic System 1984 (WGS 84), which the ICAO has adopted as the world standard.

Direction

To start to define directions on the earth, a datum needed to be selected, and the simplest datum was the direction in which the earth was rotating, which was then called East. West was then defined as the opposite of East. Then, when facing East, the pole 90° to the left was called the North Pole; the direction North was defined as the direction toward the North Pole. The pole 90° to the right, when facing East, was called the South Pole, and the direction South was defined as being opposite to North. These basic directions are known as Cardinal Points, and the directions midway between them are North-East, South-East, South-West, and North-West are known as the Quadrantal directions.

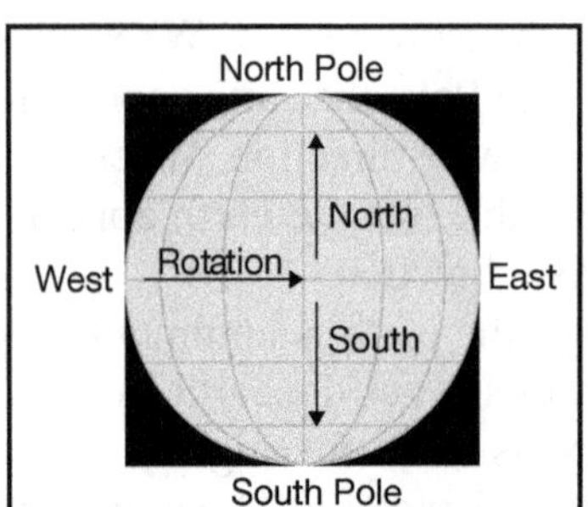

FIGURE 12-4 Basic directions.

Further subdivisions were made to give more precision, but this system does not provide the precision required in air navigation. To achieve the required precision, air navigation terminology retains the cardinal points as a reference but uses the Sexagesimal system for measuring directions.

Sexagesimal System

The Sexagesimal system utilizes the fact that a circle consists of 360° and a clockwise rotation of direction from North through the other cardinal points back to North is a circle of 360°. Therefore, North was defined as 000°, East became 090°, South became 180°, West became 270° and completing the circle North also became 360°. The use of 000° or 360° for North is acceptable. The convention for measuring direction on the earth is to measure clockwise, in degrees, from North.

In air navigation, two types of direction are generally used: True direction and Magnetic direction. True direction is the direction measured when the North datum is True North, the direction toward the North geographic pole. When specifying True direction, the abbreviation (T) is used after the number of degrees. For example, North is 000°(T), South is 180°(T), East is 090°(T), and West is 270°(T). Magnetic direction is the direction measured when the North datum is Magnetic North, the direction a suspended magnetic compass needle would point.

To avoid misunderstanding/ambiguity when receiving or transmitting directions by Radio Telephony, by convention, a three-digit group is always used (e.g., 010°(T), not 10°(T), 080°(T), not 80°(T). Receiving a direction of fewer than a three digits should be treated as possibly erroneous. For example, 10° could be a corruption of 010° or 100°, or even 110°, 210°, 310°, 101°, etc.

Earth Reference Systems

Navigation is the process of directing an airplane from one point on the surface of the earth to another. To be able to do this, all points or positions

FIGURE 12-5 Compass rose.

on the surface of the earth need to be defined accurately and unambiguously.

When defining positions on a flat surface (such as a piece of paper), Cartesian co-ordinates are generally used, where a position is defined by the distances (x, y) from two mutually perpendicular axes, X and Y.

For a spherical surface, i.e., the earth, two mutually perpendicular axes, known as longitude and latitude, are still required, but the distances (x, y) are replaced by angular distances. Lines drawn on a sphere are circles or parts (arcs) of a circle and on the earth, there are two types of circle of special interest—Great Circles and Small Circles.

Circles on the Earth

Great Circles and Small Circles and their properties are defined in the following ways:

Great Circle:

A circle on the surface of the earth, whose center and radius are those of the earth, itself is called a Great Circle. It is called "great" because a disc cut through the earth in the plane of the Great Circle would have the largest area that can be achieved.

The shortest distance between two points on the earth's surface is the shorter arc of the Great Circle joining the two points. Given two points on the earth's surface, there will be only one Great Circle joining them (unless the points are diametrically opposed).

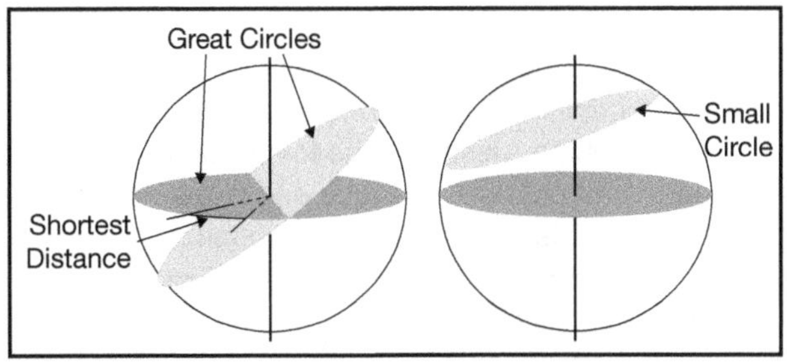

FIGURE 12-6 Great circle and small circle.

Small Circle:

A circle on the surface of the earth, whose center and radius are not those of the earth, is called a Small Circle.

The main small circles of relevance to position on the surface of the earth are the Parallels of Latitude.

The circles on the earth that are used in the position reference system are:

The equator:

The Great Circle whose plane is at 90° to the axis of rotation of the earth (the polar axis) is called the equator. It lies in an East-West direction and divides the earth equally into two hemispheres. For the definition of position on the earth, the equator is the datum for defining latitude and is the equivalent of the X axis of the Cartesian system.

The Meridians:

Meridians are semi-Great circles joining the North and South poles. All meridians indicate True North-South direction. Every Great Circle passing through the poles forms a meridian and anti-meridian.

The Prime Meridian:

The meridian passing through Greenwich is known as the Prime Meridian. The Prime Meridian is the datum for defining longitude and is the equivalent of the Y axis of the Cartesian system.

Parallels of Latitude:

The Parallels of Latitude are small circles on the surface of the earth whose planes are parallel to the equator. They lie in an East-West direction.

Graticule:

The network formed on a map or the surface of a globe by the Prime Meridian, the meridians, the equator and the parallels of latitude is called the Graticule.

The Graticule

Any position on the surface of the earth can be defined accurately and unambiguously using the Graticule. In the diagram below, the position of point A could be defined as being on a meridian *x* miles east of the Prime Meridian and on a Parallel of Latitude *y* miles north of the equator. Although the use of linear units (miles) is valid, angular measurements are preferred in navigation. Thus, the position of A is defined by the angle $\alpha°$ north of the equator and the angle $\beta°$ east of the Prime Meridian. This is the foundation of the latitude/longitude system. The intersection of the equator and the Prime Meridian are considered to be the origin of the latitude/longitude system (in the same way that the intersection of the *x* and *y* axes is the origin (0, 0) of the Cartesian system).

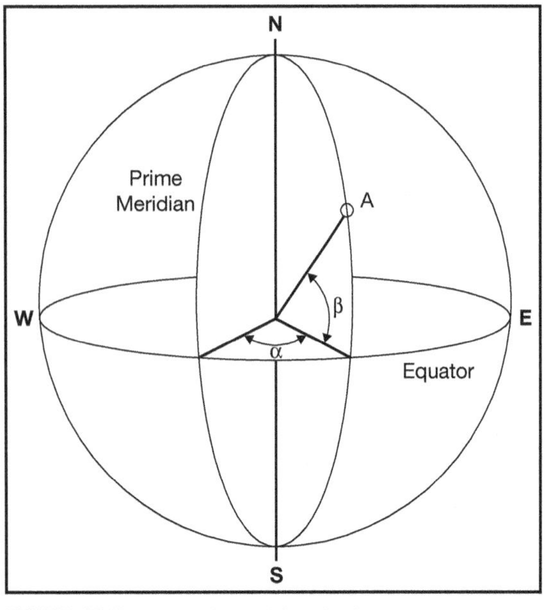

FIGURE 12-7 Latitude and longitude.

The unit used for angular measurement is the degree (°), which is defined as the angle subtended at the center of a circle by an arc equal to the 360th part of the circumference (hence, 360° in a circle). A degree can be sub-divided into 60 minutes of arc (') for more accuracy, and a minute can be further subdivided into 60 seconds of arc (") for even greater accuracy. By using seconds of arc, position on the earth can be defined to approximately 100 foot accuracy. For even greater accuracy, decimals of seconds can be used.

The above units are used for angular measurement of position, but for angular measurement of direction, it is more common to use degrees and decimals of degrees.

Latitude

The latitude of any point is defined as the distance expressed in arc (angular distance) along the meridian between the equator and the point. It is expressed in degrees, minutes, and seconds of arc measured at the center of the earth and is annotated North or South according to whether the point lies North or South of the equator.

The range of latitude values covers the equator (0°N/S) to the geographic poles, i.e, the North Pole is 90°N and the South Pole is 90°S.The definition of latitude is based on the center of the earth, namely it is the smaller angle between the line joining the point to the center of the earth and the plane of the equator. This is Geocentric Latitude. Because the earth is actually an oblate spheroid, a different definition of latitude can be used, namely, Geodetic (or Geographic) Latitude. Geodetic (or Geographic) Latitude is the smaller angle between the normal (90°) to the meridian at the point on the spheroid

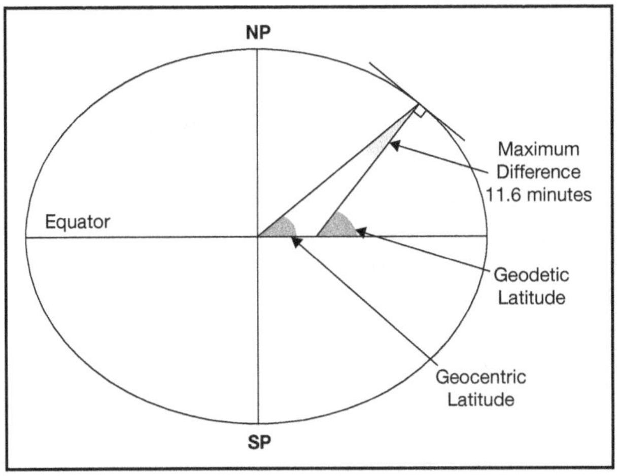

FIGURE 12-8 Geocentric and geodetic latitude.

and the plane of the equator. This normal line does not necessarily pass through the center of the spheroid as illustrated in the diagram above. (Note that the shapes used in the diagram are exaggerated for illustration purposes and the real spheroid is much closer to a sphere than shown.) The latitudes plotted on navigation charts are Geodetic (geographic) Latitudes. The maximum difference between Geocentric and Geodetic Latitudes occurs at approximately 45° N/S and is about 11.6 minutes of arc.

Special cases of Parallels of Latitude

There are several "special" cases of the parallels of latitude (other than the equator).

These are:

Arctic Circle—66½°N (NB. earth Tilt = 66½°)

Antarctic Circle—66½°S

Tropic of Cancer—23½°N (The sun is overhead this tropic on mid-summers day in the northern hemisphere.)

Tropic of Capricorn—23½°S (The sun is overhead this tropic on mid-winters day in the northern hemisphere)

Equator—0°N/S

Longitude

Longitude of any point is defined as the shorter distance in arc (angular distance) along the equator between the Prime Meridian and the meridian passing through the point. It is expressed in degrees and minutes of arc measured at the center of the earth and is annotated East (E) or West (W) depending whether the point lies East or West of the Prime Meridian.

Longitude can be measured up to 180°E or 180°W of the Prime Meridian. These two meridians are coincident and are known as the anti-Prime Meridian (APM), commonly labelled 180°E/W.

Positions in Latitude and Longitude

The latitude of New York is the arc of the meridian between the equator and New York, 41°N. The longitude of New York is the shorter arc of the equator between the Prime Meridian and the meridian through New York, 074°W. The position of New York is thus defined exactly and is illustrated in the diagram below.

When giving a position in latitude and longitude, by convention, the latitude is always given first and the longitude second, New York's position is 41°N 074°W. This can also be written as 41° 00'N 074°00'W or 4100N07400W if more accuracy is needed by using minutes of arc. If greater accuracy is required, then seconds can be included, 41° 00' 00" N 074° 00' 00" W.

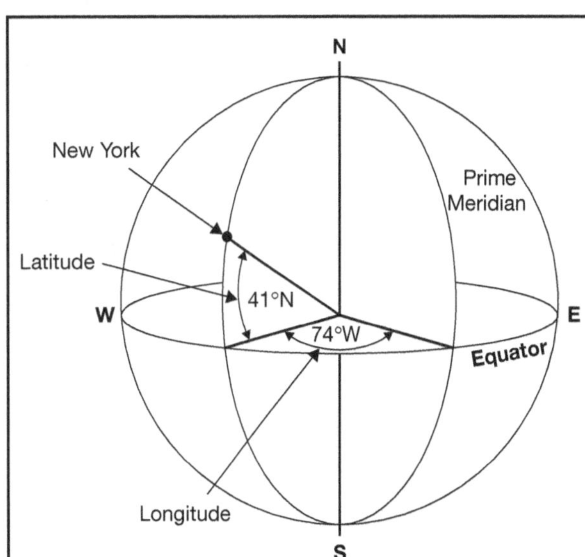

FIGURE 12-9 Latitude and longitude of New York.

Convergency

Figure 12-10 shows that all the meridians are parallel at the equator and converge toward the poles. This property of the meridians can be measured as convergency, which means the angle of inclination between the two given meridians measured at a stated latitude. Convergency is zero at the equator (the meridians are parallel at the equator) and increases to a maximum at the poles (all 360 meridians converge into a point at the poles).

The main effect of convergency is on the direction of a great circle (for example, the dashed line in Figure 12-10). The meridians define the direction of True North and because the meridians converge, the direction of the Great Circle is constantly changing. The Great circle crosses the equator in a direction of approximately 010°T and the direction steadily increases until at the end, the Great Circle direction is approximately 070°T.

Rhumb Line

Modern flight management systems can direct an airplane along a Great Circle route and manage the constantly changing direction of the Great circle, caused by convergency. However, in the early days of air and maritime navigation, compass and navigation equipment was less capable, and it was found easier to navigate along lines of constant direction—Rhumb Lines.

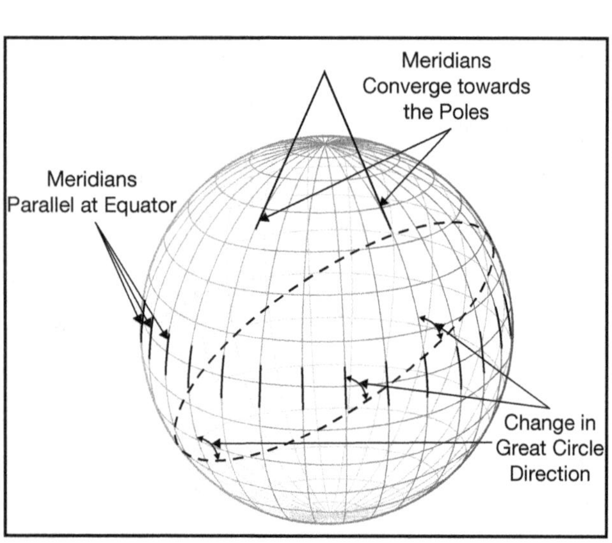

FIGURE 12-10 Convergency and great circle direction.

A Rhumb Line is a regularly curved line on the surface of the earth, which cuts all meridians at the same angle, i.e., a line of constant direction. As with Great Circles, only one Rhumb Line can be drawn between two points (unless the points are diametrically opposite). Examples of common Rhumb Lines are:

Parallels of Latitude (because they cut all meridians at 90°).

Equator (a special case because the equator is also a Great Circle).

Meridians (are also semi-great circles and the cut angle involved is 0°).

The Rhumb Line between two points will always be nearer to the equator than the corresponding Great Circle. An important property of the Rhumb Line and Great Circle joining two points is that the Great Circle and Rhumb Line directions are equal at the midpoint (the middle longitude) between the two points.

Distance on the Earth

Metric and Imperial Measures

Although we are normally interested in larger units of distance, i.e., kilometers (km), statute miles (sm), and nautical miles (nm), it is important to have an understanding of the smaller units and the relationship between metric and imperial measures.

1 meter (m) = 100 centimeters (cm) = 1000 millimeters (mm)

1 centimeter (cm) = 10 millimeters (mm)

1 meter (m) = 3.28 feet (ft)

1 foot (ft) = 12 inches (in or ")

1 inch (") = 2.54 centimeters (cm)

1 yard (yd) = 3 feet (ft)

The Kilometer (km)

The SI definition of the kilometer (km) is 1,000 meters when the meter (m) is defined in terms of the speed of light as the length of path travelled by light in a vacuum during a time interval of 1/299,792,458th of a second—this is of little use in navigation. The simpler navigation definition of the kilometer is 1/10,000th of the average distance on the earth between the equator and either pole. Thus, there are 10,000 km between the equator and either pole, and the circumference of the earth is 40,000 km. For conversions between kilometers and Imperial units:

1 Kilometer (km) = 3280 feet (ft)

1 meter (m) = 3.28 ft

The Statute Mile (sm)

The Statute Mile (sm) is defined in a Royal Statute of Queen Elizabeth I (of England) and is 5,280 feet. Although the Statute Mile is widely used on the ground (e.g., in the UK and the USA), it has limited application to aviation. Indeed, its most common use is in US Aviation Weather reports for visibility. The average distance from the equator to either pole is 6,200 sm; the circumference of the earth is 24,800 sm.

The Nautical Mile (nm)

The Nautical Mile is the most important large measure of distance used in aviation because it can be related directly to the angular measurements of the latitude/longitude graticule of the earth. Unfortunately, the compression of the earth results in slightly varying measurements for the nautical mile. The ICAO definition of the nautical mile is that it is a measure of distance of 1,852 meters. The Standard Nautical Mile is defined as a length of 6,080 feet (1,852 meters converted to feet gives a value if 6,074.56 feet, not 6,080 feet). The International Nautical Mile is 6,076.1 ft. To avoid confusion, the value 6,080 ft is generally used in any navigation calculations. The explanation below should show how differences in the definitions arise.

The full definition of the Nautical Mile is that length of arc of a Great Circle, which subtends an angle of one minute at the center of curvature of the earth's surface.

If the earth were a perfect sphere, one minute of arc of a Great Circle would be the same distance all over the earth. But, because the earth is flattened at the poles, the radius of curvature is increased, and a greater arc is required to subtend an angle of one minute at the center of curvature, hence, a Nautical Mile is longest at the poles at about 6,108 ft. The curvature of the earth is greatest at the equator, and the radius of curvature is less than elsewhere on the earth. Consequently, the length of arc required to generate an angle of one minute at the center of curvature is less at the equator than elsewhere: a Nautical Mile is shortest at the equator and measures about 6,046 ft. The average value, 6,076.1 ft, is known as the International Nautical Mile, which equates to approximately 1,852 meters. However, as stated before, for navigation, the Standard Nautical Mile of 6,080 ft should be used.

Because they all use Great Circle arcs, the definition of the Nautical Mile is similar to the definitions of latitude and longitude. Remember the following:

One minute of latitude = nautical mile (nm)

One degree of latitude = 60 minutes = 60 nm

But, One minute of longitude = 1 nm *at the equator only.*

There are 90° of latitude change between the equator and the poles, the average distance between the equator and the poles is 90 × 60 = 5,400 nm. The circumference of the earth at the equator is 360 × 60 = 21,600 nm. The circumference of the earth at the equator is also 40,000 km

Direction on the Earth

The Compass

The instrument that gives us direction is called a compass. In the previous section, True North was defined as the direction toward the North Geographic Pole and North/South would be the direction of the meridian of longitude passing through that point. Unfortunately, there is no simple compass, which will align itself and point at the North Geographic Pole to give the direction for True North. However, there is a simple magnetic compass that will align itself and point at a place known as the North Magnetic Pole.

FIGURE 12-11 A simple magnetic compass.

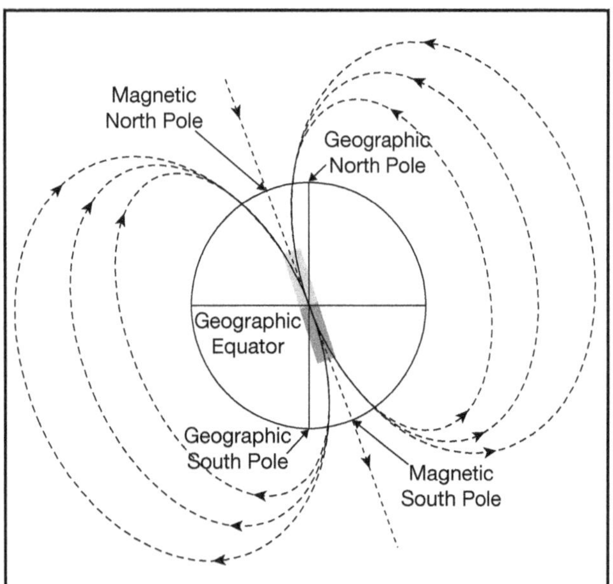

FIGURE 12-12 The earth's magnetic field and magnetic poles.

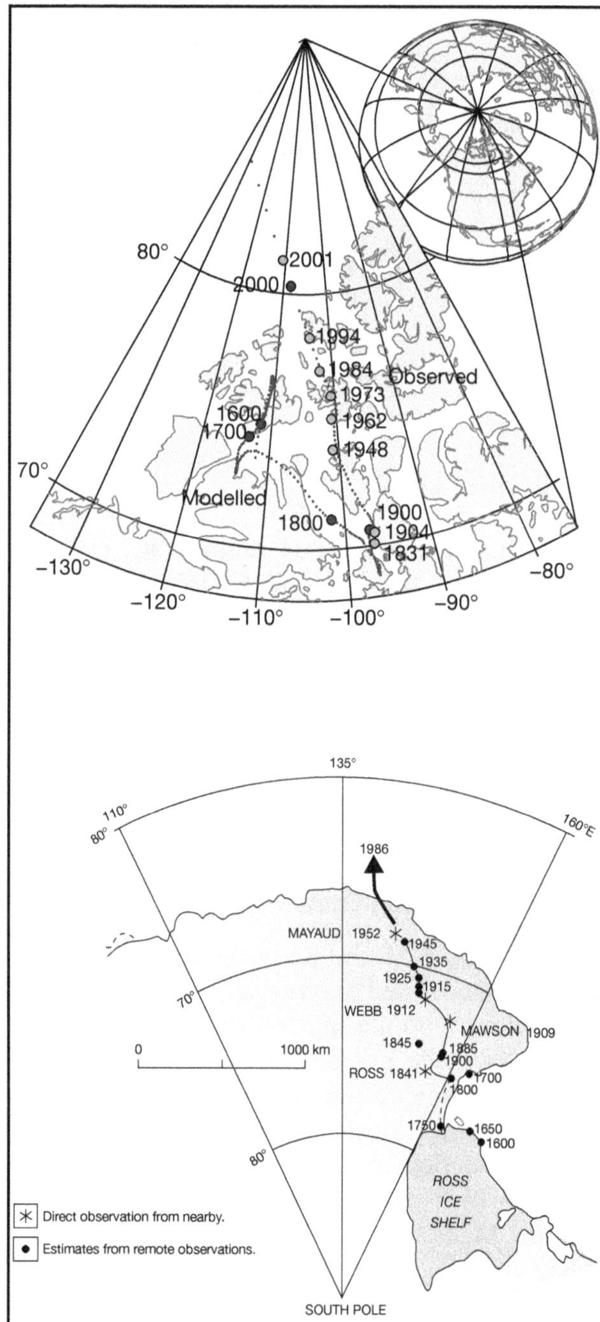

FIGURE 12-13 Changing positions of the north and south magnetic poles.

The Earth's Magnetic Poles

The earth acts as if there is a large magnet at the center producing a magnetic field around the earth's surface. Figure 12-12 shows that the positions of the magnetic poles and the positions of the geographic poles are not colocated. The poles of a magnet are labeled blue and red. For the earth, the blue pole is the North Magnetic Pole and the red pole is the South Magnetic Pole. The position of the magnetic poles changes annually (See Figure 12-13.); in 2010, the North Magnetic

Pole was 84.742°N 129.077°W, in Northern Canada, and the South Magnetic pole was 64.049°S 137.227°E, in the Antarctic.

Magnetic Variation

In aviation, navigation planning is done on aviation charts where the meridians of longitude provide a convenient datum, True North/South, for measuring direction. However, when we fly, the directional information is provided by the compass, which is based on Magnetic North. The difference between True Direction (True North) and Magnetic Direction (Magnetic North) is called variation, and the magnitude of the variation is dependent on airplane position. Refer to Figure 12-14, which is a view from above the North Geographic Pole

At position A, the compass will point West (left) of True North—Westerly Variation (or Variation West). At position B, True North and Magnetic North are in the same direction—Variation is zero. At position C, the compass will point East (right) of True North—Easterly Variation (or Variation East). Variation can be Easterly or Westerly in direction, and its value can be between 0° and 180°.

Advances in aviation technology have provided the means to store worldwide values of variation in a navigation computer database, which the airplane navigation systems automatically access. Navigation charts have the values of variation printed on them as lines of equal magnetic variation. These lines are called isogonals (or isogonic lines) and are dashed lines joining places of equal magnetic variation. The line joining places with zero magnetic variation is known as the Agonal (or Agonic) Line.

Deviation

An airplane compass is not only influenced by the earth's magnetic field, but also by the magnetic fields in the airplane caused by the airplane structure and electrical fields. These airplane magnetic fields prevent the compass from pointing to Magnetic North, and this deflection away from Magnetic North is called deviation. The deviation for a particular airplane is measured regularly and is displayed to the pilot by a Compass Deviation Card located somewhere near the compass itself.

Heading

The heading of an airplane is the direction of its longitudinal axis from aft to fore or, more simply, the direction in which the airplane is pointing (the dashed arrow in Figure 12-15). Headings may be referenced to True North (°T), Magnetic North (°M), or Compass North (°C).

Track (or Course)

If there was no wind (still air), an airplane would travel over the ground in the same direction as it is pointing. The line that the airplane follows over the ground is called the airplane's track. A still-air

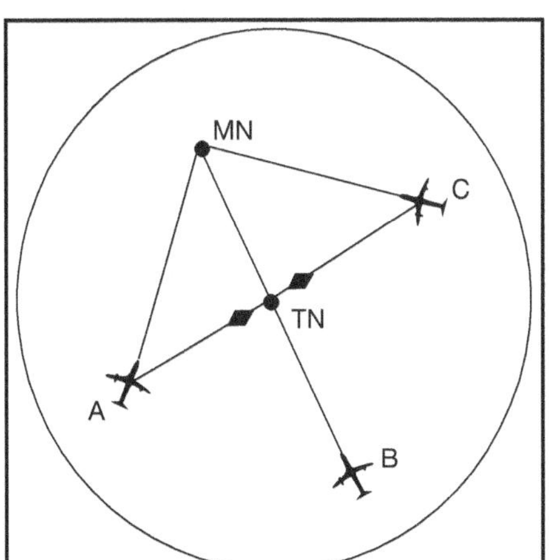

FIGURE 12-14 Variation.

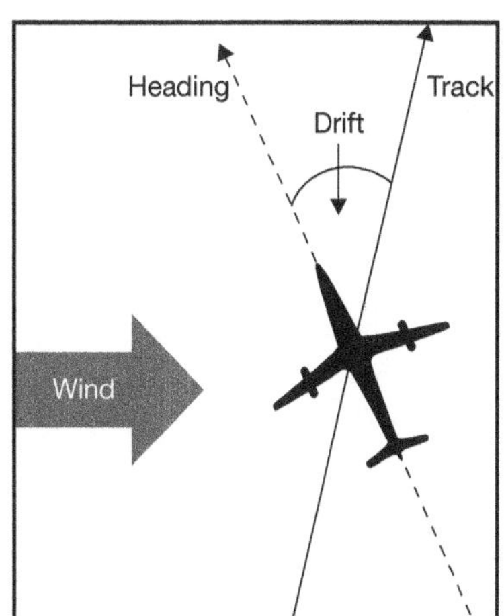

FIGURE 12-15 Heading, track and drift.

situation is very uncommon, and normally, some wind will be blowing; therefore, the airplane's track will not be the same as its heading. In Figure 12-15, the airplane is heading in one direction, the dashed arrow, but with a wind from the left, the airplane will track to the right of its heading, the solid arrow. The airplane is said to be drifting right. Drift is defined as the angle between heading and track and is measured in degrees left (port) or right (starboard) from heading. Track normally will be referenced to True North (°T) or Magnetic North (°M).

Bearing

The word *bearing* is a common term in navigation and is used to give the direction of one point with reference to another, e.g., "the bearing from A to B is 010 °T." When flyng, this would be the direction the pilot would need to fly over the ground, so this would be the airplane's track. Bearings are normally referenced to True North (°T) or Magnetic North (°M).

Relative Bearings

Pilots often need, when flying, to report the bearing of other airplanes to each other, and air traffic control needs to report the bearing of another airplane to a pilot. A very simple method was developed many years ago and is still used extensively today—the clock code. The method requires the directions around the airplane to be referenced to an analog clock face with 12 o'clock positioned at the nose of the airplane; see Figure 12-16. The bearing of another airplane can then quickly be reported, relative to the nose of the airplane, as the relevant hour on the clock.

FIGURE 12-16 Basic clock code.

An airplane straight ahead of a plane is in the pilot's 12 o'clock. An airplane directly behind the plane is in the 6 o'clock. An airplane 90° right of the nose is in the 3 o'clock. An airplane 90° left of the nose is in the 9 o'clock. Each hour of the clock represents a multiple of 30° off the nose; thus, 2 o'clock means 60° right of the nose.

The system is an excellent method of quickly directing a pilot's eyes to a particular direction. If reporting another airplane, then combining the clock code with a "high" or "low" will provide a very specific area in which to look.

CHAPTER 13

International Aviation

The growth and development of international and foreign air transportation has not paralleled that of domestic air transportation in the United States. It is necessary to understand exactly what is included in the terms international and foreign as used in air transportation. As generally used, international air transportation[1] is the conducting of air transportation by an air carrier of the United States outside the nation to another country. The class of U.S. air carrier known as International and Territorial represents this definition. Foreign air transportation,[2] on the other hand, is the operation of air carriers of nations other than the United States. In some cases foreign air carriers perform air service to and from the United States.

"Overseas" refers to a type of air route and is geographic in nature; it is not a classification of air carrier. It is a type of operation conducted by U.S. air carriers flying from the continental United States to a point outside the continental limits that is a state, territory, or possession of the United States. This type of operation can be performed by an International and Territorial Air Carrier. An example would be a flight from Miami, Florida, to San Juan, Puerto Rico, or from San Francisco, California to Honolulu, Hawaii.

One point to be noted in the study of international and foreign air transportation is that while the U.S. international air carriers are privately owned corporations, foreign air carriers for the most part are either wholly or partially owned by their governments.

Conferences and Conventions

The progression of international conferences and conventions which resulted in the evolution of the present legal framework under which international and foreign aviation operates is reviewed briefly below.[3]

Paris Conference—1910

International law presented aviation pioneers with many challenges. In 1910 it was seen that aviation was going to link different languages, cultures and legal systems to a far greater degree than sea or rail. It was also recognized that if maritime law were to be applied to aviation it would have to be expanded or, at the very least, adjusted to deal with the unique risks and demands of air travel.

A conference held in Paris in 1910 was attended by representatives from 19 European countries. It was the first attempt to reach international agreement on matters dealing with civil aviation. Those attending could not come to any basic agreements. They were divided between the concept of a freedom of the air, paralleling that of the sea; and the concept of a national sovereignty that extended into international air space. However, not all of the work of 1910 was lost. Many of the convention's less controversial draft articles were later reused in the conventions of 1919, 1926, 1928, and 1944. Their contemplated efforts were left to the Aeronautical commission of the Paris Peace Conference, which was established in March of 1919.

In the world today, communication and transportation are the two factors that support the global society. Before World War II, efforts were made to establish an international legal framework to ensure the development of international civil aviation. After World War II the potential of the airplane was recognized for international commercial purposes as well as for its future role as an instrument of better understanding among the people of the world.

Paris Convention—1919

A meeting known as the Convention Relating to the Regulation of Aerial Navigation was held on March 6, 1919. The delegation met to accomplish what was not agreed upon in 1910. The countries represented were: Belgium, Brazil, the British Empire, Cuba, France, Greece, Italy, Japan, Portugal, Romania, the Kingdom of the Serbs, Croats and Slovenes, and the United States. At their first meeting the commission agreed to produce a set of basic principles in preparing the Convention and its Annexes.

Having established the guidelines for the drafting of the Convention, the Commission established three Sub-Commissions which were legal, technical, and military. These three were aided by draft conventions submitted by France, Great Britain, and the United States. Italy submitted a draft proposal for aerial navigation laws. In actuality, the influence of the International Commission Air Navigation was limited, as well as its implementation of the convention, because during the 1920s and 1930s the airplane was not capable of world wide transport. It was not until after World War II that international air services expanded because of the improvements in the airplane to fly long distances.

This international convention for air navigation recognized that each nation had complete and exclusive control of the air space above its territory, thus preventing the air carriers of a country from flying over another country without permission. This convention:

1. Established basic principles of air sovereignty.
2. Prescribed for national registration of aircraft.
3. Placed restrictions on movement of military aircraft.
4. Established basic rules of airworthiness of aircraft.
5. Regulated pilots and others in charge of aircraft.
6. Provided for police measures.

For the purpose of the convention, the territory of a country was understood to include its national territory, that of the mother country, of the colonies, and the international waters adjacent therein.

A permanent commission was created by the Paris Convention to provide for the continued study of international aviation legal problems. This commission was known as the C.I.N.A. (Commission Internationale de Navigation Aerienne.)

The Convention was ratified by 26 countries and served as the legal document for international air transportation outside the Western Hemisphere. The United States signed the convention papers, however, it was not ratified by the United States because the convention was associated with the League of Nations to which the United States did not belong. Russia also did not ratify it, thus the two largest countries of leading importance in post World War I civil aviation never ratified the Paris Convention. The main accomplishment of this convention was the establishment of sovereignty of air space theory.

Madrid Convention—1926

The Ibero-American Convention Relating To Air Navigation was formulated at a 1926 conference by Spain, Portugal and 19 Latin American countries. The 1926 Convention largely mirrored the text and appendices of the Paris Convention of 1919. It was developed as a reaction to the unequal voting procedures and certain other features in the Paris Convention at that time. The Madrid Convention was never registered with any international body and was totally ignored in the Chicago Convention of 1944. Its lack of success was due to three factors. First, aircraft of the period were not sufficiently developed to tie together Iberia and Latin America. Second, Spain's political environment during the period was very unsettled, deteriorating into Civil War. And third, two years later the Havana Convention focused Latin American energies on North America and away from Iberia.

Pan American Convention on Commercial Aviation—1928

At Santiago, Chile, in 1923, a Pan American Conference was held to establish basic rules for operating international commercial air services. The intent was to formulate a legal document for the needs of the countries of the Western Hemisphere. It was to be related to the Monroe Doctrine of the United States. This convention reviewed the provisions proposed at the Santiago conference. It was modeled after the Paris Convention and recognized that every country had complete and exclusive sovereignty over the air space above its country and adjacent territorial waters.

Although the Paris Convention of 1919 solved most of the problems for European countries, it was not satisfactory for the countries of the Western Hemisphere. The Sixth International Conference of American States was convened in Havana in 1928.

From this conference came the Commercial Aviation Convention, properly known as the Havana Convention of 1928. The Havana Convention contains most of the basic principles established by the Paris Convention and some new ones. It differed in that it also contained some provisions relating to private international aviation law and did not provide for any continuing commission such as the C.I.N.A. This agreement was signed on February 20, 1928, by 21 countries including the United States.

The differences between the Paris Convention of 1919 and the Havana Convention of 1928 were:

1. The Havana Convention had no Annexes; all rules were contained in the treaty itself.
2. The Havana Convention was applicable only to private aircraft; government aircraft were not included.
3. Aircraft regulation was to be done according to the laws of each country. No uniformity was provided.
4. Pilots were required to demonstrate a knowledge of the regulations of each country in which they flew. This meant they were to know the rules of the twenty other countries in addition to their own.

There were no provisions for a permanent way of effecting civil aviation progress and development. Each country was expected to take the initiative in civil aviation.

The Paris Convention and the Havana Convention caused some degree of confusion in actual practice since they were two separate sets of rules. It was soon realized that there would have to be only one international organization to bring together international civil aviation developments. This did not occur until 1944.

Warsaw Convention—1929

The Warsaw Convention was the result of two international conferences held in Paris in 1925 and Warsaw in 1929. The purpose of the conferences was twofold. The first was to establish some degree of uniformity in travel/shipping documentation, i.e. tickets and waybills, etc. The second was to limit the potential liability of the air carrier in the case of accidents.

This was the first international convention pertaining to liability and has been ratified by more than sixty countries, including the United States.

The Warsaw Convention was signed in October 1929 and went into effect on February 13, 1933. It was recognized almost immediately as the most important document in international and foreign commercial air transportation. In 1934 the United States Senate gave its consent to the Convention.

The Warsaw Convention provides the rules of air carrier accident liability in international air transportation. It prescribes that the airline is liable for damages for death or injury to passengers; destruction, loss, or damage to baggage and goods; and loss resulting from delay in transporting passengers, baggage, and merchandise. It also sets standards for passenger tickets, cargo waybills, and other air travel documents.

After World War II ended many people, especially in Great Britain, suggested a revision of the Warsaw Convention. Those in the United States who urged for a revision wanted an increase in the liability limits which at that time were only $8,300 per person. These discussions continued until September 1955, when the Convention was amended. The diplomatic conference called for this event met at The Hague, Netherlands. The United States was represented, and the so-called Hague Protocol to the Warsaw Convention was signed. This Protocol made two major changes in the Convention. First it doubled the monetary limit to $16,600 as the maximum recovery for death; and second, it extended to agents of the carrier the limitation of liability now provided to the carrier. The Protocol has been signed and ratified by some 45 countries. The United States signed the Protocol on December 1, 1964. The subject of ratification of this document by the United States has been most controversial. Willful misconduct and the location of the crash can abrogate the limitation of liability.

During the early 1960s many countries expressed concern regarding the limits of liability which were doubled by the Hague Protocol of 1955. Some countries believed that limits were too low and did not relate to their economic conditions. Other countries feared that the limits were too high and imposed a financial burden on their national air carriers. In 1965, the United States denounced the original Warsaw Convention. The International Civil Aviation Organization held a special meeting in Montreal, in February 1966, in an attempt to find a solution, but no compromise was achieved.

This meeting became known as the Montreal Agreement of 1966. It was not an international agreement or a formal revision of the Warsaw Convention of 1929. It was an interim arrangement among air carriers operating passenger transportation to and from the United States. The Malta Agreement was a name given to an informal meeting of civil aviation authorities from the European governments who agreed upon the general increase in air carrier limits of passenger liability equal to that of the Montreal Agreement.

By virtue of the Montreal Agreement, the air carriers agreed to include in their tariffs a special contract permitted under the Warsaw Convention, which provided for a limit of liability for each passenger in case of death or bodily injury to the amount of $75,000. This amended the application of the Warsaw Convention as amended by the Hague Protocol of 1955. The air carriers who agreed to this arrangement must not avail themselves to any defense under the Convention, which provides that a carrier is not liable if they prove that they or their agents have taken all necessary measures to avoid the damage, or that it was impossible for them to take such measures. Thus the principle of strict liability regardless of fault had been contractually introduced by this action.

Current air carrier liability limits for international passengers range from $10,000 according to the Warsaw Convention of 1929 to a maximum of $150,000 depending upon the circumstances where the ticket was purchased.

The United States adheres to the Warsaw Convention which provides that the liability of an air carrier for a negligent act causing personal injury to one of its passengers is limited to approximately $8,300 for each passenger. If the air carrier is guilty of "willful misconduct" or if a special contract has been entered into, such limitation will not apply. The Warsaw Convention requires the injured party to bring a lawsuit within two years of the accident or the action will be dismissed. Also, the air carrier will not be liable if the carrier can prove that due care was used. This applies to injuries sustained on the airplane or while embarking or disembarking from an international flight.

The Montreal Agreement increases the liability amounts under the Warsaw Convention. Pursuant to the Montreal Agreement, many carriers agreed to increase their liability limits up to $75,000 including legal fees while also waiving their right to a defense that due care was used by the carrier. These increased liability limits apply to flights starting, ending or having a stopping place within the United States.

International Air Transportation of Goods and Litigation Under the Warsaw Convention*

On a fall day in 1929, the presidents, diplomats and even kings from throughout the world convened in Warsaw, Poland to sign a treaty relating to the regulation of claims for what was then unexplored territory. The treaty set forth the international rules relating to transportation of persons, baggage and goods by aircraft. These rules remain in effect today. The magnitude of this creation is the more amazing when one realizes that the treaty was executed only brief years after the creation of closed-cabin aircraft and immediately after the beginning of transportation of passengers on regularly scheduled airlines. More amazingly the treaty, which came to be known as the "Warsaw Convention," remains intact and vibrant to this day, with some exceptions relating to passages and baggage. This article sets forth some of the major Warsaw Convention litigation issues which shippers, freight forwarders, and cargo handlers should consider in sending their goods internationally.

To begin, the Warsaw Convention covers the claims of the sender ("consignor"), shipper and recipient ("consignee") of goods, and anyone who derives a right under them, against the international air carrier and its employees and agents. An air forwarder or customs broker may be deemed to be the agent of an air carrier. The Warsaw Convention only covers "air transportation."

Before the goods are taken, the carrier is required to have the consignor (the sender of the goods) make out and hand him a document called an "air waybill" which represents the contract of transportation. The treaty sets forth the details of what is supposed to be contained in the air waybill which includes, in part, destination, agreed stopping places, number of packages, method of packaging, particular marks, apparent condition of the goods, the value declared, who pays for the freight, and that transportation is subject to the convention. Failure to include the requirements to the air waybill and the air carrier is not entitled to avail itself of the limitation of liability provided by the treaty. On the other hand, the consignor (the person who is remanding the goods for transportation) is responsible for the correctness of the statements relating to the goods which are inserted into the air waybill. If the goods are supposed to arrive at a certain date and do not, after seven days the recipient of the goods is required to make a claim for the goods.

The sender of the goods, the consignor, is required to furnish all documents and information that may be required by the customs or police before goods are received. Any absence, insufficiency or irregularity in the information or documents entitles the air carrier to indemnification from the sender, unless the damage is due to the fault of the carrier or its agents.

The carrier is liable for damage sustained from the destruction, loss, or damage to goods if the occurrence which caused the damage took place during the transportation by air. The carrier is also liable for delay of goods. The treaty defines air transportation to include when the goods are in the custody of the air carrier at the airport. The Warsaw Convention thus does not cover when the goods are taken from the airport by independent ground transportation to the goods' final destination.

The carrier is not liable under the treaty if it can prove that it has taken "all necessary measures to avoid the damage" or that it was impossible for it to take such measures. Such measures, for example, would include the hiring of a bonded security guard to patrol at the airport warehouse.

The treaty's limitation of liability regarding goods is pegged to the 1929 gold franc (250 francs per kilogram of goods). Converted to 1998 U.S. Dollars the current amount is $25.00 per pound with one major exception. The carrier may not limit its liability if the damage was caused by the willful misconduct or gross negligence of any agent of employee as determined by the court where the claim is submitted.

A claim of hidden damage must be made within seven days of receipt of the goods, and no claim may be made after two years from the date the aircraft was to have arrived.

* This article is by Luis S. Konski, a lawyer with the law firm of Becker & Poliakoff in Miami, Florida. E-mail address is: ikonski@becker-poliakoff.com.

International air transportation may include transportation by successive carriers, some which may be entirely domestic. For example, goods coming from Germany to the U.S. Virgin Islands first have to come to Puerto Rico. Because only smaller planes can land in the Virgin Islands, the major carrier may contract with a smaller airline to transport the goods to the smaller island was deemed to be subject to the Warsaw Convention, although the ultimate carrier had no international routing.

Generally, claims under the Warsaw Convention, as a United States treaty, invoke the jurisdiction of the federal district court. To avoid federal court, and a phyrric attempt to avoid the Warsaw Convention's limitation of liability, claimants have attempted to craft claims solely under state law for negligence and breach of contract. There is a split among the district courts in this district and others whether the Warsaw Convention is an exclusive cause of action for international air carriers, thereby preempting state law causes of action such as negligence or breach of contract. Whether the suit is in state court, however, the courts uniformly agree that the Warsaw Convention sets the limit of liability whether in state or federal court. This somewhat makes the filing of a separate claim for negligence or breach of contract redundant. You can file in state court but you will be limited to Warsaw Convention damages.

If you are a shipper of goods what do you do to avoid the Warsaw Convention limitations? Simple, get insurance, and plenty of it. Make sure that the insurance is portal to portal, and includes any deviations. Make certain that the insurance includes ground transportation outside the airport.

What to do if you are an international freight forwarder? Make certain that you have blanket coverage for the same limitations, if it is available. A question, which also arises, is whether an air carrier may limit the liability under the Warsaw Convention by adding terms and conditions to the terms of the contract for transport. The general answer is no. The Warsaw Convention represents a "floor" which a carrier may not go below.

Recently, the U.S. Senate ratified the Montreal Protocol No. 4, which amends the Warsaw Convention to allow freight forwarders to process air waybills electronically (without paper). The Protocol also establishes a system of strict liability (liability regardless of fault) for cargo carriers unless the destruction, loss or damage to the cargo resulted solely from a problem inherent to the cargo, defective packaging by a party other than the carrier or its agents, an act of war or armed conflict, and an act of the public authority in charge of entry, exit or transit of cargo. Thirty nations have signed the Protocol and the effect in the short run will be to increase litigation, until the limits of the revised treaty are tested.

The Warsaw Convention, as interpreted by U.S. Courts, is an exceedingly complicated set of rules and regulations. There is no certainty, even among the courts of the same federal districts on some of the issues. While this article represents a broad summary, there are exceptions even to these generalized rules. Perhaps the best rule is to do further research.

Convention on Damage Caused by Foreign Aircraft to Third Parties on the Surface—1933

This Convention dealt with damage caused by foreign aircraft to people and property on the ground, and was signed in Rome on May 29, 1933. It imposed a strict liability on aircraft operators for sums limited by reference to the maximum permitted weight of the aircraft. Liability was secured by compulsory insurance, cash deposits, or bank guarantees. The Convention was formally adopted by very few countries, but the principles have been copied by other countries. An amended Convention in 1952 was intended to replace the original Convention of 1933. The new amendments included an increase in the liability limits and the matter of security was no longer compulsory but at the option of the individual contracting countries. The 1952 convention was revised by the Montreal Protocol of 1978. As of June 30, 1995, 38 countries have ratified the Convention of 1952.

Chicago Convention on International Civil Aviation—1944 (Chicago Conference)

With the end of World War II near, many countries foresaw the peaceful applications of international air transportation. It was obvious that action would have to take place to provide for any international agreements in aviation. The U.S. Government invited the allied and neutral powers to attend an international conference in Chicago in November and December 1944. This conference came about after discussions between the United States and Great Britain in 1943. Fifty-two nations attended; only Saudi Arabia and the Soviet Union did not attend.

The purpose was to establish a basic legal frame-work for operating international civil air transportation services. An opening message by President Franklin D. Roosevelt said that the purpose of the conference was "to establish a fundamental law that would give full recognition of the sovereignty and judicial equality of all nations so that airways would serve humanity. The future development of International Civil Aviation can greatly help to create and preserve friendship and understanding among the nations and peoples of the world." This quote is from the preamble to the Convention on International Civil Aviation signed at Chicago on December 7, 1944.

The conference had three major objectives:

- The establishment of an International Air Interim Council.
- The establishment of provisional world air routes and arrangements.
- An agreement on principles for a permanent aeronautical body and a multilateral aviation convention.

Four different proposals for the organization of civil aviation were presented by the United States, Great Britain, Canada, and Australia-New Zealand. Secret talks between the United States, Great Britain and Canada finally resolved the issue. After 37 days of continuous meetings the Chicago Conference established not one but four treaties or conventions. They were:

1. The Interim Agreement on International Civil Aviation, (U.N. Treaty Series 501, Vol. 171, page 345). Its purpose was to provide for a temporary mechanism to allow for operation until the Convention would be formally ratified. It became effective on June 6, 1945 when twenty-six countries accepted it and established the Provisional International Civil Aviation Organization. Its first meeting was held on August 15, 1945. The PICAO functioned well until it was replaced by the permanent ICAO on April 4, 1947.
2. The main Chicago Convention on International Civil Aviation (U.N. Treaty Series 102, Vol. 15, page 295), setting up the International Civil Aviation Organization (ICAO) as a permanent organization.
3. The International Air Services Transit Agreement, (U.N. Treaty Series 252, Vol. 84, page 389) whereby contracting nations mutually exchange the privileges of overflight and noncommercial landing by scheduled flights, commonly referred to as the Two Freedoms Agreement. It should be noted that the Chicago Convention provides these privileges for non-scheduled flights. This agreement became effective on June 30, 1945.
4. The International Air Transport Agreement, commonly known as the Five Freedoms Agreement (U.N. Treaty Series 502, Vol. 171, page 387), whereby parties mutually exchange the additional commercial privileges of carrying passengers, mail and freight to and from the other contracting parties, as well as beyond to other countries. This multilateral agreement is by far the least successful of the group because it ignores the capacity to be provided and the amounts to be charged for the service, the two most important issues.

It was understood that many countries would not accept this agreement. It was eventually signed by 19 countries, which reciprocally applied its provisions. Many countries denounced the agreement and only a handful remain a party to it.

The International Civil Aviation Organization as a permanent organization began on April 4, 1947.

The Five Freedoms presented in the International Air Transport Agreement entered into by members of the treaty are summarized below.

> ***Freedom One.*** *The right to fly over territory of a foreign country without landing.*
>
> ***Freedom Two.*** *The right to make a landing in a foreign country for technical reasons but not to pick up or discharge traffic.*
>
> ***Freedom Three.*** *The right to carry traffic from the home country of an air carrier into a foreign country.*
>
> ***Freedom Four.*** *The right to carry traffic from a foreign country back to the home country of an air carrier.*
>
> ***Freedom Five.*** *The right to carry traffic between two or more foreign countries on flights coming from, or going to, the home country of an air carrier.*

In addition to these recognized Five Freedoms other conditions have been expressed as freedoms and are presented below:

> ***Freedom Six.*** *The right to carry traffic between two foreign countries through the home country. This a combination of Freedoms Three and Four which has resulted from a bilateral agreement for exchanging air traffic rights.*
>
> ***Freedom Seven.*** *The right to carry traffic between two foreign countries by an air carrier operating entirely outside of its own country. This is a combination of the Fifth Freedom rights.*

Freedoms Six and **Seven** are essentially supplementary freedoms and have no official recognition.

Cabotage is an air carrier carrying traffic between two points in a foreign country and is usually prohibited. The term Grand Cabotage is used if these points are located in separate parts of the same foreign country.

> ***Freedom Eight.*** *The right to carry traffic between two points within a foreign country, therefore acting as a domestic carrier in a foreign country.*

Outcomes of Chicago Conference—1944

Bilateral Agreements

One of the three major objectives of the Conference concerned air routes. At first it was the desire of the leaders of the conference to develop a multilateral air transport agreement, by which many countries would enter into an agreement regarding air routes. However, the Multilateral Transport Agreement was rejected by a majority of the States, and all that was left to work out commercial freedoms were bilateral negotiations.

As a result of this rejection the Final Act of the Chicago Convention includes a "Form of Standard Agreement for Provisional Air Routes." One of the first of these bilateral agreements, and by far the most significant, was the Bermuda Agreement, which will be discussed later in this chapter.

The current system of bilateral agreements, which started back in the 1940s, severely restricts competition in international markets. History has amply demonstrated that competition improves efficiency, productivity, and worldwide economic growth. At the present time, DOT is attempting to create a more competitive international aviation environment for the U.S. airlines through the continuing expansion of open-skies agreements. DOT is also assessing the pros and cons of modifying cabotage constraints, modifying seventh freedom rights,[4] and increasing foreign investment in U.S. air carriers from the current 25 percent.

During the last eight years, the Administration has achieved 89 new and expanded bilateral agreements, 50 of which are open-skies. An analysis conducted by DOT showed that from 1996 to 1999, on routes connecting interior cities in both the United States and European countries, the average decline in fares was 20 percent where open-skies was in effect. In markets where there were no open-skies agreements, the average decline in fares was only 10.3 percent.

In 2000, new bilateral agreements were reached with Colombia and open-skies agreements were signed with four countries from Asia, one each from South America and Europe, and four countries from Africa. Included in the open-skies agreements was the first multilateral "Open-Skies" agreement. The agreement included Brunei, Chile, New Zealand, and Singapore and permits unrestricted international air service for all flights among the five countries for these countries' carriers.

Discussions concerning the liberalization of markets are also proceeding with other countries throughout the world. The expansion of these agreements over the next several years could significantly increase the level of activity of the more efficient U.S. carriers vis-a-vis foreign flag carriers.

The industry is expected to continue toward globalization, through the use of code-sharing agreements and alliances. Four large alliances have formed and are continuing to add members and network connections. The four are SkyTeam (Delta-Air France), Star Alliance (United-Lufthansa), Oneworld (American-British Airways), and Northwest-KLM. The alliances have been able to reduce costs through economics of scale. They have also increased revenues and passenger traffic by expanding the reach of the networks and providing seamless travel for their passengers.

Provisional International Civil Aviation Organization

An interim organization was established by an agreement on June 6, 1945. Most of its structure patterned along the lines of what the permanent organization was to be: provisional Assembly, provisional Council, and a provisional Secretariat. It was to have advisory powers only and to operate until the permanent organization (ICAO) was created. The PICAO functioned for twenty months. This provisional organization was replaced by the permanent organization when the Chicago Conference was ratified by the necessary 26 States on April 4, 1947.

In the short life of the PICAO the foundation for an international organization devoted to the needs of civil aviation was laid and fifty contracting states took concerted action to provide and maintain the facilities and services necessary for the operation of air services across national borders.

Permanent International Civil Aviation Organization (ICAO)

At the invitation of the government of Canada, the city of Montreal was chosen as the site for the headquarters of the organization.

Membership in the ICAO comes under three classes of States: (1) signatory States which include all those States who were first to ratify the Conference; (2) members of the United Nations other than original members; and (3) other States which were those who were at war with the Allies (Italy, Austria, Finland, Japan, and Germany). The ICAO has obtained the status of a United Nations specialized agency.

The organization of ICAO is divided into three main bodies which are the Assembly, the Council and the Secretariat.

The Assembly of the Organization is much like the General Assembly of the United Nations. Each of the States has one vote, and decisions in most cases are taken by a simple majority vote.

International Air Services Transit Agreement

This is the Two Freedoms agreement which is referred to as the technical freedoms. The first of these freedoms is the right of an aircraft to fly over the territory of another country who has agreed to it. This would allow an airline to fly from New York to Rome, passing over such other countries as Great Britain, France, and Switzerland. The second of these freedoms conveys the right of an airplane to land for technical reasons, such as refueling or repairs, but not for commercial purposes, such as picking up or discharging passengers. For example, the flight to Rome from New York could land at London to refuel.

This Agreement is one of the most important agreements in world aviation. If this Agreement had been the only result of the Chicago Conference the effort would have been worthwhile. The United States surrounded by two oceans and two friendly neighbors does not truly appreciate the significance of the Transit Agreement.

International Air Transport Agreement (Five Freedoms Agreement)

This agreement was signed by 16 States and consisted of the two technical freedoms plus three more commercial freedoms.

The first of these commercial freedoms is the right to carry traffic from the plane's country of origin to another country, for example, an airline carrying passengers from New York to Paris. The second commercial freedom refers to the right to pick up traffic in other countries and return to the home country of the airline, such as a flight originating in Rome and carrying passengers to New York. The third commercial freedom is the right for an airline to carry traffic between countries outside its own country.

By the end of 1994, 191 nations had deposited instruments of either ratification or adherence to the Chicago Convention and joined ICAO. Of these, 100 nations had accepted the International Air Services Transit Agreement, but only 11 accepted the International Air Transport Agreement.

The two technical freedoms and the three commercial freedoms are referred to as the "Five Freedoms of the Air." Today few countries are willing to exchange these "Freedoms" on a multilateral basis, but all accept the concept of exchanging them bilaterally. By the terms of Article 80 of the Convention on International Civil Aviation, popularly known as the Chicago Convention of 1944, all contracting States agreed to give notice of denunciation of the Paris Convention of 1919 and the Havana Convention of 1928 upon the coming into force of the Chicago Convention of 1944. Since the Convention on International Civil Aviation came into force on April 4, 1947, the Paris Convention of 1919 and the Havana Convention of 1928 are rendered ineffective except for historical purposes.

Bermuda Agreement—1946 (U.N. Treaty Series 36, Vol. 3, Page 253)

The Bermuda Agreement was not the first bilateral agreement involving civil aviation entered into by two countries. The Agreement between the U.S. and Spain was signed on December 2, 1944; Denmark and Sweden was signed on December 16, 1944; Iceland was signed on January 27, 1945; Canada was signed on February 7, 1945; Switzerland was signed on July 13, 1945; and Norway was signed on July 13, 1945. Additionally, many bilateral agreements existed under both the Paris and Havana Conventions.

Although not the most complicated bilateral air agreement in existence it was the "father" of modern bilateral air service agreements. One observer stated that these agreements could no longer be regarded as merely temporary arrangements without any general significance but together with the Chicago Convention and the Air Transit Agreement they formed a well delineated law of international air transport.

If the Chicago Convention of 1944 represents the farthest advance toward multilateral accord, then the Bermuda Agreement of 1946 represents a triumph of bilateral achievement in influencing multilateral behavior.

The two great protagonists at Chicago were the United States and the United Kingdom. At that time the United States had the world's only fleet of longrange transports, and a history of separation of business and government. At Chicago the United States had campaigned for "Open Skies," i.e., unregulated air services.

On the other hand, the United Kingdom had only a few transports—each with limited range. The United Kingdom also had a long history of government participation in commercial aviation. At Chicago the United Kingdom had campaigned for "Order in the Air" or regulated air services. The United States had economic strength, but the United Kingdom had position. London was the hub of the English speaking world, and also in those days a necessary refueling stop on the way to Europe.

In 1946, diplomats from the two nations met in Bermuda to iron out their bilateral differences. The standard format in the Final Act of the Chicago Convention was a starting point but it ignored the main issues. In the end the United States accepted the United Kingdom concept that the airlines could mutually set the tariffs to be charged (fares and rates), subject to prior review and acceptance by the two governments. For its part the United Kingdom accepted the United States concept that the individual airlines could unilaterally select capacity, i.e., its own aircraft size and service frequency, with only an "ex-post-facto" review to guard against unfair practices.

In addition to the issues of capacity and tariffs, the Bermuda Agreement also reinforced the concepts that (1) Routes were to be mutually allocated by common consent on the basis of a fair and equitable exchange, and (2) the Fifth Freedom shall likewise be allocated by common consent on a fair and equitable basis. And most importantly, the governments also agreed that this new agreement would form the model for all the future bilateral agreements each would negotiate. As a result, the Bermuda Agreement replaced the Chicago Form as the world's standard Air Service Agreement.

Though it might seem that multilateral conventions provide all that could be desired in international aviation agreements, there are limitations. There are practical, international, political, and economic relations which prevent, or deter, some nations from entering into general multilateral agreements. At the same time, various combinations of nations can work out agreements on a bilateral basis. Therefore, existing concurrently with, and in some instances complementary to multilateral agreements, the United States has entered into bilateral agreements concerning Air Navigation; Air Transport; Airworthiness of Export Aircraft; and Reciprocal Recognition of Pilot Licenses. Of particular note among these agreements are those following the pattern suggested by Resolution No. VIII—Standard Form of Agreement for Provisional Air Routes resolved at the Chicago Conference of 1944, and those following the provisions of the Bermuda Agreement. These agreements are sometimes referred to as the Chicago-Type Agreement and the Bermuda-Type Agreement.

In addition to the general multilateral and bilateral aviation agreements referred to above, there are numerous multilateral and bilateral agreements concerning Double Taxation, Radio Communications, Sanitary Regulations, and Postal Arrangements.

Rights in Aircraft Convention

The Convention on the International Recognition of Rights in Aircraft was signed in Geneva in June of 1948 and entered into force on September 17, 1953. Through this Convention the recovery of seized aircraft is assured. In ratifying this Convention the Mexican Government reserved "the rights belonging to it to recognize the priorities granted by Mexican Laws to fiscal claims and claims arising out of work contracts over other claims."

Both the United States and Mexico became Contracting Parties to this Convention but the United States does not recognize Mexico's reservation. As a result, over the next two decades it was nearly impossible for United States owners to recover aircraft stolen for the drug trade and confiscated in Mexico.

As of the end of 1994, 61 nations have deposited their instruments of ratification or adherence to this Convention. Of these only the Netherlands and Sweden have adopted the United States position regarding Mexico.

Ground Damage Convention

The Convention on Damage Caused by Foreign Aircraft to Third Parties on the Surface was signed in Rome in October of 1952 and entered into force on February 4, 1958.

As of the end of 1994, 38 nations have deposited their instruments of either ratification or adherence to this Convention. The United States was not as yet a Contracting Party.

Air Offenses Convention

The Convention on Offenses and Certain Other Acts Committed on Board Aircraft was signed in Tokyo in September of 1963 and entered into force on December 4, 1969. This Convention insures that offenses committed on board aircraft will be placed under the jurisdiction of at least one State, i.e., the one of aircraft registration. The Convention also contains provisions relating to the duties of States, the extradition of an accused offender and the powers of the aircraft commander.

As of the end of 1994, 149 nations deposited their instruments of either ratification or adherence to this Convention. The United States became a Contracting Party on December 4, 1969.

Hijacking Convention

Due to the increased incidents of seizure of aircraft for various purposes in recent years, many nations of the world met at The Hague in 1970.

As a result, the Convention for the Suppression of Unlawful Seizure of Aircraft was completed on December 16, 1970. The terms of the Convention were to become effective thirty days after the ratification by the tenth country. With the acceptance by the United States Senate and the President, ratification by the Senate was completed on September 14, 1971 and proclaimed by the President on October 18, 1971. The Convention entered into force on October 14, 1971; as of the end of 1994, 149 nations had become Contracting Parties.

Sabotage Convention

The Convention for the Suppression of Unlawful Acts Against the Safety of Civil Aviation entered into force on January 26, 1973. The law gives effect to the Hijacking Convention by providing the President of the United States with authority to suspend air service to any country which he determines is encouraging hijacking in opposition to the Hijacking Convention.

The law orders the Federal Aviation Administration to provide regulations requiring that all air carrier passengers and their carry-on baggage be screened by detection devices. Also, it requires a federal air transportation security force to be stationed at major airports under the auspices of the Federal Aviation Administration.

As of the end of 1994, 151 nations have deposited their instruments of either ratification or adherence to this Convention. The United States became a Contracting Party on November 1, 1972.

International Air Transportation Competition Act of 1979

A new law was enacted by Congress on February 15, 1980, which directs the future of the United States in international aviation affairs. This Act amended the Federal Aviation Act of 1958 in areas of international aviation matters.

The United States government extended its attitude concerning competition in the air carrier industry, formulated by the Airline Deregulation Act of 1978, into international air transportation.

The purpose of the Act is to amend the Federal Aviation Act of 1958 in order to:

- Promote competition in international air transportation.
- Provide greater opportunities for United States air carriers.
- Establish goals for developing United States international aviation negotiating policy.

The Act contains a Declaration of Policy directed to the Civil Aeronautics Board containing factors for interstate, overseas, and foreign air transportation. The Board is directed to consider the following in accordance with the public convenience and necessity:

1. Maintenance of safety as the highest priority.
2. Prevention of any deterioration of safety.
3. Availability of a variety of adequate, economic, efficient, and low-priced services by U.S. and foreign air carriers.
4. Maximum reliance on competitive market forces.
5. Development and maintenance of a sound regulatory environment which is responsive to the need of the public.
6. Encouragement of air service in major urban areas of the United States through secondary airports.
7. Prevention of unfair, deceptive, and anticompetitive practices in air transportation, and avoidance of unreasonable industry concentration.
8. Maintenance of a system of continuous scheduled interstate and overseas airline service for small communities.
9. Encouragement of an air transportation system relying on competition to provide low prices.
10. Encouragement of new carriers and existing air carriers into new markets, and strengthening of small air carriers to assure a competitive air carrier industry.

11. Promotion of civil aeronautics and a privately owned U.S. air transport industry.
12. Strengthening of the position of U.S. air carriers to assure competitive equality with foreign air carriers.

The above policy declarations are reflected in the addition to Section 1102, International Agreements, of the Federal Aviation Act by establishing "Goals for International Aviation Policy." These goals state that in formulating United States international air transportation policy Congress intends that there shall develop a negotiating policy which emphasizes the greatest degree of competition. This includes, among other things:

1. Strengthening of the competitive position of U.S. air carriers to at least assure quality with foreign air carriers including opportunities to increase their profitability in foreign air transportation.
2. Freedom of all air carriers (U.S. and foreign) to offer rates which correspond with consumer demand.
3. Minimizing restrictions on charter air transportation.
4. Maximizing the degree of multiple and permissive international authority for U.S. air carriers so that they will be able to respond quickly to shifts in market demand.
5. Elimination of operational and marketing restrictions to the greatest extent possible.
6. Integration of domestic and international air transportation.
7. Increasing the number of nonstop United States gateway cities.
8. Increasing access opportunities for foreign air carriers to United States cities on a reciprocal basis for U.S. air carriers.
9. Elimination of discrimination and unfair competition faced by U.S. air carriers in foreign air transportation.
10. Promotion of civil aeronautics and a privately owned United States air transport industry.

International Deregulation

Deregulation began with air cargo in 1977 as an experiment in the United States. The following year passenger deregulation came about with the passage of the Airline Deregulation Act. The purpose of deregulation was to provide for open competition in the marketplace and to influence the future economic and business decisions of the air carriers. This reform was limited to fares, rates, and charges. Aviation safety and air traffic control were not affected; they remained the responsibility of the government.

The application of deregulation in international civil aviation is a much different concept than domestic deregulation in the United States. This is obvious because international deregulation means the involvement of many different governments and has brought about a condition whereby more government involvement in airline activities has taken place.

In Europe it was recognized that a competitive condition was desirable, but that it would come about at a slower rate than it occurred in the United States because of the public service aspect of air carrier operations. Many countries of the world view their air carriers as an important instrument of their country designed to operate for the public good. This is especially true in the smaller developing nations.

Due to many problems, deregulation of the air carrier industry in the United States has not brought about the results that were expected. The industry has become more efficient while becoming more competitive, but not necessarily more profitable. The U.S. airline industry has become more concentrated with three air carriers capturing more than 60% of the domestic market.

As a result of deregulation in the United States, other countries have undertaken efforts to create a more competitive environment. The results in Canada were similar to the effect of deregulation in the United States. Since 1984 the Canadian air carrier industry had become highly concentrated before the National Transportation Act became effective in 1987. International deregulation evolves slowly, particularly where the major air carriers are owned wholly or in part by their governments.

In April 1986, a decision by the European Court of Justice said that air transportation agreements are subject to the competition rules of the Treaty of Rome, signed by six European countries in 1957. This meant that the air carriers could be prosecuted for violations unless permission had been extended to them by the Council of Ministers. In 1987, this Council of Ministers legislated rules and took the first step toward liberalizing the air transportation industry in Europe.

The European Civil Aviation Conference was formed consisting of 33 European countries. They developed two International Agreements concerning intra-European fares and capacity sharing. Soon other regulatory changes took place providing for more freemarket access. The European trend toward concentration of airline activities and their market share will no doubt continue in the future as a result of these and more measures to be put into effect.

International air carriers continue to find new ways to increase their strength in the market place by mergers, buying other aircarriers, and several types of marketing agreements and cooperative ventures. An important tool is the worldwide computer reservation system which has been created by groups of airlines in different parts of the world. This is playing an increasing role in marketing and selling air transportation and is among the most important forces influencing international civil aviation developments in the future. As time passes more will be heard concerning international air carrier deregulation.

The United States Policy on International and Foreign Air Transportation

Prior to World War II, the United States had no policy on international and foreign air transportation except to develop flag carriers through airmail payments. Almost all these operations were conducted by Pan American World Airways. Pan American's position as the only carrier in international competition resulted in its achieving a dominant position and always seeming to be in the position of the best qualified carrier to receive air mail contracts. Congress encouraged the concentration of one airline effort, since our laws made no such provision. Pan American's monopoly policy was a result of its own initiative. Since the close of World War II the Board has taken the position that it is in the public interest to have more than one carrier, whenever possible, of the United States operating on international routes. As a result, many Trunk Air Carriers were awarded international routes as extensions of their domestic routes.

After World War II, the United States adopted a policy which was that there should be adequate incentives to stimulate efficiency in operations; also, that by reducing costs and rates, the full economic potential of the industry would be realized. Our policy was that the United States should continue to develop and use the newest equipment and best procedures; also, to ensure that the U.S.-flag carriers would carry passengers proportional with the importance of the United States as a transportation market, and that healthy financial conditions would ensure an influx of adequate capital.

During the 1950s U.S.-flag carriers carried the largest percentage of passengers in international flights. Today, the percentage is dwindling steadily, and it appears the trend will continue. U.S. maritime shipping has declined until only a fraction of the world's shipping is done by U.S. shipping interests. This must not be allowed to happen to U.S. international air carriers.

Many international problems have come from smaller countries who feel their air carriers cannot compete directly with U.S. air carriers. All too often these countries have attempted to restrict U.S. air carriers by influencing their schedules and service. Our government has usually been unwilling to protest or request arbitration. These smaller countries, also, are reluctant to furnish statistical data regarding fifth-freedom traffic. The United States, if it is going to regain some of its losses, must engage in more realistic negotiations for route exchanges.

After World War II, the policy changed to one of maximum development of international air transportation. It became policy that no foreign-flag airline seeking part of the industry would be excluded.

The United States' policy was described in the "International Air Transport Policy" of the 79th Congress (1945–1946). This policy was fivefold:

1. United States aviation would continue to be progressive in the development of new equipment and operating procedures.
2. Adequate incentives would compel efficiency in operation.
3. Full economic potentialities of air transport would be realized by widening the market.
4. Healthy financial condition would assure the in-flow of adequate capital.
5. United States airlines would carry a volume of world traffic proportional with the importance of the United States as a market for air transport services.

The International Air Transportation Policy of the United States has undergone several changes since 1946. Richard Nixon, while President, presented his policy which superseded President John F. Kennedy's policy adopted on April 24, 1963. In September 1976, President Gerald Ford changed the policy once again.

United States Policy for the Conduct of International Air Transportation Negotiations (August 21, 1978)

Introduction

United States international air transportation policy is designed to provide the greatest possible benefit to travelers and shippers. Our primary aim is furthering the maintenance and continued development of affordable, safe, convenient, efficient and environmentally acceptable air services. Our policy for negotiating civil air transport agreements reflects our national goals in international air transportation. This policy provides a set of general objectives, designed particularly for major international air markets, on the basis of which United States negotiators can develop specific negotiating strategies.

Maximum consumer benefits can be best achieved through the preservation and extension of competition between airlines in a fair marketplace. Reliance on competitive market forces to the greatest extent possible in our international air transportation agreements will allow the public to receive improved service at low costs that reflect economically efficient operations. Competition and low prices are also fully compatible with a prosperous U.S. air transport industry and our national defense, foreign policy, international commerce, and energy efficiency objectives.

Bilateral aviation agreements, like other international agreements, should serve the interest of both parties. Other countries have an interest in the economic prosperity of their airline industries, as we do in the prosperity of ours. The United States believes this interest is best served by a policy of expansion, of competitive opportunity rather than restriction. By offering more services to the public, in a healthy and fair competitive environment, the international air transport industry can stimulate the growth in traffic which contributes to both profitable industry operations and maximum public benefits.

Goals of U.S. International Air Transportation Policy

The U.S. will work to achieve a system of international air transportation that places its principal reliance on actual and potential competition to determine the variety, quality, and price of air service. An essential means for carrying out our international air transportation policy will be to allow greater competitive opportunities for U.S. and foreign airlines and to promote new low cost transportation options for travelers and shippers.

Especially in major international air transport markets, there can be substantial benefits for travelers, shippers, airlines, and labor from increasing competitive opportunities and reducing protectionist restrictions. Increasing opportunities for U.S. flag transportation to and from the United States will contribute to the development of our foreign commerce, assure that more airlift resources are available for our defense needs, and promote and expand productivity and job opportunities in our international air transport industry.

Translating Goals into Negotiating Objectives

U.S. International Air Transportation Policy cannot be implemented unilaterally. Our objectives have to be achieved in the system of international agreements that form the basic framework for the international air transportation system.

Routes, prices, capacity, scheduled and charter rules, and competition in the marketplace are interrelated, not isolated problems to be resolved independently. Thus, the following objectives will be presented in negotiations as an integrated U.S. position:

1. Creation of new and greater opportunities for innovative and competitive pricing that will encourage and permit the use of new price and service options to meet the needs of different travelers and shippers.
2. Liberalization of charter rules and elimination of restrictions on charter operations.
3. Expansion of scheduled service through elimination of restrictions on capacity, frequency, and route and operating rights.
4. Elimination of discrimination and unfair competitive practices faced by U.S. airlines in international transportation.
5. Flexibility to designate multiple U.S. airlines in international air markets.
6. Encouragement of maximum traveler and shipper access to international markets by authorizing more cities for non-stop or direct service, and by improving the integration of domestic and international airline services.
7. Flexibility to permit the development and facilitation of competitive air cargo service.

Explanation of Objectives

1. *Pricing*. The U.S. will develop new bilateral procedures to encourage a more competitive system for establishing scheduled air fares and rates. Charter pricing must continue to be competitive. Fares, rates, and prices should be determined by individual airlines based primarily on competitive considerations in the marketplace. Governmental regulation should not be more than the minimum necessary to prevent predatory or discriminatory practices to protect consumers from the abuse of monopoly position, or to protect competitors from prices that are artificially low because of direct or indirect governmental subsidy or support. Reliance on competition and encouragement of pricing based on commercial considerations in the marketplace provides the best means of assuring that the needs of consumers will be met and that prices will be as low as possible given the costs providing efficient air service.
2. *Charters.* The introduction of charters acted as a major catalyst to the expansion of international air transportation in the 1960s. Charters are a competitive spur and exert downward pressure on the pricing of scheduled services. Charters generate new trafficand help stimulate expansion in all sectors of the industry. Restrictions which have been imposed on the volume, frequency, and regularity of charter services as well as requirements for approval of individual charter flights have restrained the growth of traffic and tourism and do not serve the interests of either party to an aviation agreement. Strong efforts will be made to obtain liberal charter provisions in bilateral agreements.
3. *Scheduled Services.* We will seek to increase the freedom of airlines from capacity and frequency restrictions. We will also work to maintain or increase the route and operating rights of our airlines where such actions improve international route systems and offer the consumer more convenient and efficient air transportation.
4. *Discrimination and Unfair Competitive Practices.* U.S. airlines must have the flexibility to conduct operations and market their services in a manner consistent with a fair and equal opportunity to compete with the airlines of

other nations. We will insist that U.S. airlines have the business, commercial, and operational opportunities to compete fairly. The United States will seek to eliminate unfair or destructive competitive practices that prevent U.S. airlines from competing on an equal basis with the airlines of other nations. Charges for providing airway and airport properties and facilities should be related to the costs due to airline operations and should not discriminate against U.S. airlines. These objectives were recognized by the Congress in legislation enacted in 1975, and their attainment is required if consumers, airlines, and employees are to obtain the benefits of an otherwise competitive international aviation system.

5. *Multiple Airline Designations.* The designation of new U.S. airlines in international markets that will support additional service is a way to create a more competitive environment and thus encourage improved service and competitive pricing. Privately owned airlines have traditionally been the source of innovation and competition in international aviation, and it is, therefore, particularly important to preserve for the U.S. the right of multiple designation.
6. *Maximum Access to International Markets.* Increasing the number of gateway cities for nonstop or direct air service offers the potential for increasing the convenience of air transportation for passengers and shippers and improving routing and market opportunities for international airlines. In addition, enhancing the integration of U.S. airlines' domestic and international air services benefits both consumers and airlines.
7. *Cargo Services.* We will seek the opportunity for the full development of cargo services. Frequently demand for such services requires special equipment and routes. Cargo services should be permitted to develop freely as trade expands. Also important in the development of cargo services are improved facilitation, including customs clearance, integration of surface and air movements, and flexibility in ground support services.

Negotiating Principles

The guiding principle of United States aviation negotiating policy will be to trade competitive opportunities, rather than restrictions, with our negotiating partners. We will aggressively pursue our interests in expanded air transportation and reduced prices rather than accept the self-defeating accommodation of protectionism. Our concessions in negotiations will be given in return for progress towards competitive objectives, and these concessions themselves will be of a liberalizing character.

Proposed bilateral agreements which do not meet our minimum competitive objectives will not be signed without prior Presidential approval.

Interagency Committee on International Air Transportation Policy

Chaired by:	Department of Transportation
Member Agencies:	Department of State
	Department of Justice
	Department of Defense
	Department of Commerce
	Council of Economic Advisors
	Council on Wage and Price Stability
	Domestic Policy Staff—The White House
	National Security Council
	Office of Management and Budget

The principle of liberal competition for the United States domestic air carriers brought about by Deregulation has permeated into the United States government's attitude with respect to its international air carrier operation and the negotiating of agreements with other countries. Parallel to domestic Deregulation, international agreements having to do with competitive actions by air carriers have become more important.

The most competitive bilateral agreement includes multiple designations, dual disapproving pricing, equal access to marketing outlets, the right to self handle, and an open charter regime. From air carrier management's perspective, such an agreement is an invitation to exercise marketing judgments to meet the public demand for service. Management has responded to this new freedom, and the effect on the industry and the public has been beneficial.

The benefits of a more world-wide competitive environment are that competition

1. Provides substantial new benefits to customers,
2. Forces the airlines to operate more efficiently, whether they are privately or publicly owned, and
3. Does not hinder the ability of airlines to earn a normal return on investment.

On August 21, 1978, President Carter announced a pro-competitive policy on international aviation. The principles contained in it were subsequently adopted by Congress in the International Air Transport Competition Act of 1979. The President stated:

"Our policy for negotiating civil air transport agreements reflects our national goals in international air transportation. This policy provides a set of general objectives designed particularly for major international air markets on the basis of which United States negotiations can develop specific negotiating strategies."

The United States' international policy has caused much controversy. Many foreign government-owned air carriers and some foreign governments have been critical of it. The criticisms may not stand up under close examination because they may be intended to protect their special interests. They fall into six areas:

1. That the United States is attempting to foist its approach on other nations which may have different national goals and is ignoring the basic principles of sovereignty.
2. That Deregulation will quickly lead to the loss of the regular high-frequency service that is required by the world business community.
3. That open competition is inconsistent with sound energy policy which may require severe restrictions in leisure travel to assure maintenance of service for business travelers.
4. That although competition may look good during the boom of a business cycle, history has shown that excessive discount pricing during growth periods has led to overcapacity and financial disaster during downturns.
5. That careful regulation and capacity control can achieve lower fares than can open entry and competition.
6. That it is foolish to talk about open skies when the industry is so dependent upon external constraints such as limited airport capacities, air traffic control, tourists' organizations, and so forth.

The foreign criticism of the United States' policy of international competition is that the United States violated the spirit of international cooperation and equality which was established at the Chicago Conference in 1944. This Conference was a milestone in the history of international aviation; it resulted in cooperation on many aspects of international civil aircraft operation.

The one area where no multilateral agreement could be reached, however, was the subject of how commercial air carriers would be permitted to compete. Presently, as it did in 1944, the United States advocates a system of essentially free entry and price flexibility. Some countries favor more regulation as a solution. No multilateral agreement has ever been reached, so the establishment of these relationships is dependent upon bilateral negotiations between participating countries.

The United States' International Air Transport Policy is a policy of commitment to creating an efficient system that can provide greater service and price options to travelers and shippers. Basically it is a return to the principles of equal competitive opportunities which the United States proposed at the Chicago Conference in 1944. The policy also provides that bilateral agreements should be designed to create a market environment where competition can flourish.

What the United States challenges is the presumption that the international system can function only if air carriers are allowed to agree on rates and if seat capacity is strictly controlled.

Between 1968 and 1977 the average load factor for U.S. air carriers on the North Atlantic route never exceeded 57%. In 1979 it rose to 69%; by 1991 it had increased slightly to 69.5%. The North Atlantic route has always been a highly seasonal market. Load factors have varied 15% to 20% between winter and summer.

Under Deregulation the domestic U.S. air carriers have adjusted their operations to accommodate for seasonal highs and lows. This flexibility is obviously beneficial and can be utilized by the international and foreign air carriers.

The public benefits of competition are not limited to greater access for leisure travelers. Business travelers also gain by the expansion of nonstop gateways to many destinations and increased frequency of service to many cities.

The result of the liberal competition policy was seen immediately as indicated by the traffic statistics for the year ended June 30, 1980.

- Traffic to Europe increased 12%
- Traffic to France increased 19%
- Traffic to Italy increased 3%
- Traffic to the Far East increased 20%
- Traffic to Korea increased 166%
- Traffic to Singapore increased 76%

Competition benefits air carriers in several ways. The potential for growth in international air transportation is directly connected to the development of non-business travel. The competitive process is better suited to developing low fares than the old system of regulation-controlled discriminate fares. Competition forces all air carriers in the international and foreign market to operate at their greatest efficiency. There is little doubt that all future bilateral negotiations will reflect the policy of competition.

A recent Gallup survey indicated that 68% of the United States population would travel abroad if they had the time and the money. In the United Kingdom the figure was 74%, and for the population of France it was 84%. This survey showed that the large majority of people want to travel. A competitive air transportation system will provide economic incentives to the air carrier industry to find methods to serve this relatively untapped market.

In April 1980 the Civil Aeronautics Board issued a tentative opinion in the International Air Transportation Association Show-Cause Proceeding approving IATA traffic conferences. Highlights of the opinion were:

1. The IATA traffic conference system was approved and granted anti-trust immunity for two years.
2. U.S. air carriers may not participate in rate conferences for travel over the North Atlantic between the United States and Europe.
3. Approval of the conferences will be effective only as long as the Civil Aeronautics Board receives assurance from IATA that in markets touching the United States, innovative fare proposals will not be subject to approval or delay by other IATA members.

The Civil Aeronautics Board affirmed its commitment to achieving a competitive international environment with individual air carrier initiative as the primary rate-setting mechanism.

Individual Civil Aeronautics Board members supporting the opinion indicated that it did not go far enough. They felt the participation of U.S. air carriers in traffic conferences should have been prohibited.

The foundation for the Civil Aeronautics Board's analysis of the International Air Transport Association traffic conference procedures is the Sherman Anti-trust Act which provides that any contract that restrains trade or commerce among the States or with foreign countries is illegal in the United States. The Board has concluded that the traffic conference procedures substantially reduce competition because they are designed primarily to establish the rates to be charged by competing air carriers.

The International Civil Aviation Organization at its Assembly meeting in September–October 1980, stated that the industry grew appreciably faster than the general economy during the 1970s and that the growth is expected to continue. World scheduled international air traffic grew at a 10% annual rate compared to the world's gross product growth rate of 3.8%. An average yearly decline in rate levels of 3.4% would appear to have contributed to this growth. Future growth will be affected by world economic growth which was predicted to be 4% a year. In light of this, yearly passenger increases were forecast to be 9%.

The effect of deregulation in the United States has spread to other countries. England has passed a law that makes it mandatory for their Civil Aviation Authority to consider the beneficial effects of competition when awarding routes. The British Civil Aviation Act became law in November 1980. It legally enabled the British government to sell all or part of British Airways to private investors.

The new law strengthens the powers of the British CAA in granting new air carrier routes by eliminating the power formerly held by the trade minister to give guidance to the CAA in determining what routes to award to which air carrier.

The entire world is watching the United States' system of air carrier deregulation. If successful, deregulation may spread to areas where the economic regulation is still being practiced.

The United States will offer to negotiate 'open skies' agreements with all European countries willing to permit U.S. carriers essentially free access to their markets. Discussions will be limited to European countries since Europe is moving toward the free flow of passengers and goods. After reexamining its policy, the U.S. has decided it should be willing to offer open skies agreements to countries who agree with our free market principles.

The U.S. has already taken an important step in this direction with its underserved cities program, opening U.S. gateway cities to service by foreign airlines without insisting on prior negotiations for such rights. Foreign airlines have authority to provide service to selected American cities under the program.

During the 1980s, the U.S. developed more market-oriented agreements with its aviation partners than ever before. The more liberal environment is being challenged now by a number of countries. He noted there have been attempts by the U.S.'s trading partners to limit the rights of U.S. airlines under existing agreements. The United States places a very high value on its bilateral agreements; respect those agreements, and honor them. The same is expected from their trading partners. In the past, the U.S. has offered such agreements only to a few of its largest aviation partners.

Free Trade

The standard nation-to-nation bilateral is the present manner for international agreements. However, a new concept is being discussed and promoted by many air carriers and countries who desire the abandonment of the bilateral system and replace it with the method of international "free trade."

Free Trade advocates open market access and competitive pricing, rules to safeguard competition, no nationality limitations on airline ownership and transitional measures to ease acceptance of the new system. Small nations and their air carriers are concerned that globalization of airlines will be further supported by a liberal, multilateral system and lead to domination by a few airlines. Those favoring Free Trade countered that the bilateral system restricts competition and stifles the growth of the air carrier industry, which is critically needed for economic development.

If the industry is to respond to consumer needs it must be free of the constraints on competition and market access that bilaterals have imposed in the past. The only point of agreement is that a hybrid system of bilaterals and new styled multinational agreements will coexist for awhile. In the mean-time the free trade advocates seem to have their method for change. Many have endorsed the concept of nations joining together in a bilateral agreement and then allowing other nations to accept the liberal trading rules when they are willing and able. Thus the multilateral agreement could come into being and expanded to other nations until Free Trade becomes accepted. This process will speed up as regions join in economic groupings in order to grow. A useful product of open skies negotiations would be a model agreement that could put Free Trade on its way.

The change, if and when it comes, will require several years and much negotiation.

Bibliographic Note

One of the problems in the study of international civil aviation has been the lack of total concentrated research material. Several independent sources and their materials would have to be gathered together and then carefully searched for the particular subject matter required. This is very time consuming. Fortunately, since 1996, this is no longer the case. A very worth-while text has been prepared by Adrianus Groenewege entitled the *Compendium of International Civil Aviation.* This book is published by the International Aviation Development Corporation, 3460 Peel Street, Suite 1803, Montreal, Quebec, Canada H3A 2M1. The book costs under $200. Any up-to-date well-stocked library will have a copy available. It is highly recommended for further study of any subject matter presented in this chapter.

The text covers the major milestones in international civil aviation from 1990 through 1996 and is a general review of historical events.

A detailed analysis of the structure of international civil aviation covers the basic legal foundations of the several conferences, conventions and agreements that formulate the present legal aspect of international civil aviation.

An alphabetical listing of worldwide organizations and a description of each is included. A section presents the definition of the aviation terms commonly used throughout the industry and is a very good resource for the meaning of various technical as well as non-technical terms.

A larger section of the book is devoted to abbreviations, acronyms, world airport codes and other important research information including countries, capitals and their population, and distances between major world airports. Available are conversion tables for area, distance, pressure, speed, temperature and volume. A wealth of statistics is accessible for anyone desiring data.

This book is a valuable reference for any serious-minded aviation enthusiast or student.

Endnotes

1. The Warsaw Convention and the Chicago Convention have two slightly different definitions of what constitutes International Transportation.
2. The term "foreign air transportation" is unique only to the United States and can be offensive to some non-United States air carriers and aeronautical authorities.
3. Much of the text subject on conferences and conventions has been supplemented by Aero-Accords, Inc. of Seattle, Washington, Earl R. Scott, President. For a complete study of the International Conferences and Conventions reviewed in this chapter, the book recommended is "Compendium of International Civil Aviation" by Adrianus D. Groenewedge, published by the International Aviation Development Corporation, 3460 Peel Street, Suite 1803, Montreal, Quebec, Canada H3A 2M1.
4. Seventh Freedom rights allow a carrier to pick up passengers from a country other than its own and deliver them in a third country, also not its own, on flights that do not connect to its homeland. (Source: ALPA)

AVS 1200
Introduction to Aviation

Aviation
Acronyms, Words and Phrases

Acronyms

• ADF	• CVR	• FBO	• IMC	• PF	• UHF
• AOPA	• DGPS	• FDR	• ISA	• PNF	• UTC
• ARTCC	• DH	• FIR	• JAA	• RADAR	• VDF
• ARTS	• DME	• FMS	• NAS	• RNAV	• VFR
• ATA	• DOT	• FSS	• NASA	• SAR	• VHF
• ATC	• EAA	• GA	• NAT	• SSR	• VMC
• ATIS	• EAS	• GPS	• NDB	• SST	• VNAV
• CAA	• EFIS	• GPWS	• NOAA	• STOL	• VOR
• CAS	• EGT	• HUD	• NOTAM	• TACAN	• Z
• CDI	• ELT	• IATA	• NTSB	• TAS	
• CFIT	• EPR	• ICAO	• NWS	• TCAS	
• CG	• FAA	• IFR	• OAT	• TDWR	
• CRM	• FAR	• ILS	• PAR	• TRACON	

Words and Phrases

- Air carrier
- Air commerce
- Aircraft
- Airframe
- Airplane
- Airport
- Airship
- Air traffic
- Air traffic clearance
- Air traffic control
- Air transportation
- Alternate airport
- Altitude
- Approved
- Area navigation
- Balloon
- Brake horsepower
- Calibrated airspeed
- Category-used in respect to certification of airmen and aircraft
- Ceiling
- Class-used in respect to certification of airmen and aircraft
- Commercial operator
- Controlled airspace
- Crew member
- Critical altitude
- Critical engine
- Equivalent airspeed
- Extended over-water operation
- Fireproof
- Fire resistant
- Flame resistant
- Flammable
- Flash resistant
- Flightcrew member
- Flight level
- Flight plan
- Flight time

Words and Phrases

- Foreign air carrier & commerce
- Glider
- Gyroplane
- Helicopter
- Indicated airspeed
- Interstate air commerce & transportation
- Intrastate air transportation
- Large aircraft
- Large aircraft
- Lighter-than-air aircraft
- Load factor
- Medical certificate
- Night
- Non-precision approach procedure
- Overseas air commerce & transportation
- Pilotage
- Pilot-in-command
- Precision approach procedure
- Propeller
- Rating
- RNAV
- Rotorcraft
- Second-in-command
- Small aircraft
- Standard atmosphere
- True airspeed
- Traffic pattern
- Type-used in respect to certification of airmen and aircraft

AVS 1200
Introduction to Aviation

Man's Efforts to Fly
and
Development of the Airplane

Early Aviation Theorist

- **Daedalus & Icarus (Mythology)**
  - Feather & Wax
  - First in-flight structural failure

Early Aviation Theorist

- **Friar Roger Bacon - England (circa 1250)**
  - Heavier Than Air Concept (air support)
  - Flapping Wings
    - Ornithopters

Early Aviation Theorist

- **Leonardo Da Vinci - Italy (circa 1550)**
  - Flapping-wing Device
  - Parachute
  - Helicopter

Lighter-than-Air Ship

- **Joseph & Etienne Montgolfier - France**
  - Hot Air Balloons
    - First appearance June 4, 1783
    - First aerial passengers (sheep, rooster and duck) September 19, 1783
    - First successful test pilot deRozier November 21, 1783

Dirigible

- **Henri Giffard – France (1852)**
  - Dirigible
    - 1st steerable balloon

The Zeppelin Airship

- **Count Ferdinand Von Zeppelin – Germany (circa 1900)**
  - – Rigid Airships

Early Aviation Theorist

- **Sir George Cayley - England**
    - *"The whole problem is confined within these limits, namely to make a surface support a given weight by the application of power to the resistance of air."*
  - – Fixed-wing Device over the Ornithopter
  - – Aerodynamic Theory
    - Cayley's Papers –"On Aerial Navigation" (1809)

Early Aviation Activists

- **Horatio Philips - England (1884)**
  - – Double surface aerofoil sections

Early Aviation Activists

- **Otto Lilienthal - Germany (1896)**
  - Gliders

Early Aviation Activists

- **Octave Chanute (1896)**
  - Pratt Truss System - Gliders
  - Early Gliding experiments in the Indiana Dunes

Powered Flight

- **Samuel P. Langley - USA (1903)**
  - Aerodrome

Powered Flight

- **The Wright Brothers - USA (1903)**
  - Aerodynamic Research
  - Flyer I

Powered Flight

- **Alberto Santos-Dumont – Paris 1906**
  - First powered flight in Europe

Powered Flight

- **Glenn Curtiss - USA (1908)**
  - *June Bug*
    - Won first aviation prize offered in US
  - Curtiss Aircraft
    - Largest aircraft manufacturer in the world during World War I

Powered Flight

- **Louis Bleriot – 1909**
  - First international flight
    - Crossing the English Channel
- **Adolphe Pegoud – 1913**
  - First inverted flight and loop
- **Lincoln Beachey – 1914**
  - First US stunt pilot

AVS 1200
Introduction to Aviation

Basic Aerodynamics

The Four Forces of Flight

- **Lift**
- **Weight**
- **Thrust**
- **Drag**

Airfoil

- **Leading Edge**
- **Trailing Edge**
- **Chord**
- **Mean Camber Line**
- **Camber**
- **Relative Wind**
- **Angle of Attack**

Lift

- Bernoulli
- Newtonian
- Laminar Flow

Weight - Gravity

- Operating Empty Weight (OEW)
- Payload
- Zero Fuel Weight
- Maximum Takeoff Weight
- Ramp Weight
- Useful Load

Thrust

- Reciprocating Engines
- Gas Turbine Engines
- Rocket Engines

Drag

- Induced
- Parasitic
- Airspeed/Drag Relationships

Airfoil Stall Characteristics

- Buffet
- Reduction in Control Effectiveness
- Loss of Lift
- Recovery

The Three Axis of Rotation

- Longitudinal
- Lateral
- Vertical

Primary Control Surfaces

- **Ailerons**
  - Roll
    - About the Longitudinal
- **Elevators**
  - Pitch
    - About the Lateral
- **Rudder**
  - Yaw
    - About the Vertical

Stability

- **Dynamic**
- **Static**

Performance

- **Takeoff**
- **Climb**
- **Cruise**
- **Fuel Efficiency**
  - Range
  - Endurance
- **Descent**
- **Landing**

AVS 1200
Introduction to Aviation

World War I
And
Post-War Aviation

Early Aircraft Engines

- Antoinette Series (Radial)
- The Gnome and La Rhone (Rotary)

Aviation 1917-1918

- **World War I**
  - Rapid Development of Airframes & Engines
  - Automotive Influences Entered the Aviation Industry
    - Assembly-line processes

Aviation post WWI

- **First Atlantic Flights**
  - US Navy-Curtiss Flying Boats - May 1919
  - British Vickers Vimy (non-stop) - June 1919
    - Alcock and Brown
  - British Airship R34 (first round trip) - June 1919

Aviation post WWI

- **First Around the World (1924)**
  - Donald Douglas World Cruisers (5 started, 2 finished)
    - 175 days
    - 26,553 miles
    - 363 flight hours

Aviation post WWI

- **Barn Storming Era**
  - WWI Aircraft Surplus
  - Wide Open Spaces
  - Pilots to Fly
  - Enthusiastic Population

Aviation post WWI

- **Golden Age of Flight**
  - Pushing the boundaries of speed
  - Altitude Limits
  - Expanding the Utility of the Aircraft and its Influence

AVS 1200
Introduction to Aviation

World War II
and Post-War Aviation

World War II

- Airline operations reduced or suspended
- Some airlines provided military assistance
- Manufacturers transformed production to military aircraft

World War II

- Rapid advancement in aircraft technology
  - Radar
  - Gas Turbine Engines
  - Pressurization
  - Turbo-supercharging
  - Wright's Whirlwind and Cyclone engines
  - Pratt and Whitney *Wasp*

Postwar Changes

- Development of Helicopters
- Improved General and Commercial Aircraft
- Improved International Services
- US Airline Expansion
- Development and Conversion of Airports
- Expansion of National Airlines

International Conferences

- Chicago Conference of 1944
  - Establishes Provisional International Civil Aviation Organization (PICAO)
- Bilateral Agreements based on the Chicago Standard Form

Lagging Aviation Infrastructure

- Inadequate Route Coverage by Radar
- Sparse Precision Approaches
- Inadequate Air Traffic Control
- Inadequate Airports to Handle Larger Aircraft and Jets

Developments 1958-1978

- Aircraft Speeds Increasing
- Passenger Demand Increasing
- Size of Aircraft Increasing
- International Competition for North Atlantic and Pacific Routes Increasing

Developments 1958-1978

- Corporate Aviation Departments Increasing
- Increasing Individual Aviation
- Increasing Noise Around Airport

Introduction to Aviation

Federal Legislation
and
Deregulation

Federal Legislation

- Air Mail Act of 1925
  - The Kelly Act
    - Authorized Postmaster General to seek competitive bidding for contract mail routes
      - CAM 1 - CAM 12
- Air Commerce Act of 1926
  - Creation and maintenance of navigation aids
  - Rules and regulations for aviation safety

Federal Legislation

- Air Mail Act of 1930
  - McNary-Watres Act
  - Gave Postmaster authority over "competitive" bidding
    - "lowest responsible bidder" – flown route day and night
- Air Mail Act of 1934
  - Black-McKellar Act
  - Competitive bidding reinstated
  - Separation of airlines and aircraft manufacturers

Federal Legislation

- Civil Aeronautics Act of 1938
  - McCarran-Lea Act
  - Created
    - Civil Aeronautics Authority
    - Administrator of Aviation
    - Air Safety Board
- 1940 Amendment to Civil Aeronautics Act
  - Two new agencies from original three
    - Civil Aeronautics Board (CAB)
    - Civil Aeronautics Administration (CAA)

Federal Legislation

- Federal Airport Act of 1946
  - Provided funding for fifty percent of cost of airport construction or improvement
- Federal Aviation Act of 1958
  - Abolished CAA and created the Federal Aviation Agency (FAA)
  - FAA
    - Long-range planning and implementation
    - All air safety research and development
  - CAB
    - Aircraft accident investigation
    - All aspects of economic regulations

Federal Legislation

- Department of Transportation Act of 1966
  - Created Department of Transportation (DoT)
    - Federal Aviation Agency became Federal Aviation Administration within DoT
  - Created National Transport Safety Board (NTSB)
    - Assumed aviation investigative responsibilities from CAB
- Airport and Airway Development Act of 1970
  - Created Airport and Airway Trust Fund
    - Provide grants for improvements

Federal Legislation

- Air Cargo Deregulation Act of 1977
  - Phased out CAB control of air cargo operations
    - Eliminated CAB economic regulations
      - Market demand
- Airline Deregulation Act of 1978
  - Phased out CAB control of airline operations
    - Eliminated CAB economic regulations
      - Free-market pricing
    - Established safeguards for service to small communities

Federal Legislation

- International Air Transportation Competition Act of 1979
  - Amended Federal Aviation Act of 1958 to
    - Promote competition in international transportation
    - Provide greater opportunities for US air carriers
    - Establish goals for developing US international aviation negotiating policy

Federal Legislation

- Airport and Airway Improvement Act of 1982
  - Reestablished the FAA's airport grant program
    - Inactive since 1981
  - Renamed Trust Fund program the Airport Improvement Program (AIP)
  - Required airports serving air carriers to secure "Airport Operating Certificates"
    - FAR - Part 139

Pre-Deregulation

- CAB (1938-1985):
  - Acted as a public utility
  - Granted authority to regulate:
    - Entrants
    - Routes/service areas
    - Fares
    - Profit margins
    - Types of aircraft used
    - Even seating configurations!!

Pre-Deregulation

- Economists critical of economic regulation:
  - Pricing and entry restrictions caused restricted service and insufficient pricing competition, inflated airline costs, unsatisfactory profits
  - Jets introduced in 60's (wide-bodied in 70's):
    - Aircraft faster, more efficient, more comfortable, more reliable, therefore:
      - Increased passenger demand
      - Existing airlines unable to meet demand
      - New entrants not allowed in the club; therefore:
        - » Public disgruntlement

Deregulation

- Air Cargo Deregulation Act (1977):
  - Operate on any domestic route
  - Charge whatever market would bear
  - CAB approval:
    - No longer "public interest"
    - But "fit, willing, and able"
  - Growth of express package carriers:
    - Absorbed local incumbents ⇒
    - Hub & Spoke networks

- Passenger carriers given more freedom in setting routes
- Initially prospered and growing public support; therefore:

- Airline Deregulation Act (Oct, 1978)
  - overwhelming bi-partisan support
  - opposed by "in the club" airlines
  - CAB
    - 1981: no authority over routes
    - 1983: no authority over fares
    - 1985: no CAB:
      - Antitrust/essential air service to FAA
      - International regs (BASA/BAA) to Dept of State

- Goals of deregulation:
  - increase competition
  - improve service with reduced fares
  - stimulate economy
  - introduce innovative service offerings
  - enhance productivity and efficiency
- Rapid impact:
  - Airlines reshaped networks
  - New entrants

Effects of Airline Deregulation

- Establishes Free Market Pricing
- Airlines saw Increased Profits and Route Access (Airlines Supported Deregulation)
- New Airlines saw Ability to Enter Market
- Ten Year Essential Air Service (EAS) Program Established to Smaller Communities
- CAB ceased to exist at close of business on December 31, 1984

Changes Promoted by Deregulation

- Major Airlines Assumed Superior Position by:
  - Size of Route Structure
  - Computer Reservation System
  - Mileage Bonus Programs
  - Strong Financial Resources
  - Acquired Smaller, Two-engine, Two-pilot Aircraft
  - Negotiated Wage Concessions

Consolidation of Route Structure

- Hub and Spoke Design System
- Exclusive Feeder Lines with Commuter Carriers
- Route Structure (circa 1930)

Airline Reaction

- Airlines were forced to re-focus on:
  - Cost savings:
    - more efficient aircraft
    - personnel reductions
    - wage and benefit reductions
    - route changes
- Many went under

AVS 1200
Introduction to Aviation

Federal Aviation Administration
and
National Transport Safety Board

Prelude to
Federal Aviation Act of 1958

- **Grand Canyon Accident 1956**
  - TWA Super Constellation Collides with United DC-7
- **Las Vegas Accident 1958**
  - DC-7 and Military Jet Collide
- **Maryland Accident 1958**
  - Viscount and Military T-33 Collide

Federal Aviation Act of 1958

- Establishes the Federal Aviation Agency
- New FAA Responsible for Regulation, Certification, and Safety
- Civil Aeronautics Board Still Investigates Accidents and Sets Economic Policy

Basic FAA Responsibilities

- Regulating civil aviation to promote safety and fulfill the requirements of national defense
- Encouraging and developing civil aeronautics, including new aviation technology
- Developing and operating a common system of air traffic control and navigation for both civil and military aircraft.
- Performing research and development with respect to the National Airspace System and civil aeronautics
- Developing and implementing programs to control aircraft noise and other environmental effects of civil aviation
- Regulating U.S. commercial space transportation

Rules & Procedures

- **International - International Civil Aviation Organization (ICAO)**
  - Standards and Recommended Practices - SARPs
- **U.S. - Federal Aviation Administration (FAA)**
  - **NPRM** - Published in *Federal Register* for public comment
    - *After review proposed rule could be made into law*
- **European – European Aviation Safety Agency (EASA)**

Federal Aviation Regulations (FARs)

- **FAR Regulation Part 1**
  - Definitions & Abbreviations
- **FAR Regulation Part 21**
  - Aircraft Certification Process
- **FAR Regulation Part 23**
  - Design Requirements for Small Aircraft (<12,500lbs)

Federal Aviation Regulations (FARs)

- **FAR Regulation Part 25**
  - Design Requirements for Large Aircraft (>12,500lbs)
- **FAR Regulation Part 36**
  - Noise Standards
- **FAR Regulation Part 61**
  - Certification of Pilots & Instructors

Federal Aviation Regulations (FARs)

- **FAR Regulation Part 71**
  - Airways, Routes, Airspace
- **FAR Regulation Part 73**
  - Special Use Airspace
- **FAR Regulation Part 91**
  - General Operating & Flight Rules

Federal Aviation Regulations (FARs)

- **FAR Regulation Part 121**
  - Air Carrier Operations (Large Aircraft)
- **FAR Regulation Part 135**
  - Air Taxi & Commercial Operators (Small Aircraft)
- **FAR Regulation Part 141**
  - Pilot Flight Schools
- **FAR Regulation Part 147**
  - Aviation Maintenance Technician Schools

Accident Forces FAA to Act

- New York, 1960
  - DC-8 and Super Constellation Collide (While Under Radar Surveillance)
- Forces FAA to:
  - Increase Airways
  - Improve Radar Equipment
  - Increase Restrictions to Airspace

Department of Transportation (DOT) Act of 1966

- Federal Aviation Administration
  - Name change when FAA became part of DOT
    - *Agency* changed to *Administration*
- National Transportation Safety Board
  - Established under DoT
- Civil Aeronautics Board
  - Economic policies only

National Transportation Safety Board

- Origins of the NTSB come from the Air Commerce Act of 1926
- 1966 congress changed the agency to be independent and their role was to determine the probable cause of:
  - Highway accidents
  - Train accidents
  - Marine accidents
  - Pipeline accidents
  - Hazardous material accidents
  - AVIATION accidents

National Transportation Safety Board

- The mission of the NTSB:
  - to determine the "probable cause" of transportation accidents and to formulate safety recommendations to improve safety
- 1974 Congress made NTSB completely independent from DOT
- NTSB has no authority to regulate, fund, or be directly involved in the operation of any mode of transportation

International Aviation Accidents

- Annex 13 to the Chicago Convention (international treaty)
  - Governing structure for civil aviation throughout the world.
- NTSB is a key player
  - Charged with responsibility for compliance with US obligations
- Civil accidents outside U.S. territory
  - NTSB appoints U.S. representative to the investigation and provides advisors from U.S. industries
    - NTSB provides assistance to the authorities of the investigation.
    - NTSB may even be given full responsibility for the investigation
- NTSB also provides technical assistance in the international field
  - Such as CVR & FDR analysis

AVS 1200
Introduction to Aviation

FAA
Certification

Propulsion

- **Reciprocating Engines**
  - Normally aspirated
  - Turbo-charged
- **Turbojet**
- **Turbofan**
- **Turboprop**
- **Turboshaft**

Instrumentation

- **Electronic Flight Instrument Systems (EFIS)**
  - Flight
    - Primary Flight Display – PFD
    - Navigation Display – ND
  - Engine
    - Engine Indicating and Crew Alerting System – EICAS
      - » Beoing
    - Electronic Centralised Aircraft Monitoring – ECAM
      - » Airbus
- **Navigation**
  - Flight Management System
    - Global Positioning System - GPS
    - Inertial Reference System - IRS

Major Classifications of Airplanes

- **Airliners**
  - Large Transport Category Aircraft
- **Commuters**
  - Turboprops & Regional Jets

Major Classifications of Airplanes

- **Military**
  - Trainers
  - Airlift
  - Aerial Refueling
  - Bombers
  - Fighters
  - Battlefield Support

Major Classifications of Airplanes

- **General Aviation**
  - Training
  - Personal
  - Business
  - Air Taxi & Charter
  - Air Freight & Courier Services
  - Agricultural
  - Forestry
  - Fish & Wildlife
  - Construction
  - Aerial Photography
  - Offshore Drilling
  - Ultralights

Certification of Aircraft

- **The Certification and Registration of Aircraft**
  - **Category:**
    - Normal
    - Utility
    - Acrobatic
    - Limited
    - Restricted
    - Experimental
    - Provisional
    - Transport
  - **Class:**
    - Airplanes
    - Rotorcraft
    - Gliders
    - Lighter-than-air
  - **Type:**
    - Cessna 152
    - Beechcraft KingAir

Aircraft Certification According to Airman Qualifications

Category:
- Airplanes
- Rotorcraft
- Gliders
- Lighter-than-air

Class:
- Single-engine land
- Single-engine sea
- Multi-engine land
- Multi-engine sea
- Helicopter
- Gyroplane
- Sailplane
- Blimp
- Balloon

Type:
- Cessna 152
- Piper Cherokee
- Beechcraft KingAir
- Hughes 500

Aircraft Certificates

- Production Certificate
- Type Certificate
- Airworthiness Certificate

Pilot Certification

- **FAR part 61**
- Basic FAA Requirements for Certification
- Student
  - Required before solo
  - 16 yrs old + 3rd Class medical
  - Expires after 2yrs
- Private
  - 17 yrs old + 3rd Class medical
  - Pass both a written and a practical (flight) test

- Commercial
  - 18 yrs old + 2nd Class Medical
  - At least a private pilot certificate
  - Pass both a written and a practical (flight) test
- Instructor
  - Commercial pilot certificate
  - Instrument rating in category and class
  - Pass 2 addition written tests and a practical (flight) test

- Airline Transport
  - 1500 hrs flight time
  - 500 hrs of cross-country flight time
  - 23 yrs old
  - Instrument rating

Medical Examination

- First Class - 1 yr (<40) or 6 mth (≥40)
- Second Class - 1 yr
- Third Class - 5 yr (<40) or 2 yr (≥40)

Tests

- Written Test
- Flight Test (Practical)

Simulators / Flight Training Devices

- Reduce Costs
- Enhance Safety
- Controlled Environment

PC Based Training

- Convenience
- Affordable
- Adaptability

Introduction to Aviation

Aircraft Design

Aircraft Design Considerations

- FAR 21 - Certification of New Aircraft
- FAR 23 - Certification of Small Aircraft (under 12,500 lbs)
- FAR 25 - Certification of Large Aircraft

Airplane Configuration

- Payload - Passengers & Cargo
- Performance - Range, Speed, & Runway Capabilities
- Marketing - Profitability

Engine Configuration

- Reciprocating
- Turboprop
- Turbofan - *quiet*
- Turbojet
- Economics
- Emergency Situations
- Performance
- Noise (FAR 36)
- Maintenance

Cabin Configuration

- Economics
- Safety
- Comfort
- In-flight Services
- Convertibility to Freight

Design Phases for Aircraft

- Design Specifications
- Scale Models/Wind Tunnel Tests
- Computer Modeling & Simulation
- Prototype Aircraft - Compliance with Regulations & Verify Performance
- Flight Testing
- Production

Human Factors of Design

- Type of Information - Visual, Aural, Tactual, Combination
- Warning - Red
- Caution - Yellow
- Advisory - Green

Human-Machine Closed Loop System

- Input
- Display
- Human
- Controls
- Machine
- Output
- Feedback

Location and Direction of Movement

- Logical Flow Patterns
- Simple

Fly-by-Wire

- Computer Controlled (digital) Flight Control Systems
- Weight Reduction
- Easier Diagnostics

Materials

- Consideration given to:
  - Weight
  - Strength
  - Fatigue
- Composite materials
  - B-787 Dreamliner
    - 50+% composite for weight reduction

Computer Monitoring Systems

- Flight Management System (FMS)
- Electronic Flight Information Systems (EFIS)

Introduction to Aviation

National Airspace System (NAS)

Rules of the Air

- **Visual Meteorological Conditions (VMC)**
  - 1000 Foot Ceiling and 3 Miles Visibility
- **Instrument Meteorological Conditions (IMC)**
  - That which is less Than VFR
- **Classes of Airspace**

Purpose of Air Traffic Control System

- To provide safe separation of IFR aircraft using the National or International Airspace System

Latest ATC Equipment

- **Old system**
  - Monochrome, plain-view displays
  - Not highly flexible in event of peak traffic, severe weather
  - Little back-up protection against outages
  - Not capable of displaying real-time weather intensity
- **New system**
  - High-resolution color monitors
  - Improved flight data and weather displays
  - Better protection against outages
  - Capable of displaying real-time weather intensity

Terminal Facilities

- Clearance Delivery
- Ground Control
- Local Controller
- Central TRACON Automatic System (CTAS)
  - provides automation tools for planning and controlling arrival air traffic

En-Route Facilities – Air Route Traffic Control Centers (ARTCC)

- 21 Centers operated by the FAA
- Will Provide Separation of IFR Aircraft from Other IFR Traffic and Terrain
- Workload Permitting ARTCC will Provide
  - Separation of VFR Aircraft
  - Traffic Advisories to IFR and VFR Aircraft
  - Separation from Severe Weather to IFR and VFR Aircraft

AVS 1200
Introduction to Aviation

Airports

Airports

- Types:
  - Private
  - Military
    - 250 airports were developed for national defense
      - Especially during WW2
  - Municipal
    - Federal funding was disallowed by the Air Commerce Act of 1926

Federal Airport Act of 1946

- Authorized funding for capital improvements
- Grants and matching funds to Municipalities
  - Most of the 250 WW2 airports were declared surplus
    - Returned to Municipalities

Airport/Airway Development Act (1970)

- Jet aircraft with large payloads required improved airport and terminal facilities
  - Users were charged a Levy
    - Held in trust to provide Federal grants and matching funds
- National Plan of Integrated Airport Systems (NPIAS)
  - Developed by the FAA
  - Identified 3,000 eligible airports for funding
    - "Adequate" airport: must be located within 20 SM of 97% of U.S. population

NPIAS

- Recognizes **3** General Classes of Airports
  - Commercial Service Airports:
    - 2,500 PAX Enplanements Per Year
  - Reliever Airports:
    - General Aviation Airports within the Metro Area
  - General Aviation Airports:
    - General Aviation Airports Outside the Metro Area

NPIAS

- Commercial Service
  - 546
  - 422 are primary: > 10,000 enplanements p.a.
- Reliever
  - 260
    - Average of 28 GA based there (27% of total GA)
- GA
  - Considered by NPIAS if 10 based GA and > 20 SM from other NPIAS

NPIAS

- FAA Hub Classification:
  - Different to airline
  - City + Standard Metropolitan Statistical Area:
    - Large: at least 1% of total enplanements
    - Medium: at least 0.25% of total enplanements
    - Small: at least 0.05% of total enplanements
    - Non-hub

Airport Components

- Three
  - Airside
  - Terminal Operation
  - Landside

Airside

- Runways
- Taxiways
- Aprons
- Influences on design
  - Land area
  - Type of users (aircraft): size/weight/pax
  - Location of terminal/maintenance facilities
  - Available land area
  - Movement of ground service vehicles

Terminal Operation

- Provides the link between landside and airside
- 3 main areas:
  - Curb-side
  - Central processing area
  - Links to the airside

Terminal Operation

- Must meet the needs of all users:
  - Primary users: Pax, Mail & Cargo
  - Secondary users:
    - Airport/airline/gov't employees
    - Meeters-greeters
    - Well-wishers, etc.

Landside

- Development has generally lagged:
  - FAA does not usually provide funds
- Require
  - Entry - Exit Roads
  - Parking
  - Taxi / Shuttle
  - Hotels
  - Rental Cars

Operational Considerations

- Runway Layout
  - Local Terrain
  - Local Weather
- Other Local Airport Operations
- ATC Requirements

Social Considerations

- Local Pax Demand
- Ground Access
- Noise
- Local Environmental Impact/Land Use
- Wildlife Habitats

Cost Considerations

- Topography
- Soil
- Construction Material
- Local Services
- Land Values

AVS 1200
Introduction to Aviation

Airports -
Operations and Challenges

Residual-Cost Approach

- Airlines Share Some of the Airport's Investment Risk
- Airlines Agree to Pay Certain Costs of Operations By Sharing Financial Risk
  - Thus, The Airline Can Exercise Financial Control Over New Airport Investments
- Difficulty In Generating Excess Profits- Reduce Airline Charges

Compensatory Approach

- Airport Assumes Full Financial Risk of Operation
- Airport Has No Guarantee That Rents & Fees Will Cover Operational Costs
- Greater Flexibility & Freedom To Undertake Capital Improvements

Airport Income

- Fuel Flow Tax
- Hanger
- Landing
- Concessions
- Parking
- Leased Areas

Airport Operating Expenses

- Administration & Staff
- Maintenance of Facilities
- Ground Equipment
- Utilities
- Security
- Communications

Airport Non-operational Expenses

- Debt Service-Funds to Repay the Loan

Source of Airport Funds

- Sale of Bonds
- State Grants
- Federal Grants
- Private Financing

Airport Capacity Challenge - Congestion

- After 1983, Airlines Increased Operations While Airports Did Not Increase Capacity
- Development of Airline Hub Airports
- Proliferation of Marketing Agreements Between Larger Carriers & Commuters
- Curfew Restrictions

Results of Congestion

- Economic Delays
- Safety Margins Could Be Compromised

Increase Airport Capacity

- Build New Airports
- Build New Runways
- Make Optimal Use of Existing Resources (Airport Modification)

AVS 1200
INTRODUCTION TO AVIATION

Human Factors in Aviation

Human Factors concerns the relationship between people and equipment, procedures, the environment, and other people

Human Factors involves:

- Physiology
- Psychology
- Sociology
- Anthropometry
- Bio-mechanics
- Technology

"Human Element in Aviation"

- The most reliable
- The most adaptable
- The most valuable
- **THE MOST VULNERABLE**

ACCIDENTS

- How many?
- What causes them?
- What stages of flight are most dangerous?

Accident Rates in Perspective

- For a 1:1,000,000 risk of being killed or seriously injured
- 5 minutes on a bicycle
- 30 minutes in a car
- 5 hours travel by air

Accident Rates

- Travel by air is relatively safe
- Flight safety has improved year by year
- **Still room for improvement**

Causes – Summary (1994-2003)

- 1. Loss of Control in Flight (32)
- 2. Controlled Flight into Terrain (CFIT)(24)
- 3. Landing (16)

CFIT

- Main accident cause in the 1980s, particularly in general aviation
- Much reduced by introduction of Ground Proximity Warning System (GPWS)

FLIGHT PHASES - SUMMARY

- Landing (and approach) accounts for approx 50% of accidents but only 4% of flying time
- Take-off and climb accounts for 25% of accidents but nearly 50% of fatalities

THE SHEL(L) CONCEPT

- Illustrates the interfaces between humans (the crew) and machine systems, software, the environment , and other humans

"SHELL" Model

- S = Software (Checks, SOPs)
- H = Hardware (Machines)
- E = Environment
- L = Liveware (Human Beings)
- L = Liveware (Pilot)

HUMAN INFORMATION PROCESSING

- RECEIVE
- PROCESS

THE BASIC STAGES

Detecting – information received from ***senses*** and paid ***attention*** to

Information interpreted (***perception***)

Decisions taken

Action (***Response*** selection and execution)

Feedback (How well am I doing?)

INFORMATION PROCESSING

- Most processing involves 2 sources
- Information from outside world (senses)
- Information already stored in memory
- Our mental model is based on the senses and past experience

SENSORY STORES

ICONIC MEMORY

- 70% of information processed by man enters the visual channel

- Visual = Iconic
- Iconic memory = 0.5 – 1.0 seconds
- Rapidly decays

- Used in cine film to enable impression of smooth movement

SENSORY STORES

ECHOIC MEMORY

- Aural
- 2 – 8 seconds
- Decays more slowly

ATTENTION MECHANISMS

- When "man focuses on certain features of the environment to the exclusion of others"

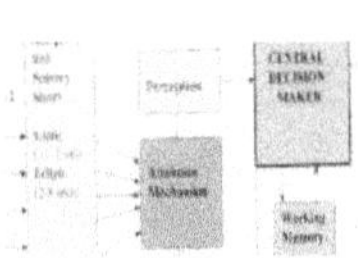

- Why do we need an attention mechanism?

- Limited number of channels
- Brain works best on single channel
- Working memory limitations (discuss later)

ATTENTION MECHANISMS

- Attention Mechanisms select the stimuli we can detect and then pass them on as a basis for perception, thought and decision making
- 2 general types
  - Selective
  - Divided

PERCEPTION

- Converting sensory inputs into meaningful mental structures
- Need mental models with which to compare
- Mental concepts are already stored from past experience
- "We can perceive only that which we can already conceive"

BUT THE BRAIN/EYE COMBINATION CAN PLAY TRICKS ON YOU!!

EXPECTATION

LEFT/RIGHT CONFLICT

Look at the chart and say the COLOUR not the word

YELLOW BLUE ORANGE
BLACK RED GREEN
PURPLE YELLOW RED
ORANGE GREEN BLACK
BLUE RED PURPLE
GREEN BLUE ORANGE

Power of the Mind

The phaonmneal pweor of the hmuan mnid, aoccdrnig to a rscheearch at Cmabrigde Uinervtisy, it dseno't mtaetr in waht oerdr the ltteres in a wrod are, the olny iproamtnt tihng is taht the frsit and lsat ltteer be in the rghit pclae. The rset can be a taotl mses and you can sitll raed it whotuit a pboerlm. Tihs is bcuseae the huamn mnid deos not raed ervey lteter by istlef, but the wrod as a wlohe.

MENTAL MODELS

- To perceive an accurate mental model, you need
- SENSORY INPUTS
- PAST EXPERIENCE (CONCEPTS)

PERCEPTION ERRORS

- EXPECTATION
- (*Paris in the the Spring*)
- CONFIRMATION BIAS
- *Seeking only information that confirms our mental model and ignoring other inputs that might disprove the model*

CENTRAL DECISION MAKER

- Input perceived
- Decision and response to be made
- Will involve memory:
  - Short term memory
  - Long term memory

SHORT-TERM MEMORY

- "Working" Memory
- 10 – 20 seconds
- Average of 7 ± 2 items retained
- Info lost by interference or replacement ....UNLESS transferred into long-term memory by rehearsal
- PERCEPTION takes place in working memory

INCREASING SHORT TERM MEMORY

- Chunking
  - *Not 367129860*
  - *but 367 129 860*
- Association

LONG-TERM MEMORY

- SEMANTIC MEMORY
- EPISODIC MEMORY
- *PROCEDURAL/MOTOR MEMORY*

SEMANTIC MEMORY

- Long term and accurate
- Stores general knowledge of the world
- Languages
- Often said when we cannot remember a word, it has not been lost but we cannot find where it is stored.

EPISODIC MEMORY

- Memory of events or 'episodes' in our lives
- Not as accurate as semantic memory
- Can be modified by desires or expectations
- Problems with eye witness accounts after accidents
- **Most amnesia (loss of memory) affects episodic memory**

CENTRAL DECISION MAKER LIMITATIONS

- Information processing limited by:

1. Short-term memory capacity
2. Single-channel capability

CENTRAL DECISION MAKER LIMITATIONS

- If limited to single channel processing only, we would not be able to carry out simultaneous tasks
- Would not be able to fly a turn and talk to the first officer simultaneously
- To satisfy multi-tasking requirements, we can use Motor Programs (Procedural Memory or Skills)

Motor Programmes (Procedural Memory or Skills)

- Some experts consider Motor Programs as third part of Long Term Memory
- Diagram shows separately for clarity only

FEEDBACK

- Feedback mechanism continually in use
- Even when motor programmes in use, feedback required for adjustments

BEHAVIOUR

- Rasmussen's **SRK** model useful for explaining pilot's behaviour

- **S** = Skill-Based Behaviour
- **R** = Rule-based Behaviour
- **K** = Knowledge-based Behaviour

SKILL-BASED BEHAVIOUR

- Skill acquired by practice (rehearsal) until it becomes a motor program
- Stored as "non-declarative knowledge"
- After "switch-on" no further reference to the Central Decision Maker is required
- Often un-monitored
- Accident Prone

ERRORS OF SKILL-BASED BEHAVIOUR

- Action Slip
- Environmental Capture
- Much more common in experienced pilots than novices
- More likely to occur when pilot is tired, over- or under-aroused

RULE-BASED BEHAVIOUR

- Uses the experience of others
- Stored as sets of rules in long-term memory but triggered by the Central Decision Maker
- Highly reliable (practice/rehearsal)
- Written Checklists reduce recall errors
- Taught most thoroughly in simulators

ERRORS OF RULE-BASED BEHAVIOUR

- Error of Commission most common.
  - e.g. Using the wrong procedure after misidentification of problem
- Departure from the rules
  - e.g. Valid GPWS warning ignored

KNOWLEDGE-BASED BEHAVIOUR

- Unusual circumstances not covered by the rules, SOPS, checklists.
- Requires thinking & reasoning
- Making a hypothesis (what's wrong here?)
- Making an action decision

ERRORS OF KNOWLEDGE - BASED BEHAVIOUR

- Incorrect Mental Model - ambiguous data
- Over-Confidence
- Lack of Situational Awareness
- Confirmation Bias
- Frequency Bias

HUMAN ERROR

- ***"It is in the nature of man to err"***
- ***" Only the fool perseveres in error"***
- ***Anyone can make a mistake but only a fool makes the same mistake twice***

SOME SOURCES OF PILOT ERROR

- Liveware/Software
- Liveware/Hardware
- Liveware/Environment
- Liveware/Liveware

MINIMIZING OCCURRENCE OF ERRORS

- Situational Awareness
- Avoid Expectancy
- Know Automation Problems
- Stay Motivated
- Professionalism
- Improved Management Culture

MINIMIZING OCCURRENCE OF ERRORS

- Equipment Design
- *Murphy's Law " If something can go wrong, then it is only a matter of time until it does go wrong"*
- Cross Monitoring
- Authority Gradient
- Controlled Intervention

Introduction to Aviation

The Human-Machine System:
Crew Resource Management
(CRM)

Crew Resource Management (CRM)

- Efficient Use of People, Equipment, and Information
- Crew Selection Process
- How Much Information Can a Human Absorb? - Time Frame
- Automation to Reduce Workload

Negative Aspects of Automation

- Neglecting Skills
- Inattention
- Mismatched Hardware to Liveware (Humanware)

Crew Communication

- Brief/Debrief
- Inquiry
- Assertion
- Advocacy
- Clarification
- Feedback (Crew Self-Critique)

Decision Making

- Choice Selection
- Conflict Resolution

Crew Environment

- Leadership
- "We Climate"
- "You Climate"
- Concern for Operation
- Informing
- Followership

PARENT (P), ADULT (A), CHILD (C) TRANSACTIONS

- High Authority to Low Authority = P-C / A-C
- Equal Authority = A-A
- Most flight deck transactions should be A-A
- Captain promotes adult responses and open 2-way communication
- P-C on flight deck can result in an angry or subdued First Officer/crew
- If an authoritarian Captain uses P-C mode, crew should respond in A mode to avoid either domination or conflict

TEAMWORK/LEADERSHIP

- 'Leaders are born, not made ' ??
- 'Leaders are made, not born' ??
- Probably a little bit of both

TEAMWORK/LEADERSHIP(2)

- Captain is appointed, so foundation of leadership role on the flight deck is established for us

EFFECTIVE LEADERSHIP

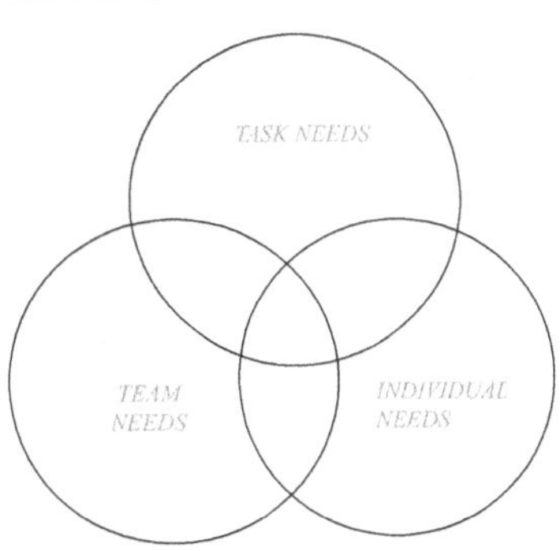

P/G Personality Model

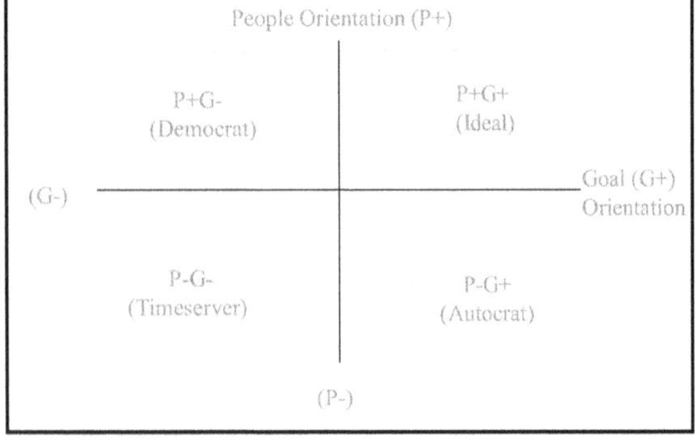

P+,G+ IS BEST

- P+,G+ is best because the job gets done but the crew are considered, consulted and involved
- When leading a team, need to be aware of some risk factors affecting group decision
- Risky Shift
- Conformity
- Compliance

RISKY SHIFT

- Also called Group Polarisation
- Occurs when a group have a tendency to a particular attitude
- Group position 'shifts' towards an extreme
- A group of careful people will 'shift' to an extremely careful decision – the 'risk' is it may be too careful
- A group of adventurous will 'shift' toward a more adventurous decision – the 'risk' is it may be too adventurous
- Flight crew tend to be in the latter category

CONFORMITY

- If the majority of the group state an opinion, an individual may conform and not state a different opinion.
- If you disagree, SPEAK UP
- Keep an open mind

COMPLIANCE

- Similar to conformity and occurs when people want to be helpful
- Often occurs when decision is taken by a person of higher status ,e.g., a Captain
- Validate the decision in your own mind
- If you don't agree, SPEAK UP!

FACTORS AFFECTING FLIGHT DECK RELATIONSHIPS

- 3 dominant factors:

- STATUS
- ABILITY
- ROLE

STATUS

- Leadership/followership interaction often determined by rank.
- Seniority seen as symbol of status
- Easy for Captain to question the FO's actions
- More difficult for FO to question Captain

- Need a reasonable 'AUTHORITY GRADIENT'

AUTHORITY GRADIENTS

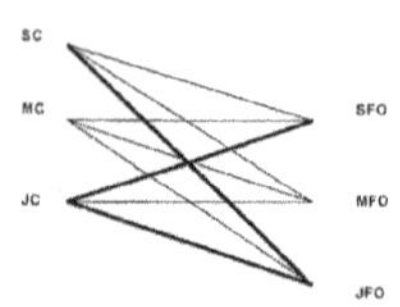

AUTHORITY GRADIENTS(2)

- SC to JFO – too steep
- *JFO over-awed and may not question a wrong decision*
- JC to SFO- wrong way
- *SFO may resent JC 'authority*
- JC to JFO – shouldn't happen
- *Too inexperienced*

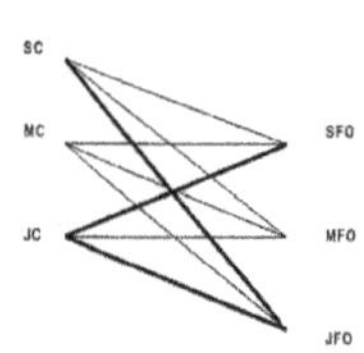

ROLE

- Who is the 'handling pilot'?
- Who is the legal commander?

ABILITY

- How good at their job are people perceived to be?
- A 'good' Captain may be perceived to be a poor pilot
- Most dangerous when an autocratic Captain is a weak pilot
- Democratic people may tend to trust others' ability too much

Workload Management

- Automation
- Distribution

Situational Awareness

- Planning
- Vigilance
- Preparation
- Sterile Cockpit
- Distraction
- Avoidance
- Discuss Issues
- Analyze

Professionalism

- Exists only when a service is performed for someone, or for a common good.
- Achieved only after extended training and preparation.
- True performance is based on study and research.
- Posses the ability to reason logically and accurately.
- Ability to make good judgments not limited to standard patterns and practices.
- Must be true to yourself and to those you serve.

The 8 Step Judgment Process

- Vigilance
- Problem Discovery
- Problem Diagnosis
- Alternative Generation
- Risk Analysis
- Background Problem
- Decision
- Action

AVS 1200
INTRODUCTION TO AVIATION

Principles of
Navigation

The Solar System

The Solar System

- Eight Planets orbiting the Sun
  - Mercury
  - Venus
  - EARTH
  - Mars
  - Jupiter
  - Saturn
  - Uranus
  - Neptune

Kepler's Laws

- First Law states
  - each planet moves in an ellipse and the Sun is at one of the foci
- Second Law states
  - as the planet moves around the Sun it sweeps out an equal area in equal time

The Earth Orbit

Earth closest to the Sun at the PERIHELION (3 January)

Earth farthest from the Sun at the APHELION (3 July)

The Earth Orbit

- The Earth's axis is inclined (tilted) at an angle of 66½° to the orbital plane.
- The inclination (tilt) is sometimes stated as 23½° between the plane of the Equator and the plane of the Earth's orbit around the Sun.

The Earth

The Poles

The Poles are defined as the extremities of the axis about which the Earth spins.

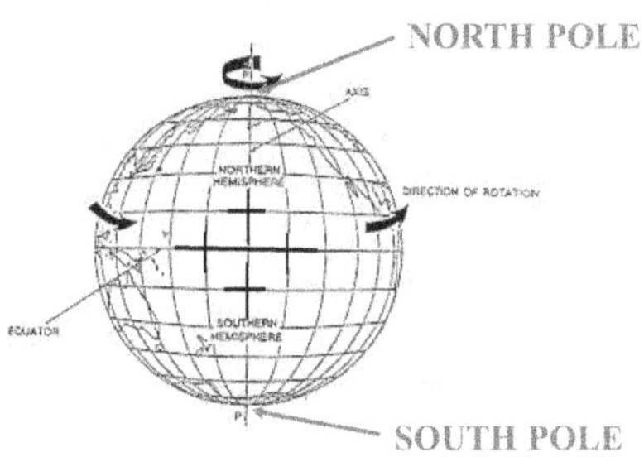

Shape of the Earth

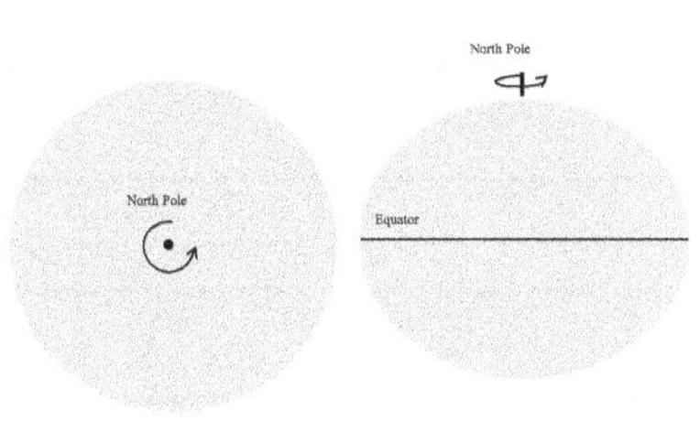

Oblate Spheroid

Earth Compression

The Earth's polar diameter is 42 km (23 nm) less than its equatorial diameter.

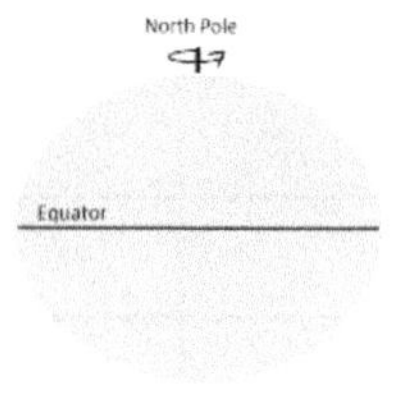

Basic Direction on the Earth

EAST is defined as the direction the Earth is spinning.
WEST is defined as the opposite to East.
NORTH and SOUTH are define by the position of the poles.

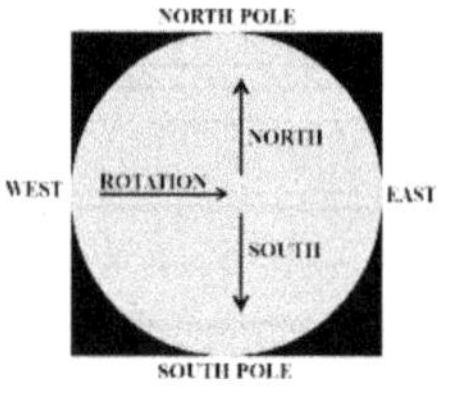

Basic Direction on the Earth

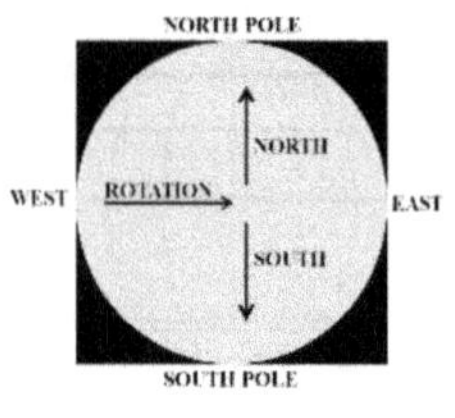

- The North Geographic Pole is the datum used for measuring TRUE direction (True North)

Sexagesimal System

Direction is measured CLOCKWISE from North in a circle (360°)

North is **000°** or **360°**

West is **270°**

East is **090°**

South is **180°**

Position on the Earth and Change of Latitude/Longitude

Circles on the Earth

A circle on the surface of the earth whose center and radius are those of the earth itself is called a GREAT CIRCLE

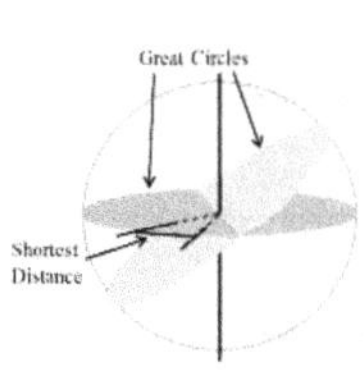

The shortest distance between two points on the Earth's surface is the short arc of the Great Circle joining the two points.

Circles on the Earth

A circle on the surface of the earth whose center and radius are NOT those of the earth itself is called a SMALL CIRCLE

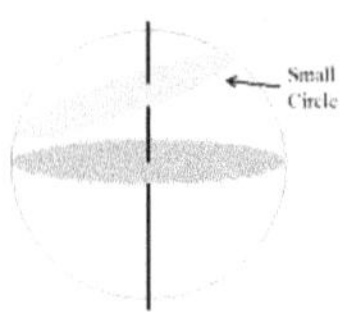

Circles on the Earth

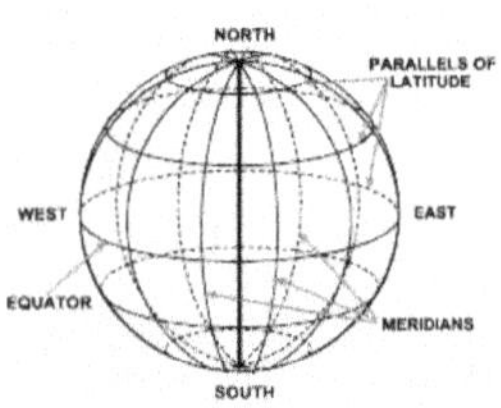

EQUATOR – The Great Circle whose plane is 90° to the axis of rotation of the Earth.

MERIDIANS – Semi-Great Circles joining the North and South poles.

Circles on the Earth

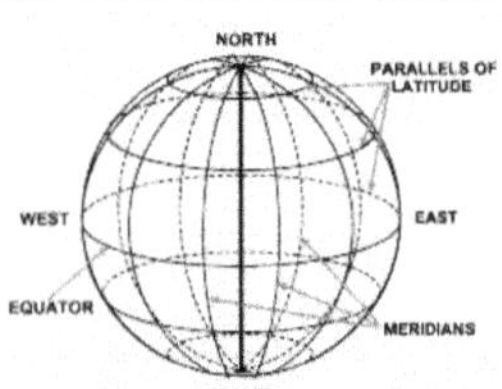

PRIME (or GREENWICH) MERIDIAN – Meridian passing through Greenwich.

PARALLELS OF LATITUDE – Small Circles whose planes are parallel to the Equator (lie in an East-West direction).

Earth Reference System

GRATICULE – Network formed on surface of the earth by the Prime Meridian, the Meridians, the Equator, and the Parallels of Latitude

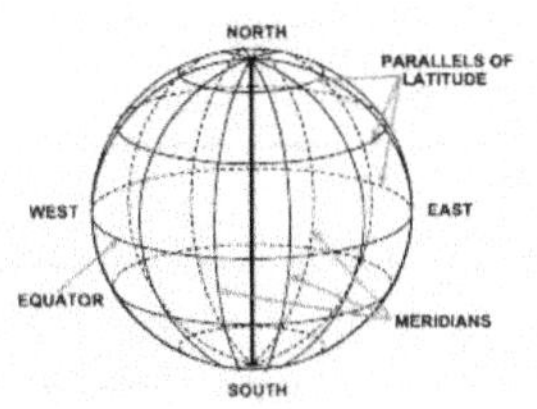

Earth Reference System

The earth reference system LATITUDE and LONGITUDE is based on angular measurement (degrees °, minutes ', seconds ") measured at the center of the earth.

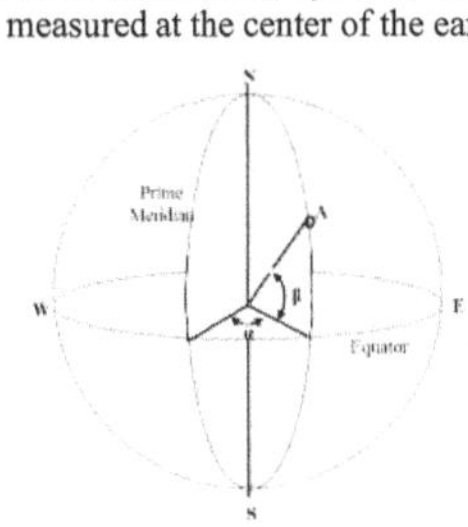

Latitude

LATITUDE – is the arc (angular distance) measured along the meridian through the point from the Equator to the point.

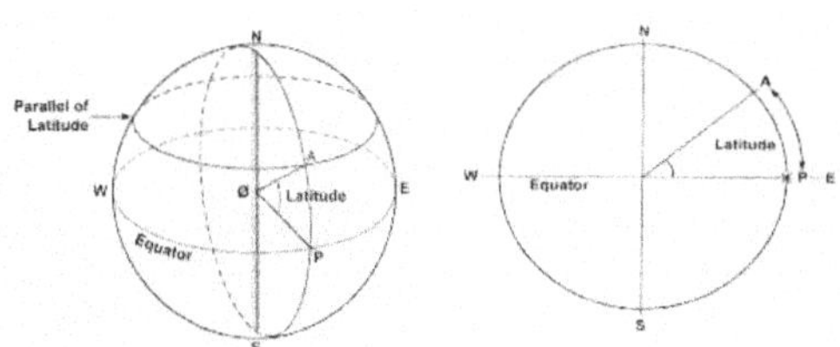

Examples - 40°N 40°00'N 4000N

Geocentric and Geodetic Latitude

GEOCENTRIC – measured at the center of the earth.
GEODETIC – measured using normal (90°) to meridian.

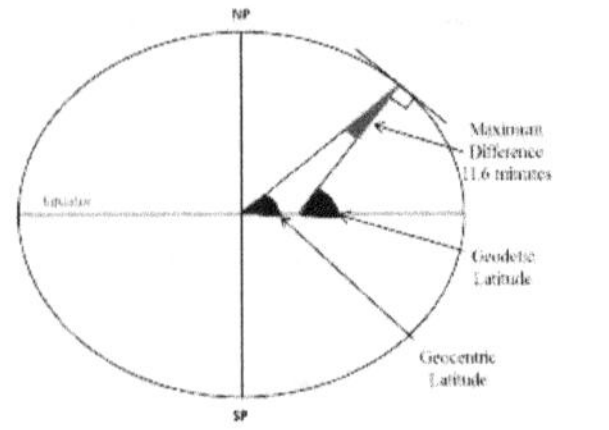

Longitude

LONGITUDE – is the shorter distance of the arc along the Equator between the Prime Meridian and the meridian through the point.

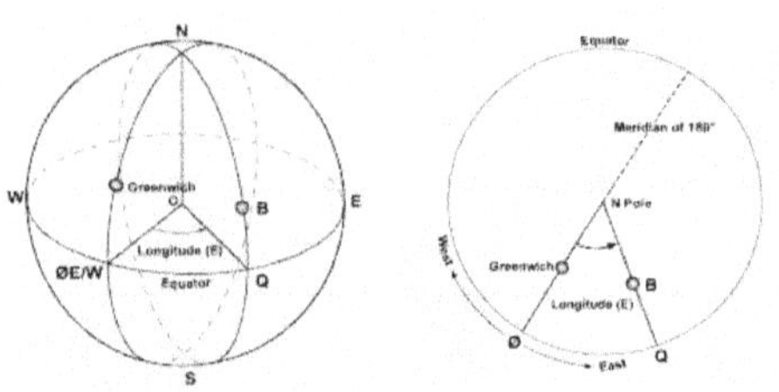

Examples - 040°E 040°00'E 04000E

Latitude and Longitude Conventions

Latitude 2 DIGITS FOR DEGREES - 05°N, 09°S, 90°N

Longitude 3 DIGITS FOR DEGREES - 005°E, 025°W, 160°E

Latitude FIRST - 41°N 074°W or

41°00'N 074°00'W or

4100N 07400W

Change of Latitude (ch. lat)

CH.LAT – Angular difference between 2 latitude angles

SAME hemisphere SUBTRACT

OPPOSITE hemisphere ADD

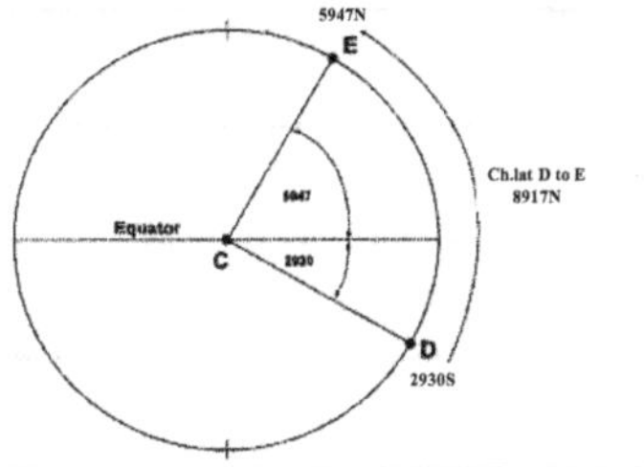

Change of Longitude (ch. long)

CH.LONG – Angular difference between 2 longitude angles

SAME hemisphere SUBTRACT

OPPOSITE hemisphere ADD

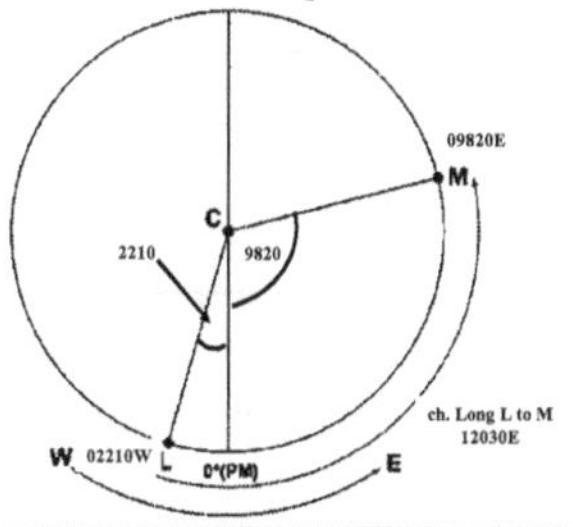

Convergency

CONVERGENCY – angle of inclination between 2 given meridians measured at a stated latitude.

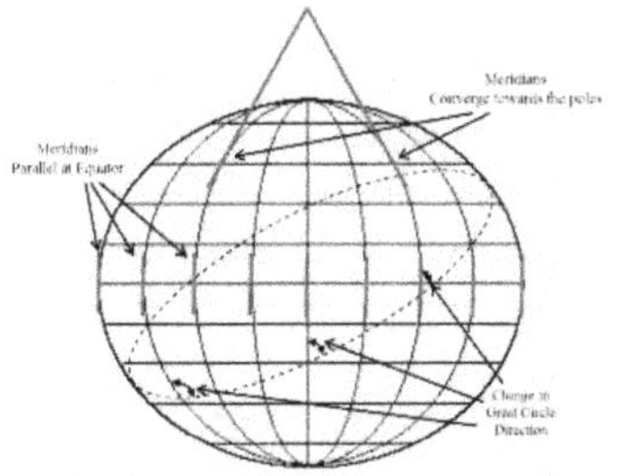

Rhumb Line

RHUMB LINE – crosses all meridians at same angle (constant direction)

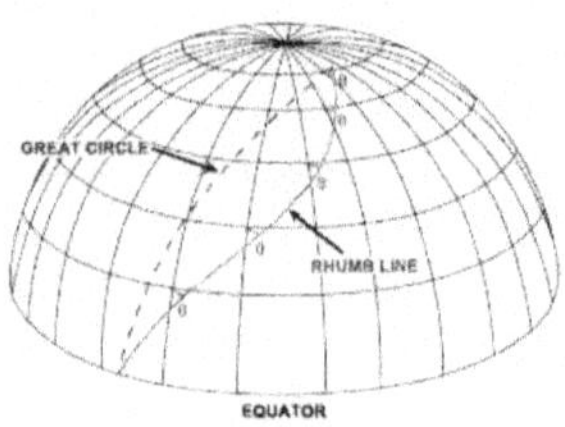

Rhumb Lines

PARALLELS OF LATITUDE – cut all meridians at 90°

EQUATOR – also a Great Circle

MERIDIANS – also semi-Great Circles with cut angle 0°

All Rhumb Lines (EXCEPT PARALLELS OF LATITUDE) end at either Pole.

Great Circle

GREAT CIRCLE – direction is constantly changing.

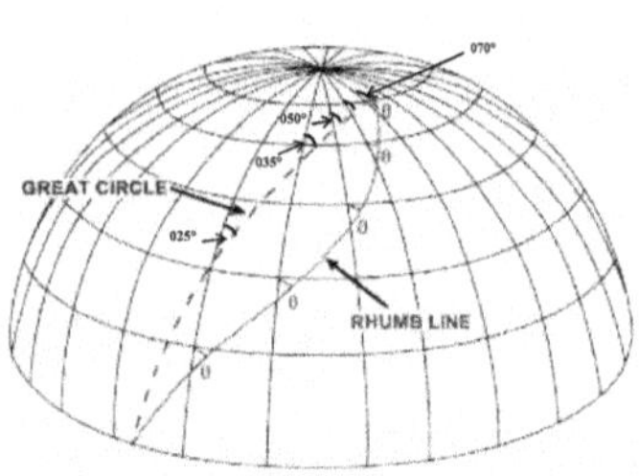

Convergency

On the SAME latitude

Convergency = ch. long x sin(lat)

On DIFFERENT latitudes

Convergency = ch. long x sin(mean lat)

Direction on the Earth

Direction

- True North
  - Direction measured with reference to the north geographical pole
  - True North is ALWAYS along a meridian
  - By convention, direction is measured clockwise from north in the horizontal plane from 000° to 360°
  - Measured in degrees and always expressed as a three-figure group e.g. 090 °, 270 °, etc

Direction

- Magnetic North
  - The direction a freely suspended magnetic needle will point towards
  - Geographic and magnetic poles are NOT coincident, separated by approximately 900 nm

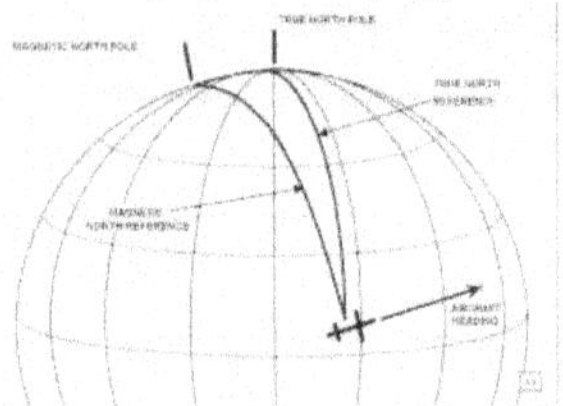

Direction

- Variation
  - The angular difference between the direction of True North and Magnetic North is called Variation
  - Magnetic North direction may lie either to the West or East of the True North direction
  - Measured in degrees and is named West or East accordingly

Direction

- Isogonals
  - Lines joining points of equal variation

Distance on the Earth and Great Circle Distance

Units of Distance

- Kilometre (km)
  - Metre (m)
  - Centimetre (cm)
  - Millimetre (mm)
- Statute Mile (sm)
  - Foot (ft)
  - Inch (in)
- Nautical Mile (nm)

Kilometer

- 1/10,000 of the average distance from equator to nearest pole
- Meter
  - length of platinum-iridium kept in a vault in Paris
  - distance travelled by light in a vacuum in 1/299,792,458 sec
- 1 km = 1000 m
- 1 m = 100 cm
- 1 m = 1000 mm

Statute Mile

- Statute of the English Parliament in reign of Elizabeth 1
- 5,280 ft = 1 sm
- 1 ft = 12 in

Nautical Mile

- Historically defined as a "minute of arc along a great circle of the Earth"
  - Earth not a perfect sphere
    - 1 nm at equator = 1843 m
    - 1 nm at poles = 1862 m
- International Nautical Mile is defined as 1852 m
  1852 m = 6076.1 ft
- Standard Nautical Mile is defined as 6080 ft

Nautical Mile

Arc on a GREAT CIRCLE

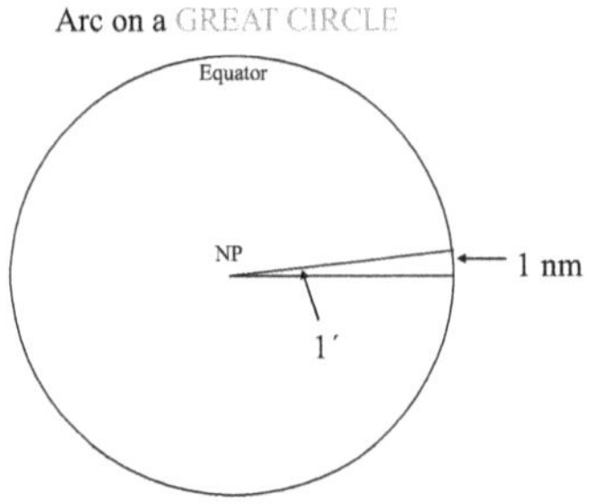

1° = 60′ = 60 nm on a GREAT CIRCLE

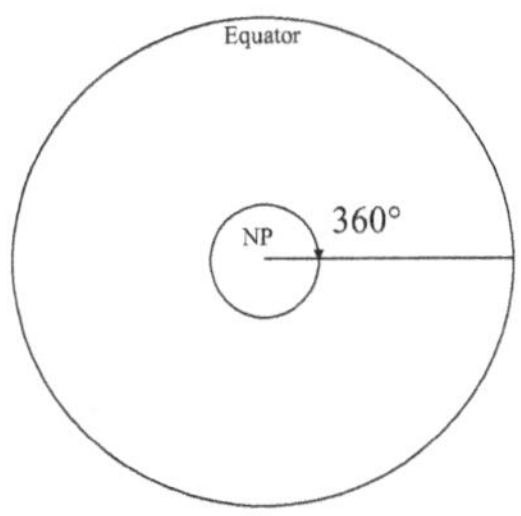

1° = 60 nm

360° = 360 x 60 = 21600 nm

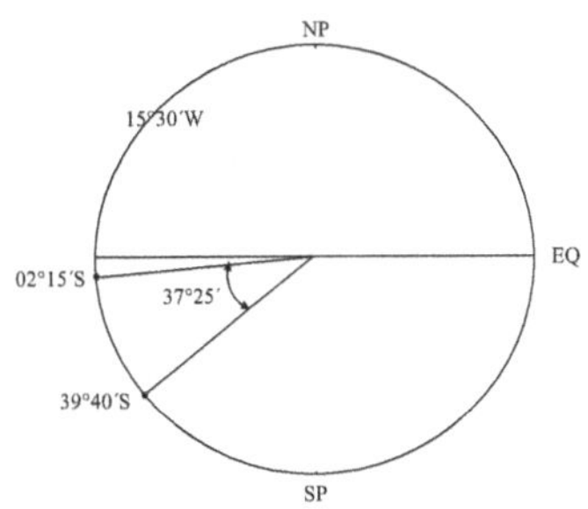

37°25′ = 37 x 60 + 25 = 2220 + 25 = 2245 nm

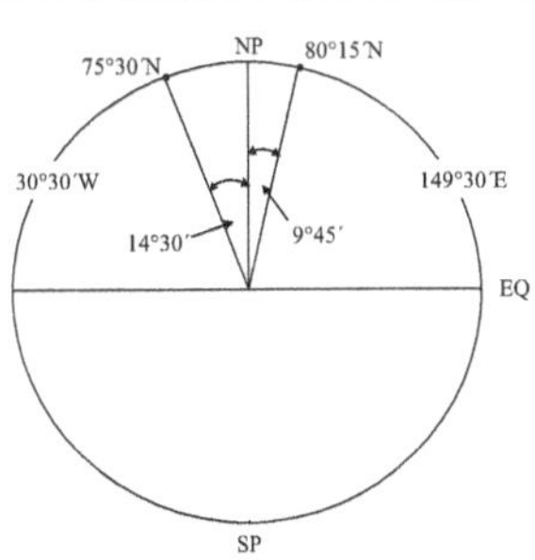

14°30′ + 9°45′ = 24°15′ = 24 x 60 +15 = 1455 nm

Distance Conversions

- 54 nm = 62 sm = 100 km
- 1 nm = 1.15 sm = 1.852 km
- 1 nm = 6080 ft = 1852 m
- 1 sm = 5280 ft = 1.6 km
- 1 km = 3280 ft
- 1 ft = 12 in
- 1 in = 2.54 cm

AVS 1200
Introduction to Aviation

International Air Transportation

International Agreements and Organizations

International Agreements

- **Paris Convention 1919**
  - "Sovereignty Principle"
    - each nation has complete sovereignty over the airspace above its territories and territorial waters.

International Agreements

- **Warsaw Convention 1929**
  - The liability of the Operator (and Agents) of aircraft specifically relating to matters of loss of life or injury to passengers, loss caused through delay etc..
  - $100 000 (as updated by the Guatemala City Protocol 1971 and Montreal 1999). If claim is more than $100 000, airline has the right to offer defence. Lesser amounts for other loss.
  - A contract exists when the operator issues a ticket. If no ticket issued, Operator cannot rely on Warsaw to limit compensation. Montreal 1999 cover the implications for a *ticketless* airline

International Agreements

- **Rome 1933/52**
  - Concerns the liability of an owner/operator of an aircraft which causes **damage to persons or property on the surface of the earth.**
  - The operator is liable
    - but the liability is limited to a sum proportional to the mass of the aircraft
    - The Montreal Conference 1978 has subsequently adjusted the amount of liability

International Agreements

- **Chicago Convention 1944**
  - Focused discussion on economic issues
  - United Kingdom
    - Desired to restrict access
  - United States
    - Desired to open access
  - Established general rules which international air transport could operate
  - Provisional International Civil Aviation Organization (PICAO) established

International Agreements

- **Tokyo 1963**
  - Dealt with **offenses and acts committed onboard an aircraft**. Its object is to ensure that such offenses and acts do not go unpunished (penal offenses)
    - Empowers aircraft commander and others, to prevent the commission of such acts, and to disembark the person concerned
    - An aircraft commander has the authority to
      - Restrain a suspect
      - Require or authorise crew members to restrain a person
      - Request or authorise passengers to restrain a person

International Agreements

- **The Hague 1970**
  - Diplomatic conference convened in order to adopt the Convention for the Suppression of **Unlawful Seizure of Aircraft** (hijacking)
    - There is a definition of such seizure, and an undertaking by member states, to make such offenses punishable by severe penalties

International Agreements

- **The Montreal Convention 1971**
  - Contains similar provisions on **jurisdiction, custody, prosecution and extradition of an alleged offender** as the Hague Convention
    - Amended by extensions of definitions and provisions on jurisdiction
    - Dealt with acts committed on the ground affecting international civil aviation

World Trade Organization (WTO)

- Formed by 1994 (Uruguay) round of trade talks
  - Rules of trade through agreements (GATT and GATS)
  - Objectives
    - Free trade flow
    - Liberalization of markets
    - Dispute settlement

General Agreements on Tariffs and Trade (GATT)

- Multilateral trading system for merchandise trade
  - Eliminate customs duties
  - No technical barriers
  - No preferred suppliers
  - No unreasonable import quotas
  - No unfair subsidies

General Agreement on Trade in Services (GATS)

- Multilateral trading system for service sector
  - No favoritism
  - Transparency
  - Market access
  - National treatment

Airline Agreements

- **The Hague 1919**
  - International Air Traffic Association (IATA)
    - Founded by European Airlines to make air transportation convenient and more acceptable to the traveling public
    - limited to a European dimension until 1939 when Pan American joined.
- **Havana, Cuba 1945**
  - International Air Transport Association (IATA)
    - Initially 57 members from 31 nations, mostly in Europe and North America.
    - Today it has some 240 members from 126 nations in every part of the globe.

Provisional International Civil Aviation Organization (PICAO)

- Formed Network of Air Routes
- Formed Global Air Navigation Infrastructure

International Civil Aviation Organization (ICAO)

- 96 articles
  - Accepted by "contracting states"
    - Adopt the standards
    - Article 38 – inform ICAO of differences
- 18 Annexes
  - Specified the establishment of "Standards and Recommended Practices" (SARPS and PANs)

International Civil Aviation Organization

- Aim:
  - To ensure safety, regularity and efficiency of International Civil Aviation operations.
- Contracting States
  - Countries which are members of ICAO

International Civil Aviation Organization

Annex 1 Personnel Licensing	Annex 10 Telecommunications
Annex 2 Rules of the Air	Annex 11 ATC
Annex 3 Meteorology	Annex 12 SAR
Annex 4 Maps and Charts	Annex 13 Accident Investigation
Annex 5 Units of Measurement	Annex 14 Aerodromes
Annex 6 Operation of Aircraft	Annex 15 Aeronautical Information Service
Annex 7 Registration	Annex 16 Environment
Annex 8 Airworthiness	Annex 17 Security
Annex 9 Facilitation	Annex 18 Dangerous Goods

Freedoms of the Air

- International Air Services Transit Agreement:
  - **First Freedom**:
    - To fly through the airspace of a Contracting State
  - **Second Freedom**:
    - To land in a Contracting State for maintenance

Technical Rights

1. Right to Overfly without landing

Technical Rights

2. Right to Land for Technical Purposes

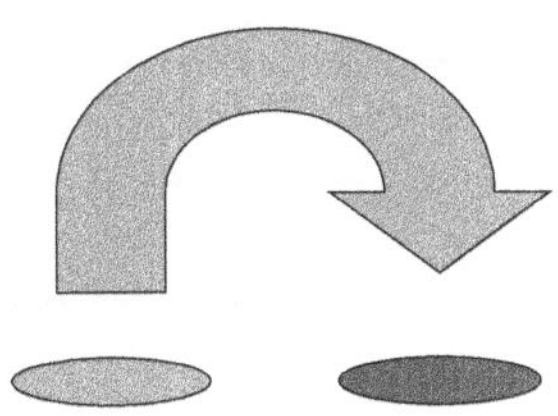

- International Air Transport Agreement:
  - **Third Freedom**:
    - To disembark 'traffic' in a Contracting State
  - **Fourth Freedom**:
    - To embark 'traffic' in a Contracting State
  - **Fifth Freedom**:
    - To embark 'traffic' in a Contracting state and disembark the 'traffic' in another Contracting State without transiting through the State of Registration

- TRAFFIC is defined as passengers, mail or cargo

Traffic Rights

3. Right to put down Passengers, Cargo, & Mail

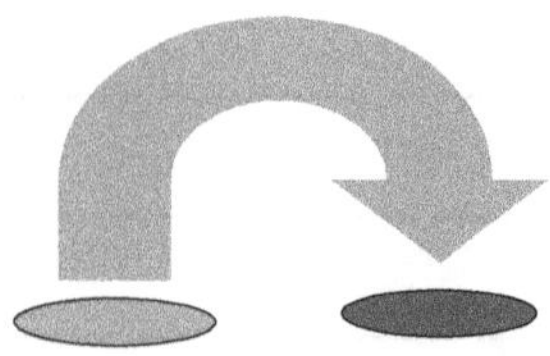

Traffic Rights

4. Right to pick up Passengers, Cargo, & Mail

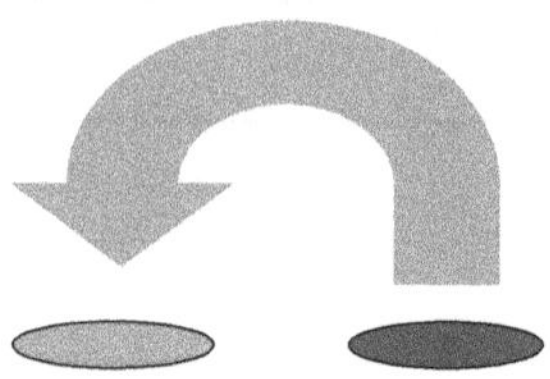

Traffic Rights

5. Right to operate between 2 other countries

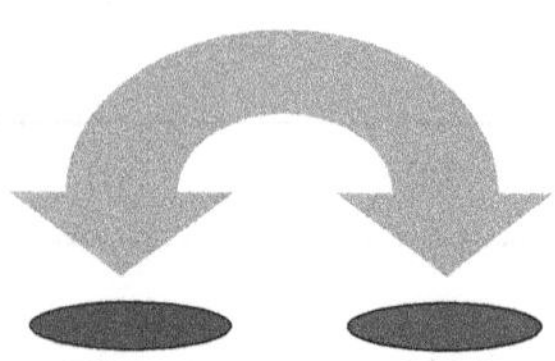

International Operations

Preparation

- Can be categorized into
  - Advanced Preparation
  - Route Specific Preparation

  Or
  - Aircraft/Crew/Passengers
  - Acquisition and Study of Flight Information and Regulations
  - Route Planning
  - Flight Planning
  - En Route and Destination Arrangements
  - Special Equipment Acquisition, Weather briefing and Flight Plan filing
- Checklists ensure that ***ALL*** planning and preparation is done

Aircrew

1. Current in international operations and regulations
2. Adequate rest-pilot's discretion*
3. Trip itinerary* - include stops along the way
4. Passenger manifest* -including full names, citizenship, dates of birth, current addresses, next of kin, any drugs property labeled and documented
5. List of U.S. embassy locations
6. FAA Airmen's Certificates
   a. Pilot's license
   b. Current medical certificate
7. FCC radiotelephone license
8. Current passports (county clerk)
9. Visas or tourist cards* (information on acquisition provided by the IFIM or professional planning service)
10. Second form of identification with photo (besides passport)
11. Proof of citizenship
    a. Birth certificate
    b. Naturalization papers
    c. Social security card
12. Immunization records
13. Traveler's checks
14. Credit cards
    a. Internationally accepted
    b. Recommendation from professional planner
15. Currency for every country*
16. Customs forms
    a. U.S. Customs Form 178
    b. General declaration
    c. List of hours of customs operations
    d. List of inspections required
    e. Copy of "U.S. Customs Guide for Private Flyers'
17. Current charts*
    a. Enroute charts, approach charts (obtain through Jeppesen, NOS or DMA)
       i. High attitude enroute charts
       ii. Approach chartsb.
    b. Current flight and entry regulations of country
18. All luggage properly marked
19. IBAC aircrew cards
20. List of emergency phone numbers
    a. Police
    b. Lawyers
    c. Next of kin

Aircraft

1. Airworthiness certificate
2. Registration (Note: Aircraft may not fly Internationally with a temporary registration or a copy-must be the original)
3. Radio license (FCC-check expiration date and be sure HF radio frequencies are permitted)
4. Current aircraft operations manual with weight and balance
5. Minimum equipment list (FAA)
6. Metric conversion tables (with pre-converted aircraft size and weights)
7. Airframe and engine logs
8. Certificates proving insurance coverage (from insurance company)*
9. Aircraft maintenance manuals
10. Import papers for aircraft of foreign manufacture
11. Liquor sealing device(s)
12. ETP/Wet footprint calculations*
13. Foreign meteorology reference book*
14. HF radio telephone operating manual and documentation
15. Copy of IFIM/AIP
16. Battery (age and time since deep cycle)*
17. Emergency quick reference guide*(tailored to suit needs of the trip planned)
18. Checklists
19. Notarized letter of authorization to use aircraft if borrowed, leased, or rented
20. 12" Registration marks and external ID plate are needed if ADIZ is penetrated

International Flight Planning Services

- IFPS do most of the work involved in acquiring planning information and making arrangements
- Several organizations offer this service
- The sort of items these services arrange include
    - Crew Transportation and Hotel Reservations
    - Slot Times
    - Discount Fuel
    - Acceptable International Credit Cards for Fuel Purchase
    - Overfly Permits
    - Ground Service and Agents
    - Flight Planning and Flight Logs

North Atlantic Route System

- There are a few VOR/NDB routes in the North Atlantic – **BLUE SPRUCE ROUTES**
- Other than Blue Spruce Routes little navaid coverage at low altitudes
- Radio communications limited for VHF (line of sight)
- HF is recommended (required for IFR – **above FL060**)

Minimum Navigation Performance Standards (MNPS)

- MNPS airspace extends from 27ºN to 90ºN and from FL280 to FL420
- Only available to aircraft and crews that are
  - Qualified
  - Equipped
  - Licensed to use it

Extended Time (Range) Operations (ETOPS)

- "Engines Turning Or Pilots Swimming"
- ETOPS are regulations dealing with over water operations
- Only applicable to part 121 operations
- Before 1985, only 4-engined aircraft
- After 1985, 2-engined aircraft with certain restrictions
- Single engine still air cruise speed restriction
  - 60 min, 75 - 120 min, 180 min

Organized Track System (OTS)

- Takes into account passenger demand, time zone differences and airport noise restrictions
- Two flows
  - Westbound from Europe in the morning
  - Eastbound from North America in the evening
- Concentrates most traffic uni-directionally
- Peak traffic flow at 30°W
  - Westbound between 1130Z and 1900Z
  - Eastbound between 0100Z and 0800Z

OTS

- Separation constraints are
  - 60 nm laterally and 10-15 minutes longitudinally
  - *Economically* FL310-390
- Organized every 12 hours close to minimum cost paths
- "Opposite Direction" flight levels can be utilized - FL350, 390 east and FL330, 370 west
- Several hours gap between termination of one flow direction to commencement of next

Composite Route System

- Established between Anchorage and Tokyo
- Composite lateral/vertical separation at and above FL280 (50 nm lateral and 1000 feet vertical, on the same route 2000 feet vertically or 20 minutes longitudinally)
- 5 ATS routes R-220, R-580, A-550, R-591 and G-344
  - R-220 westbound: FL290, 310, 330, 350, 370, 410, 430
  - R-580 westbound: FL280, 300, 320, 340, 360, 380, 400, 420
  - R-590 eastbound: FL290, 310, 330, 350, 370, 390, 410, 430
  - R-591 eastbound: FL280, 300, 320, 340, 360, 380, 400, 420
  - G-344 eastbound: FL290, 310, 330, 350, 370, 390, 410, 430
- There can be exceptions e.g. R-580 and R-590 bidirectional

Flexible Track System (FTS)

- Permanently established between Orient and Hawaii
- Times
  - Tokyo FIR – 1200z to 1700z
  - Oakland FIR – 1300z to 1900z
- All IFR flight levels at and above FL290
- One or 2 routes, if 2 routes separated by at least 100 nm laterally
- North FTR and South FTR
- Geographic coordinates transmitted by NOTAM at approximately 0500z daily by Tokyo ACC

AVS 1200
Introduction to Aviation

Air Carriers

Problems

- Since 2000 U.S. airlines have dealt with:
  - impacts of 9/11
  - bankruptcy of five network carriers
    - US Airways (02-03,04-05), United (04-06), Northwest (05-07), Delta (05-07)
    - Now American
  - record high fuel prices
  - most serious economic downturn since the Great Depression
  - heightened concerns about a pandemic that turned into reality in 2009

Recession

- The recession led to:
  - an erosion of wealth
  - double-digit unemployment
  - declining corporate travel budgets
  - close-fisted consumers
- All of which contributed to a softening of demand for air travel

Industry Response

- Previous recessions (in 1991 and 2001) carriers stimulated passenger demand by reducing fares sharply
- Initial response to the current economic downturn was to modestly cut fares and to better match supply (seats) and demand (passengers)
  - carriers reduced flying to hold the line on costs

- eliminated unprofitable routes
- grounded older, less fuel efficient aircraft
- charged separately for services historically bundled in the price of ticket
- initiated new services which customers were willing to purchase
- exhibited capacity discipline

- Their focus on additional revenue streams bolstered the industry to profitability in 2010 (for the first time since 2007)

System Capacity (Available Seat Miles)

- ASM is the overall yardstick for how busy aviation is both domestically and internationally
  - 7.4 percent decrease during 2009
  - 0.5 percent decrease in 2010
  - 3.7 percent increase in 2011
  - grow at an average of 3.1 percent per year through 2032

Aircraft Size

- Average size of domestic aircraft declined by 0.3 seats in FY 2010 to 121.6 seats, but average size of domestic aircraft increased in FY 2011 and is expected to increase by 0.2 seats in FY 2012 to 122.8 seats
  - network carriers continue to reconfigure their domestic fleets
  - demand for 70-90 seat aircraft continues to increase
  - the number of 50 seat regional jets in service is expected to fall

2012 FAA Forecast

- Lackluster activity in the near term, with a return to growth over the long term
  - not expected to snap back to levels published in the previous FAA forecast
- The shaky global economy that took hold in the latter part of 2008 expected to continue its squeeze on air travel demand through 2012

Commercial Aviation

- Mainline carriers will continue to drive down their costs by
  - better matching flight frequencies
  - aircraft gauge with demand
  - delaying deliveries of newer aircraft
  - grounding older aircraft
  - pressuring regional affiliates to accept lower fees for contract flying

General Aviation

- The downturn in the economy has dampened the near-term prospects for the general aviation industry
  - long-term outlook remains favorable.
- Growth in business aviation over the long term
  - driven by a growing U.S. and world economy.

AVS 1200
Introduction to Aviation

Air Cargo

Economics

- US $50 billion business
- Transports 35% of the value of goods traded internationally
- A critical part of the airline business
- Economic uncertainties growing but current cargo market conditions positive

- Revenues have been growing strongly
- In many regions, cargo profitability is back to pre-recession levels
- International air freight markets in July 2010 rose to a level 4% higher than pre-recession peak
- However, the financial position of the industry took a dip in 2011

- despite some apparent recovery in April, the most important markets — North America, Europe and Asia-Pacific — have seen a general decline in 2011

Economic Future

- Growth has now moved into a slower phase
  - more evenly spread geographically
  - still above trend
- All key indicators for air freight are pointing to slower growth in the rest of 2011 and 2012
- Economic growth is likely to continue to be very unbalanced

Revenues and Yields

- Cargo yields have rebounded faster than passenger yields
  - But have not recovered to pre-recession peaks
- Cargo revenues are expected to be up over a quarter this year as both volumes and load factors rise sharply

Capacity

- Capacity is now coming back into the market as a result of
  - the expanding passenger fleet
  - existing freighter fleet better utilized
- Load factors now back to pre-9/11 levels and aircraft numbers stabilizing
- Significant numbers of aircraft returning to service from storage
- New deliveries continued at a rate of around 100 aircraft a month.

Competition

- Ocean container freight rates up 80%+ while air freight yields rise 25%
- Containerized freight was slower to pick up than air freight
  - speediness of air overcame the cheapness of sea at the start of the economic upturn

Profitability

- Cargo profitability – among the US cargo airlines – remaining close to the levels of 2010 fourth quarter.
- Net profit margins of around 4% and operating margins of 6% are close to pre-recession levels in 2006 and 2007.
- Asian airlines are likely to have seen a sharper improvement in profitability
  - region where volumes, load factors and yields have risen the most in the past year.

Security

- The air cargo industry is experiencing
  - increasing security costs
  - inconsistency of terms, definitions and national requirements
  - unhelpfully diverse opinions regarding security controls
- Air cargo security is mainly funded by industry and its customers

- Estimates for compliance with US 100% screening mandate range from US$ 3.5 billion over 10 years to USD $4.5 billion in year 1 alone
- Whatever the true cost, it is huge and growing
- Unless costs can be contained/reduced they are unsustainable

Cargo Facilitation

- Customs authorities around the world require advance submission of electronic cargo information for risk and assessment purposes.
- Not all the requirements and systems are aligned with international standards.
- Cargo Facilitation provides a standard approach to comply with government regulations requiring the provision of cargo information.

Unit Load Devices

- With the increasing number of wide-bodied aircraft now in operation, Unit Load Devices (ULD) have become a key element of high efficiency in air transport.
- Currently ULD management is an under-recognized activity within the air cargo industry
- The right and serviceable container in the right place at the right time is critical to the running of the cargo operation

Dangerous Goods

- Ensuring that undeclared dangerous goods do not get on board an aircraft is one of many key objectives of a dangerous goods program.
- Standards for documentation, handling and training has produced a very high degree of safety in dangerous goods transport.

Live Animals

- Animals have been transported by air since the early 1930's.
- Carriage of live animals by air is considered the most humane and expedient method of transportation over long distances.
- Both safety and animal welfare are addressed in all regulatory issues pertaining to transportation of live animals by air

Perishable Goods

- Perishable goods were among the first commodities carried by air.
- Airlines have developed very effective handling techniques for chilled and frozen products
  - provide shippers with optimum, cost-efficient packaging methods for these time and temperature sensitive commodities.

AVS 1200
Introduction to Aviation

General Aviation

History

- One of the first mass produced General Aviation aircraft was
- The *Moth*
  - Produced by De Havilland Company of Great Britain
  - First flown in 1925
    - 2 seat biplane (powered by Cirrus engine)
  - Selected as the training aircraft for the first 5 Royal Aero Club flying schools

  - *Gypsy Moth* produced in 1928 (powered by De Havilland Gypsy engine)
  - By 1929 estimated that of every 100 aircraft flown in UK, 85 were *Moths*
- Used by military of 21 countries around the world in 1930's

- **Travel Air Manufacturing Company**
  - One of the leading United States aircraft manufactures founded in 1925 by
    - Walter Beech
    - Clyde Cessna
    - Lloyd Stearman
  - Attained international notoriety for establishing several speed and distance records

- **Taylor Brothers Aircraft Corporation**
  - Founded in 1927 by Clarence and Gordon Taylor
  - William T. Piper purchased bankrupt company in 1930
  - Renamed **Piper Aircraft Company** in 1937 and *Piper Cub* was built
  - The Cub was the first inexpensive training aircraft produced in the U.S. in large numbers

- **Business Travel**
  - Lockheed JetStar (1957)
  - First dedicated business jet to enter service
  - Rockwell Sabreliner (1958)
    - Initially developed for military training
  - Grumman G-1 (1959)
    - Twin turboprop business aircraft, first flew on August 14, 1958
    - In August 2006 a total of 44 Grumman Gulfstream I aircraft remain in service

- British Aerospace BAe 125 (1962)
  - The world's best selling mid-size corporate jet
- Dassault Falcon 20 (1963)
  - Pan American contracted to distribute the Mystère 20 (later became *Falcon 20*) in the western hemisphere
- Learjet (1963)
  - Founded in late 1950s by William Powell Lear Jr. as **Swiss American Aviation Corporation**
  - Renamed the **Lear Jet Corporation** in 1963
  - Original Learjet 23 was a six to eight seater and first flew on October 7, 1963

General Aviation Characteristics

- All aviation activity other than
  - Scheduled commercial air transport, and
  - Military
- Scheduled air transport is larger in terms of passenger numbers
- GA is larger in the number of flights (and flight hours, in the U.S.)
  - GA carries 166 million passengers each year
  - more than any individual airline, though less than all the airlines combined

General Aviation Structure

- General Aviation Manufacturers Association (GAMA)
  - National trade association representing over 50 American aircraft manufactures
  - Over 300,000 General Aviation aircraft in the world (over 200,000 in the USA)

- Aircraft Owners and Pilots Association (AOPA)
  - Founded on May 15, 1939
    - Goal of keeping general aviation fun, safe and affordable
  - Membership consists mainly of general aviation pilots in the United States
  - "AOPA Pilot" and "AOPA Flight Training".

- National Association of Flight Instructors
  - 16,000
- Executive
  - 40,000
- Business
  - 134,000
- Personal Use
  - 88% are single engine airplanes

Structural Changes in General Aviation

- Declining number of aircraft manufacturers during 1980's
  - Volume of exports plummeted from 3,395 units in 1979 to 440 in 1986
  - Declined from 100% of SE aircraft sold in US in 1980 to less than 70% today (mainly due to lower production by US manufactures)
- Declining number of pilots
- Emergence of the kit planes

Product Liability

- By the 1980s the world's insurance underwriters began to withdraw product liability insurance coverage from U.S. general aviation manufacturers
- Industry-wide liability premiums went from about $24 million in 1978 to $210 million in 1985
  - Product liability ranged from $70,000 to $100,000 per unit built and shipped during 1987

- Private/personal aircraft involved in a number of accidents (mostly related to pilot error)
  - 80 percent of the cases won by manufactures
- Manufactures increasingly unable/unwilling to deal with the enormous legal costs
- Reduced production
- Increased costs

Limited Liability

- In 1994 Congress limited liability for aircraft manufacturers
- Cessna starting to build single engine airplanes again
  - 172, 182 and 206
- Piper recovering from bankruptcy re-commenced manufacturing in 1995

Innovations

- Experimental
  - Experimental Aircraft Association (EAA)
    - EXPERIMENTAL placard
      - Modified or built from scratch
- Composites
  - Most common composite materials, including fiberglass, carbon fiber, and kevlar
    - At least two parts, the substrate and the resin
  - Airframes
    - Tails, wings, fuselages, propellers
    - Boeing 787 Dreamliner – over 50% composites

- Cockpit Displays
  - EFIS (Glass Cockpit)
  - Saab 2000, Airbus 320
- Navigation Systems
  - GPS
    - Not restricted to navigational uses.

Corporate Aviation

- **National Business Aviation Association** (NBAA)
  - Founded in 1947 and based in Washington, DC
  - Represents more than 8,000 companies that rely on general aviation aircraft
- Use of Aircraft for Business
  - Professional Crews
  - Transport Corporate Staff
  - Transport Customers
  - Transport Light Freight

Corporate Advantages

- Flexibility of Schedule
  - Take-off at anytime
- Flexibility of Airports
  - Land anywhere
- Increased Management Productivity
  - Reduced travel time
  - Face-to-face meetings

Corporate Aircraft Competition

- e-mail
- Fax
- Tele-conference
- Computers

Pilot Training

- College & Universities
  - Aviation degrees
- Flying Clubs
- FBOs

AVS 1200
Introduction to Aviation

General Aviation
- Corporate/Business Aviation

Need

- Companies represent many different professions and locations, but all need
  - fast, flexible, safe, secure and cost-effective access to destinations across the country and around the world.

Benefits

- Employees can meet, plan and work with each other aboard business aircraft, productivity en-route is greatly enhanced.
- May be only transportation option available, to link population centers and manufacturing facilities.
- Studies have shown that business aviation contributes greatly to local economies across the country.
- Employees can stop at several locations, then return to headquarters the same day reducing fatigue and cost.

Economics

- Business aviation contributes $150 billion to U.S. economic output and employs more than 1.2 million people.
- General aviation activities – including sales of new and previously owned airplanes, as well as maintenance and other operational support – generates substantial financial benefits for every state in the nation.
- The vast majority of GA aircraft used for business purposes worldwide are manufactured, operated, serviced and maintained in the U.S.

- Airplanes manufactured outside the U.S. often are "completed" (outfitted) in the U.S. with American-made avionics, electronics, systems, engines, paint, interiors and other aircraft components.
- Companies that utilize business aircraft outperform non-aviation users in annual earnings growth, stock and dividend growth, total share price, market capitalization and other financial yardsticks.
- Only 3 percent Fortune 500 companies
- 97 percent by organizations, including governments, universities, charitable organizations and businesses – large, medium and small.

Airports

- Business aviation reaches 10 times the number of U.S. airports (over 5,000 public-use facilities) than the airlines do.
  - Majority of U.S. airline flights only go to and from 70 major airports, and the total number of U.S. destinations served by air carriers has declined.
- Business aircraft flights account for just 4 percent of the total traffic at the busiest airports used by the commercial airlines.

Who

- Small companies operate the majority of business aircraft.
- Most companies (59 percent) operating business aircraft have fewer than 500 employees
- Seven in 10 have less than 1,000 employees.

Company Employees

- More productive aboard business aircraft than they are even in their own offices.
- Productive collaboration among company employees occurs nearly eight times more often than aboard commercial aircraft.
- Productive collaboration with customers occurs nearly seven times more often than on commercial aircraft.

Humanitarian Flights

- Following Hurricane Katrina in 2005, business aircraft brought 130,000 pounds of relief supplies to communities across the Gulf Coast, including those with no airline service.
- In 2009, companies that flew humanitarian missions, 69 percent flew between one and five trips, 13 percent between six and 10 trips and another 16 percent more than 10 trips that year.
- In early 2010, after the earthquake in Haiti business aircraft conducted more than 700 flights, transporting more than 3,800 passengers and delivering more than 1,400,000 pounds of critical supplies to relief groups on the island.

Safety

- Business aviation has achieved a safety record that is comparable to that of the major airlines.
- Besides complying with stringent government safety and security regulations, business aircraft operators participate in a variety of voluntary programs designed to enhance safety and security.
- Numerous federal officials, including the inspector general of the Department of Homeland Security, have found that general aviation "does not present a serious homeland security vulnerability."

AVS 1200
Introduction to Aviation

Careers in Aviation

PILOT

- Agricultural Pilot
  - Specially designed aircraft (including helicopters)
    - Apply herbicides, insecticides, seeds and fertilizers on crops, orchards, forests, fields and swamps
    - Aerial surveys of cattle and crops
    - Fighting forest fires
  - Fly low level with heavy loads, avoiding trees, power lines and fences

PILOT

- Air Taxi or Charter
  - Flies fare paying passengers 'anywhere, anytime' but usually for short trips over varying routes in single engine or light twin engine planes.
  - Greater the number of flying hours and the more complex the flying skills the more varied the opportunities for advancement as a pilot

PILOT

- Corporate Pilot
  - Fly aircraft owned by business and industrial firms
    - Transport company executives to branch plants and business conferences
      - Arrange in-flight meals, transportation at destinations
      - Supervise servicing and maintenance of aircraft
  - Fly in any weather into unfamiliar airports in aircraft ranging from light twin engine to corporate jet

PILOT

- Airline Pilot
  - Start as a First Officer on a regional carrier
    - Gain experience and flight hours
    - Low pay, long hours, many sectors per flight
  - Major carriers generally require 2000+ hours and some require college degrees

Engineer

- Airlines
  - Airfield facilities
  - Deep maintenance facilities
- Corporate
  - Routine maintenance of company owned aircraft
- Maintenance Facilities
  - Routine maintenance, refurbishment
  - Individuals, firms, airline contracts

Administration

- Airline
  - Operations
    - Flight Dispatch
    - Flight Scheduling
- Airports
  - Management
    - Operations
    - Security
    - Safety

Website

- http://www.avjobs.com/careers

Collegiate Organizations

- Alpha Eta Rho
  - Alpha Eta Rho - WMU Theta Chapter
  - Alpha Eta Rho International Aviation Fraternity

www.ingramcontent.com/pod-product-compliance
Lightning Source LLC
Chambersburg PA
CBHW081124180926
51611CB00004B/46
* 9 7 8 1 4 6 5 2 4 8 4 2 8 *